The Chemistry of Colour Application

R.M. Christie
Reader in Colour Chemistry, Heriot-Watt University

R.R. Mather
Senior Lecturer in Colour Chemistry, Heriot-Watt University

R.H. Wardman
Senior Lecturer in Colour Science, Heriot-Watt University

Blackwell
Science

Editorial Offices:
Osney Mead, Oxford OX2 0EL
25 John Street, London WC1N 2BL
23 Ainslie Place, Edinburgh EH3 6AJ
350 Main Street, Malden
MA 02148 5018, USA
54 University Street, Carlton
Victoria 3053, Australia
10, rue Casimir Delavigne
75006 Paris, France

Other Editorial Offices:

Blackwell Wissenschafts-Verlag GmbH
Kurfürstendamm 57
10707 Berlin, Germany

Blackwell Science KK
MG Kodenmacho Building
7–10 Kodenmacho Nihombashi
Chuo-ku, Tokyo 104, Japan

First published 2000

Set in 10/12.5pt Times
by DP Photosetting, Aylesbury, Bucks
Printed and bound in Great Britain by
MPG Books Ltd, Bodmin, Cornwall

DISTRIBUTORS

Marston Book Services Ltd
PO Box 269
Abingdon
Oxon OX14 4YN
(*Orders:* Tel: 01235 465500
Fax: 01235 465555)

USA
Blackwell Science, Inc.
Commerce Place
350 Main Street
Malden, MA 02148 5018
(*Orders:* Tel: 800 759 6102
781 388 8250
Fax: 781 388 8255)

Canada
Login Brothers Book Company
324 Saulteaux Crescent
Winnipeg, Manitoba R3J 3T2
(*Orders:* Tel: 204 837 2987
Fax: 204 837 3116)

Australia
Blackwell Science Pty Ltd
54 University Street
Carlton, Victoria 3053
(*Orders:* Tel: 03 9347 0300
Fax: 03 9347 5001)

A catalogue record for this title is available from the British Library

ISBN 0–632–04782–8

Library of Congress
Cataloging-in-Publication Data
Christie, R.M.
The chemistry of colour application/
R.M. Christie, R.R. Mather, R.H. Wardman.
p. cm.
Includes bibliographical references and index.
ISBN 0-632-04782-8
1. Dyes and dyeing. 2. Coloring matter.
I. Mather, R.R. II. Wardman, R.H. III. Title.
TP910.C37 1999
667′.3–dc21 99-42716
CIP

For further information on
Blackwell Science, visit our website:
www.blackwell-science.com

Contents

Preface	vii
1 Introduction	**1**
Bibliography	4
2 An Introduction to Dye and Pigment Molecules	**5**
2.1 Dyes and pigments	5
2.2 Intramolecular and intermolecular forces	7
2.3 Properties of dyes and pigments: colour, affinity and fastness properties	10
2.4 The classification of dyes and pigments	12
Bibliography	30
3 Introduction to Substrates	**31**
3.1 Introduction to polymers	31
3.2 Textiles	33
3.3 Paper	42
3.4 Leather	43
3.5 Plastics and rubber	44
3.6 Ceramics, cement, concrete and glass	48
Bibliography	50
4 Surface Coatings: Paints and Printing Inks	**51**
4.1 Paint	52
4.2 Printing inks	69
Bibliography	82
5 Application of Dyes	**83**
5.1 Selection of dye	83
5.2 Cellulosic textiles	84
5.3 Protein textiles	98
5.4 Synthetic and cellulose acetate textiles	108
5.5 Dyeing fibre mixtures	114
5.6 Levelling agents	115
5.7 Paper	117
5.8 Leather	119

5.9 Application of dye mixtures 121
Bibliography 122

6 Pigments and their Application **123**
6.1 The functions of pigments in their application 124
6.2 Requirements of pigments for application 128
6.3 The chemical structures and properties of pigments 132
6.4 The manufacture of pigments 142
Bibliography 144

7 Textile Printing **146**
7.1 Introduction 146
7.2 Printing pastes 147
7.3 Printing styles 150
7.4 Printing processes 157
7.5 Carpet printing 164
7.6 Drying and fixation 166
7.7 The flash-ageing technique 166
7.8 Washing off 166
7.9 Pigment printing 167
7.10 The future of textile printing 169
Bibliography 172

8 Introduction to the Physical Basis and Measurement of Colour **173**
8.1 Introduction 173
8.2 Electromagnetic radiation 174
8.3 Absorption of radiation 178
8.4 Quantitative aspects of light absorption 180
8.5 Quantitative analysis of dyes in solution 184
8.6 Fluorescence and phosphorescence 185
8.7 Specification of surface colours 188
8.8 The C.I.E. system of colour specification 196
8.9 Instruments for colour measurement 206
8.10 Uses of colour specification 209
Bibliography 226

9 Machinery for Dyeing Textiles **227**
9.1 Introduction 227
9.2 Basic principles 229
9.3 Dyeing in the loose fibre form 231
9.4 Yarn dyeing 232
9.5 Dyeing of woven and knitted fabrics 237
9.6 Garment dyeing 247
9.7 Future developments 251
Bibliography 252

10 Pigment Dispersion Technology **253**
10.1 The mechanism of pigment dispersion 253
10.2 Forces involved in pigment dispersion 257
10.3 Industrial dispersion equipment 258
Bibliography 265

11 Other Applications of Colorants **266**
11.1 Introduction 266
11.2 Colour photography 266
11.3 Food colorants 270
11.4 Hair colorants 271
11.5 Biological applications 273
11.6 Electrical/electronic applications 276
11.7 Titration indicators 278
Bibliography 278

12 Environmental and Toxicological Implications of Coloration **279**
Bibliography 282

Index 283

Preface

Colour plays a major part in our everyday lives. It influences our moods and emotions and generally enhances the way in which we enjoy our surroundings. Natural colours are all around us, in the earth, the sky, the sea, the animals and in the vegetation, in the trees, leaves, grass and flowers. In addition, there is a wealth of examples of synthetic colours, products of the chemical manufacturing industry, which we tend to take so much for granted nowadays. Synthetic colours are applied in the clothes we wear, in paints, plastic articles, in a wide range of multicoloured printed material such as posters, magazines and newspapers, in photographs, cosmetics, ceramics, and on television and film. Colour is introduced into these materials using dyes or pigments.

There are numerous textbooks on the market dealing with a variety of topics in colour science. For example, the Society of Dyers and Colourists, which is the professional organisation principally involved with dyes and their application to textiles, have produced an excellent series of textbooks at a reasonably advanced level dealing with various aspects of textile coloration. Similarly, the Oil and Colour Chemists Association offer a range of publications dealing with pigments, surface coatings and printing inks. At the Scottish Borders Campus of Heriot-Watt University in Galashiels, we have a long history of involvement in teaching colour chemistry at undergraduate and postgraduate level. From our experience, we have perceived for some time the need for a single textbook which deals at a reasonably introductory level with the physical and chemical principles underlying the application of colour. This encompasses both dyes and pigments and their application to a wide range of substrates, including traditional substrates such as textiles, surface coatings and plastics, as well as some of the newer non-traditional electronics, reprographics and biomedical applications of colour.

It is that need which is addressed in this textbook. It provides a broad, yet stimulating introduction to the science of colour application which will prove useful to undergraduate students. In addition, the treatment of the topics is at a level which is appropriate to those in industrial and academic organisations who require an overview of the subject. At the end of each chapter, the reader is directed to other texts which deal with individual aspects of colour application in greater depth and also with topics, such as the chemistry of the synthesis of dyes and pigments, which are not specifically addressed in this book.

We are indebted to The Society of Dyers and Colourists for permission to reproduce the diagrams shown in Figures 7.1–7.3, 7.5, 7.6, 9.4, 9.6, 9.8–9.10, 9.12 and 9.13. We are also grateful to Ray Tremlett and Ian Knowles for general advice on the preparation of Chapters 3 and 5.

R.M. Christie
R.R. Mather
R.H. Wardman

Chapter 1

Introduction

Colour vision arises from the interpretation by the brain of the electrical impulses produced by the interaction between light and the photoreceptors in the eye and transmitted to the brain along the optic nerve. The part of the electromagnetic spectrum which interacts with the photoreceptors is referred to as the visible region and for most organisms, including humans, this region covers a well defined band of wavelengths. The ability of animals to perceive colour is important in a variety of ways. For many animals, colour is essential for survival and, in the course of evolution, organisms have developed specific coloristic properties, for example for the purpose of camouflage. Whilst some species have developed highly conspicuous coloration as a means of drawing attention to themselves, others have developed coloration which enables them to blend into their natural habitat so that they are camouflaged effectively from predators.

Humans are particularly interested in colour, but for a number of reasons which extends far beyond that of straightforward survival. At least in the developed countries, day-to-day survival is not an immediate concern, but humans use colour for many purposes. Naturally, humans are interested in their appearance and so the style and the colour of the clothes they wear are important in this respect. Indeed this very interest in style and fashion, of which an important aspect is colour, is the basis of a major industry worldwide. It is one of the industries to which the content of this book has relevance. Beyond the choice of an individual about appearance, there lies the use of style and colour in garments for corporate identity purposes. Here the market is vast because modern lifestyles require us to identify groups of people, whether they work for one's favourite airline or play football for one's local team. The most historic form of the corporate wear market is that of the military forces, where identification of friends and foes in battle is a fundamental requirement.

The use of colour for identity has great importance to commercial organisations, where customer expectation demands consistent colour schemes and design of the products they purchase from a given manufacturer. This feature applies across wide ranges of consumer products and the principle applies equally to the appearance of petrol station forecourts as it does to the appearance of shampoo containers. The colours and colour schemes used in packaging have a direct influence on the image of a product. Thus whilst light, pastel colours might be used successfully to sell soft toilet tissues, it is often necessary to use a dark colour or black for products where an image of solidity, robustness or toughness is to be projected. Humans also use colour to assess the quality of certain products which may deteriorate over a period of time. A good example of this use is in the assessment of the quality of food, particularly red meat, where the customer has a perception of the shade of red which symbolises quality. If the meat on view at the counter is slightly off-shade, for whatever reason, the customer is deterred from buying.

In addition to selecting the colours for the garments they wear, humans also can control the colours of their surroundings. It is well known that soft, pastel colour schemes contribute to tranquil moods and in the workplace can increase work efficiency. In the opposite sense, colour schemes using vivid or highly contrasting colours tend to generate excitement and enthusiasm in mood. The way in which people decorate their homes depends on their individual personality; the types of decoration used in public buildings, such as schools, hospitals, shops, offices, etc. are selected to create moods appropriate to the activities carried out in the particular situation. This decorative market for colour, which involves application of surface coatings (i.e. paints) in addition to textiles, is also a large and lucrative market, and one of relevance to this book.

Colour is also used extensively as a means of communication, through devices such as liquid crystal displays, neon signs and visual display units, all of which are taken for granted in everyday life. However these three examples illustrate the fact that the colours which are perceived can be generated by various physical processes. Most of the objects we view around us appear the colours they do because the wavelengths of light from a source (for example the sun in the case of daylight) have been absorbed to various degrees by the dye or pigment present in them. The colours we observe are due to the combined effect of the light not absorbed and therefore reflected to our eyes. These are termed surface colours because we are viewing the light reflected from surfaces.

Most of this book is concerned with surface colours and the technology by which coloured objects are produced. Coloured effects are produced either by dyeing, whereby dyes actually diffuse into the interior of substrates such as textile fibres, or by the application of dyes or pigments to the surface of substrates using coating techniques such as painting or printing. Surface application can be used for a wide range of substrate types, including paper, textiles, metal and plastic. The difference between dyes and pigments is a simple one. Dyes are soluble in water and are therefore easily applied to fibres from hot aqueous solution. It is not essential that the dye solution is hot, but the transfer of dye from solution into the fibres occurs at much faster rates at higher temperatures. Many application classes of dyes become attached to fibres as a result of intermolecular forces of attraction of which there are many types (see Chapter 2). One particularly important class, reactive dyes, react chemically to form a covalent bond with the fibre during the application process.

Pigments are insoluble in water and indeed insoluble in most organic solvents. This makes it impossible to apply pigments to textile fibres in the same way as dyes. Plastic materials and a small amount of synthetic fibres are coloured by a process called mass pigmentation, in which the pigment powder is dispersed into the polymer while it is in a liquid phase. Commonly, pigments provide colour by application to the surfaces of substrates, either by painting or by printing. In order to facilitate application and to bind the pigment particles to the surface of the substrate, pigments are formulated in a suitable medium, which on drying after application, forms a hard, durable film on the substrate. The chemistry of the application of these so-called surface coatings is dealt with in Chapter 4.

This simple distinction between dyes and pigments, which is based on their solubility, is a useful one, but is not without some exceptions which can lead to confusion. There exists a class of dyes called disperse dyes, which have only a low solubility in water and therefore have a character which is intermediate between dyes and pigments as defined above. They are still referred to as dyes because they can be satisfactorily applied to textile fibres by the traditional dyeing methods involving water. Another class of dyes, called vat dyes, are totally insoluble in water and to all intents and purposes are pigments according to the above definition. However these dyes are special in that they are capable of being chemically modified by reduction to a water soluble form called the 'leuco' form, which is soluble in water. These dyes are therefore applied in their leuco form by a traditional dyeing process from water. When the transfer of dye from the solution to the fabric is complete, the fabric is treated with an oxidising agent which converts the dye back to its insoluble 'parent' form. In this instance the final result is that the insoluble dye particles are uniformly distributed through the cross-section of the fibre. Needless to say, because the dye particles are insoluble, the wet fastness properties of such dyed fabrics are excellent. Yet another class of dyes, the azoic dyes are also insoluble in water. In this case, the dyes are azo compounds actually formed in the fibres. The process is carried out by treating the fabric to be dyed firstly with a solution of a water soluble coupling agent and then drying it. The fabric is then passed through an ice-cold solution of a diazonium salt, wherein the insoluble azo dye forms. As with the vat dye types, the dye molecules are trapped within the fibres so dyeings of excellent wash fastness properties are possible.

As will be obvious from the above, the applications of dyes or pigments to substrates are essentially chemical processes. The purpose of this book therefore is to provide an introduction to the processes of coloration and to explain the underlying chemical principles involved. The approach which has been adopted is to deal firstly with the colorants themselves, in terms of their fundamental chemistry (Chapter 2). This is followed by a discussion of the chemistry of the various types of polymeric substrates which are coloured including textiles, plastics, ceramics, glass and concrete (Chapter

3). In dealing with the colorants and the substrates, connections are made to enable an understanding to be developed of the reasons why certain colorants are used for certain substrate types. Chapter 4 deals with some of the fundamental principles of the technology of surface coatings. Chapters 5, 6 and 7 then deal with the processes by which the colorants are applied to traditional substrates, Chapter 5 dealing with the dyeing of textiles, Chapter 6 with pigment applications and Chapter 7 with the specific topic of textile printing. In Chapter 8, the theory of colour is explained, together with an account, in some detail, of the way in which colour is specified. This subject is an important one because whatever product is being made for retail, whether it be butter, a shirt, paint or lipstick, the precise required colour must be produced, batch after batch. The ability to measure and define colour and colour differences in quantitative terms is central to the quality management of the coloration process. Other colour application technologies, such as those used in a range of electronic applications, modern reprographics and biomedical applications have emerged in recent years and are rapidly becoming mature processes. Many of these non-traditional uses of colour have the potential for high added value and are therefore very important areas of innovation commercially. Chapter 11 is devoted to a selected range of these technologies, with the aim of giving the interested reader the background to their underlying principles. Finally, Chapter 12 covers an issue of tremendous relevance to all industrial activities nowadays, that of the environment. Like any other chemical process, the manufacture of colorants and the coloration application methods employed potentially lead to pollution due to waste products. The ways in which the industry has responded to the awareness of environmental considerations are dealt with in Chapter 12.

The book therefore covers many different aspects of coloration and will be of interest to students of textile dyeing as well as those interested in surface coatings technologies. In addition to science and technology based students, it will also be of interest to students following design courses who require an understanding of the background technology of coloration processes.

Bibliography

Colour for Textiles. A User's Handbook, Wilfred Ingamells, The Society of Dyers and Colourists, Bradford (1993). ISBN 0-901956-56-2.

Color Chemistry: Synthesis, Properties and Applications of Organic Dyes and Pigments, H. Zollinger, VCH, Weinheim & New York (1987). ISBN 0-89533-421-4.

Chapter 2

An Introduction to Dye and Pigment Molecules

2.1 Dyes and pigments

Colour may be introduced into manufactured articles, for example textiles and plastics, or into a range of colour application media, for example paints and printing inks, for a variety of reasons but most commonly the ultimate purpose is to enhance the appearance and attractiveness of a product and improve its market appeal. The desired colour is generally achieved by the incorporation into the product of coloured compounds referred to as dyes and pigments. The term *colorant* is frequently used to encompass both types of colouring materials. The aim of this chapter is to provide an insight into the chemical structures of dye and pigment molecules. With this in mind, it incorporates a discussion of the methods used to classify dyes and pigments, an outline of the properties which dyes and pigments are required to exhibit in the course of their application, and finally a survey of the typical structural features of a selected range of dyes and pigments which aims to illustrate how the molecules are designed to fit their specific purpose. A more detailed discussion of the chemistry of these materials, particularly in relation to their application, is contained in later chapters.

For an appreciation of the chemistry of colour application, it is of fundamental importance that the distinction is made between dyes and pigments as quite different types of colouring materials. Dyes and pigments are both commonly supplied by the manufacturers as coloured powders. Indeed, as the discussion of their molecular structures contained in this chapter will illustrate, the two groups of colouring materials may often be chemically quite similar. However, they are distinctly different in their properties and especially in the way they are used. Dyes and pigments are distinguished on the basis of their solubility characteristics: essentially, dyes are soluble, pigments are insoluble. A comparison of the general characteristics of dyes and pigments is contained in Table 2.1.

The traditional use of dyes is in the coloration of textiles, a topic covered in considerable depth in Chapter 5. Dyes are almost invariably applied to the textile materials from an aqueous medium, so that they are generally required to dissolve in water. Frequently, as is the case for example with

Table 2.1 A comparison of the general characteristics of dyes and pigments.

Characteristic	Dyes	Pigments
Solubility	Required to be soluble or capable of solubilisation especially in water	Required to resist dissolving in any solvents which they may contact
Traditional applications	Textiles, leather, paper	Paints, printing inks, plastics
Method of application	Applied to textile fibres from an aqueous dyebath solution	Dispersed into a liquid medium which subsequently solidifies
Main chemical types	Organic only: azo, carbonyl and arylcarbonium ion	Organic types: azo, phthalocyanine and carbonyl; inorganic types
Application classification	For textile applications, the subdivision into dye application classes is of considerable importance	Pigments are multi-purpose materials and application classification is relatively unimportant
Colour properties	Full range of coloured species	Include coloured, white and metallic species

acid dyes, direct dyes, cationic dyes and reactive dyes, (some of the application classes of textile dyes discussed later in this chapter) they dissolve completely and very readily in water. This is not true, however, of every application class of textile dyes. Disperse dyes for polyester fibres, for example, are only sparingly soluble in water and are applied as a fine aqueous dispersion. Vat dyes, an important application class of dyes for cellulosic fibres, are completely insoluble materials but they are converted by a chemical reduction process to a water-soluble form which may then be applied to the fibre. There are also a wide range of non-textile applications of dyes, many of which have emerged in recent years as a result of developments in the electronic and reprographic industries (see Chapter 11). For many of these applications, solubility in specific organic solvents rather than in water is of importance.

In contrast, pigments are colouring materials which are required to be completely insoluble in the medium into which they are incorporated. The principal traditional applications of pigments are in paints, printing inks and plastics, although they are also used more widely, for example, in the coloration of building materials, such as concrete and cement, and in ceramics and glass. The chemistry of pigments and their application is discussed in more detail in Chapter 5. Pigments are applied into a medium by a dispersion process, which reduces the clusters of solid particles into a more finely-divided form, but they do not dissolve in the medium. They remain as solid particles held in place mechanically, usually in a matrix of a solid polymer. A further distinction between dyes and pigments is that while dye molecules are designed to be attracted strongly to the polymer molecules which make up the textile fibre, pigment molecules are not required to exhibit such affinity for their medium. Pigment molecules are, however, required to be attracted strongly to one another in their solid crystal lattice structure in order to resist dissolving in solvents. The discussion of the nature of intramolecular and intermolecular forces which is found later in this chapter (section 2.2) is highly relevant to an understanding of these properties.

2.2 Intramolecular and intermolecular forces

All matter is held together by forces of attraction. Individual molecules owe their stability to forces which hold together the constituent atoms in an appropriate arrangement. These forces operating within molecules are termed intramolecular forces. In addition, molecules are attracted to one another by a set of forces which are termed intermolecular forces. Inevitably these forces involve electrons, the tiny negatively-charged particles which are considered to orbit the positively-charged nuclei of atoms. An understanding of the origin and nature of these forces is essential to explain many features of the chemistry of textile dyeing and pigment application. This section contains an outline of the principles of the most important forces of attraction, illustrated as appropriate in Fig. 2.1, and which can be divided into the following five types:

(1) Covalent forces
(2) Ionic forces
(3) Dipolar forces
(4) Van der Waals forces
(5) Hydrogen bonding

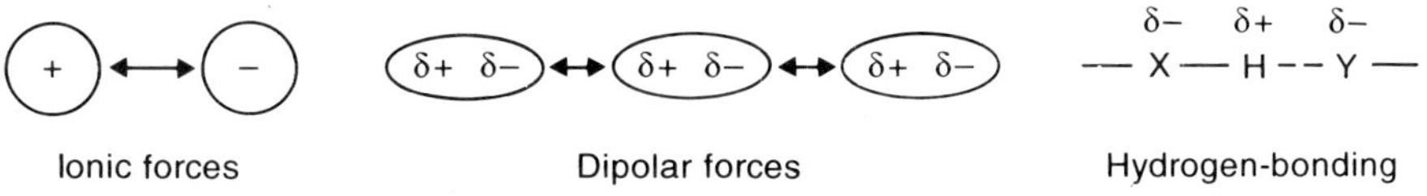

Fig. 2.1 Ionic, dipolar and hydrogen-bonding forces.

Covalent forces

Covalent forces, the strongest of all forces, result from a sharing of electrons between atoms. Molecules are held together by a system of covalent bonds between the constituent atoms. An especially important use of covalent bonding in colour application technology is the application of reactive dyes to cellulosic and protein textile fibres. These dyes react chemically with the fibre to form a covalent bond which combines the dye and the polymer into a single molecule.

Ionic forces

Ions are produced when electrons are transferred from one atom to another to form a positively-charged species (*cation*) and a negatively-charged species (*anion*). Strongly electropositive atoms (metals) readily form cations, while strongly electronegative atoms, notably oxygen and the halogens, readily accommodate the negative charge of anions. One of the most fundamental rules in chemistry is that ions of opposite charge are strongly attracted to one another by an ionic (or electrostatic) force, while ions of the same charge repel one another. Many, although not all, textile dyes are ionic in nature. For example, acid, direct and reactive dyes are anionic in nature, while basic dyes are cationic. The ionic nature of these types of dyes is an important factor in providing their water-solubility, a useful property for their application to textiles. Ionic forces of attraction between dye and fibre molecules are of importance in a number of textile dye systems, particularly in the acid dyeing of protein fibres and in the basic (cationic) dyeing of acrylic fibres. Another example of the importance of ionic forces in colour technology may be found in some inorganic pigments. The intermolecular forces which hold the crystal lattice of many inorganic pigments together are ionic in nature.

Dipolar forces

In many textile applications of dyes, dipolar forces may make a contribution to the forces of attraction between dye and fibre molecules. A molecule is described as being *polar* if there is a considerable degree of charge separation throughout the molecule leading to centres of partial negative ($\delta-$) and partial positive ($\delta+$) charge. Occasionally, the terms *hydrophilic* and *hydrophobic* are used to describe polar and non-polar character in molecules, respectively. A covalent bond in a molecule will have a polar character if the two atoms involved have different electronegativities, leading to an unequal sharing of the electrons. Dipolar intermolecular forces involve the attraction of the positive centres in one polar molecule for the negative centres in another, as illustrated in Fig. 2.1. As a result of these forces, polar molecules are generally attracted to each other more strongly than are non-polar molecules of comparable molecular size. However, as the charges are only partial, dipolar forces are weaker than ionic forces.

Two further types of force which may be included in this category are *ion–dipole* and *dipole–induced dipole* forces. Ion–dipole forces, as the name implies, involves the attraction of ions of one charge for a centre of partial charge of opposite sign due to a dipole in a molecule. Ion–dipole forces between dye ions and water molecules are largely responsible for the water-solubility of ionic dyes. Dipole–induced dipole forces of attraction are caused by the presence of a dipole in one molecule inducing oppositely-oriented dipoles in neighbouring molecules so a weak attraction between the molecules results.

Van der Waals forces

Van der Waals or *dispersion forces* are the forces of attraction involved between non-polar molecules. They are the weakest of intermolecular forces. Although in a non-polar molecule there is no overall charge distribution, the electrons are in constant motion so that at any instant small dipoles will be present. These instantaneous dipoles in turn induce oppositely-oriented dipoles in neighbouring molecules and a weak attraction between the molecules results. The reader should be aware that there is no real consistency in the literature concerning the use of the term van der Waals force, since it is commonly also used to encompass dipolar forces and hydrogen bonding.

Hydrogen bonding

In hydrogen bonding, a hydrogen atom serves as a bridge linking two other atoms. Strong hydrogen bonding only occurs when the two atoms involved are highly electronegative (most importantly oxygen, nitrogen, and the halogens, fluorine and chlorine). Hydrogen bonding of this type is often considered as a particularly strong form of dipolar interaction resulting from the polarisation of the covalent bond between the hydrogen and the electronegative atom. The hydrogen atom, as a result of its partial positive charge, is strongly attracted to the other electronegative atom which carries a partial negative charge (Fig. 2.1). One of the most notable examples of a molecule which readily participates in hydrogen bonding is water. The extensive intermolecular hydrogen bonding in the structure of liquid water is responsible for many of the unique properties of this familiar substance.

Many of the desirable properties of dyes and pigments may be attributed at least in part to hydrogen bonding, both intermolecular and intramolecular. For example, intermolecular hydrogen bonding is encountered in the crystal lattice of a number of organic pigments, and this gives rise to the desirable property of high insolubility. In addition, many dyes and pigments owe their good technical performance, especially lightfastness, to strong intramolecular hydrogen bonding. Several examples of this will be found in section 2.4. Intermolecular hydrogen bonding between functional groups on dye molecules and polymer molecules is thought to play a part in many interacting dye–fibre systems. A more unusual type of hydrogen bonding between the π-electron system of an aromatic ring, of the type invariably found in dye molecules, and hydrogen-donor groups such as the hydroxy (OH) group is also possible.

2.3 Properties of dyes and pigments: colour, affinity and fastness properties

Colour

The most obvious requirement of a dye or pigment is that it must have an appropriate colour. The physical basis of colour is discussed in detail in Chapter 8. A few of the basic concepts which relate particularly to the molecular structures of dyes and pigments and their influence on colour are outlined in this section. A dye owes its colour to the selective absorption by dye molecules of certain wavelengths of visible light. The remaining wavelengths of light are either transmitted or reflected, thus giving rise to the observed colour. The absorption of visible light energy by the molecule promotes electrons in the molecule from a low energy state, or *ground state* to a higher energy state, or *excited state*. The dye molecule is therefore said to undergo an electronic transition during this excitation process. The energy difference, ΔE, between the electronic ground state and the electronic excited state is given by Planck's relationship

$$\Delta E = \mathrm{h}\nu$$

where h is a constant (Planck's constant) and ν is the frequency of light absorbed. Alternatively, the relationship may be expressed as

$$E = \mathrm{hc}/\lambda$$

where c is the velocity of light (also a constant) and λ is the wavelength of light absorbed. Thus there is an inverse relationship between the energy difference between the ground and excited states and the wavelength of light absorbed. As a consequence, for example, a yellow dye which absorbs short wavelength (blue) light, requires a higher excitation energy than, say, a green dye which absorbs long wavelength (red) light.

The classical theory relating the colour of dyes and pigments to their molecular structure involves the concepts of *chromophores* and *auxochromes*. In this approach, chromophores are regarded as the groups of atoms principally responsible for colour of the molecule. The most important chromophores defined in this way are the *azo* (—N=N—), *carbonyl* (C=O), *methine* (—CH=) and *nitro* (NO_2) groups. However, dye molecules may also require the presence of functional groups referred to as auxochromes. Auxochromes are rather more loosely defined as groups of atoms in the molecule which are essential to augment and enhance the colour of the dye. These groups normally increase the intensity of the colour and shift the absorption to longer wavelengths of light. Commonly encountered auxochromes include *hydroxy* (OH) and *amino* (NR_2) groups. In dye molecules, chromophores and auxochromes are linked to one another through a conjugated system, i.e. a system of alternate double and single bonds. Commonly this conjugated system takes the form of six-membered rings which are termed aromatic. Aromatic rings, which confer particular stability on organic molecules, are an essential feature of all dye and organic pigment molecules. The numerous examples of chemical structures which follow in later sections of this chapter will illustrate the many ways in which chromophores, auxochromes and conjugated aromatic systems, together with other structural features designed to confer particular application properties, are incorporated into dye and pigment molecules.

The chromophore and auxochrome theory, which was first proposed more than 100 years ago, still retains some merit today as a simple method for explaining the origin of colour in dye molecules, although it lacks rigorous theoretical justification. It may be applied to most chemical classes of dye, including azo, carbonyl, methine and nitro dyes, but for some classes, for example the phthalocyanines, it is not appropriate. Nowadays, modern theories of chemical bonding, based on either the valence bond or the molecular orbital approaches, are capable of providing a much more sophisticated account of colour and constitution relationships. In fact, a mathematical approach using computer-based molecular orbital methods, for example the PPP-MO method, allows some aspects of the colour of dye molecules to be calculated and this provides a powerful tool for colour chemists in the design of new dyes. In this approach, a set of data associated with all the atoms and bonds in the molecule, including bond angles, bond lengths, electron affinities and valence state ionisation potentials, are used to calculate electronic transition energies and hence the wavelength of light absorbed. This therefore allows the hue of the dye to be predicted using Planck's relationship. In essence, it is no longer absolutely essential to synthesise a new dye with a particular chemical structure to establish what its colour properties might be. A computer program can provide a reasonably accurate prediction.

Affinity

The term 'affinity' is used here to describe, in qualitative terms, the intermolecular forces of attraction between colorant molecules and their substrate. This qualitative use of the term is useful in descriptions of the organic chemistry of colour application and should not be confused with the term 'standard affinity' which has a precise thermodynamic meaning in the physical chemistry of textile dyeing (derived from measurements of the relative quantities of dye absorbed by the fibre and remaining in solution when the dyeing process has reached equilibrium). In textile dyeing, the term 'substantivity' is commonly preferred to describe the suitability of a dye for a particular fibre, as a subjective assesment of the uptake of dye by the fibre from a dyebath in a reasonably short dyeing time. Dyes must exhibit affinity, or substantivity, for textile fibres in order to fix themselves to the fibre initially, and to resist removal by washing afterwards. Dyes owe their affinity for textile fibres in the main to the strong intermolecular forces between individual dye molecules and individual molecules of the polymer which makes up the fibre. A dye designed for a particular fibre therefore contains certain structural features which provide the appropriate forces of attraction for the fibre concerned. In contrast, pigments are dispersed as solid particles into their polymeric application medium and are held there mechanically. They are therefore in contact with this medium only at the surface of the pigment particles. The intermolecular forces of attraction between the pigment and the application medium are generally rather weak. Pigment molecules, however, are tightly bound to one another within the crystal lattice which makes up the particles by strong intermolecular forces. As a result, they do not have a tendency to dissolve in their application medium.

Fastness properties

All colouring materials are expected to exhibit a range of fastness properties in their application, ensuring that they withstand the effects of the environment in which they are placed both in processing and in their useful lifetime. Most of these fastness properties are assessed as the ability to resist a change in colour when exposed to some external effect, such as exposure to light, heat, solvents and other chemicals. Each particular application will have its own requirements and clearly some will be more demanding than others. Fastness to light, the ability to resist fading as a result of exposure to direct sunlight, may be of critical importance for many applications. This is so in the case of a pigment used in a paint applied to a car body, or for a dye in the textile fabric used in car upholstery but it is of lesser importance for applications where prolonged external exposure is not envisaged. Textile dyes may require to be fast to a variety of treatments to which the textile materials might be subjected, including washing, perspiration, rubbing and sublimation, and pigments to solvents and heat. Fastness properties, like affinity, are frequently dependent on the nature and strength of both intramolecular and intermolecular forces. Examples of ways in which the molecular structure of dyes and pigments can influence fastness properties will be found later in this chapter.

2.4 The classification of dyes and pigments

Colorants may be classified usefully in two separate ways, either according to chemical structure or according to the method of application. In the next two sections the chemical classification and the application classification of colorants will be discussed separately. The most important reference work dealing with the classification of dyes and pigments is the *Colour Index*, which is produced by The Society of Dyers and Colourists. This series of volumes provides a comprehensive listing of the known commercial dyes and pigments. Each colorant is given a C.I. Generic Name, which includes the application class, the hue and a number. This system of nomenclature for dyes and pigments is universally accepted and is frequently used throughout this book. The *Colour Index* gives information for each dye and pigment on the methods of application and on fastness properties. The volumes also list the companies which manufacture the products, together with trade names, and give the chemical constitutions where these have been disclosed by the manufacturer.

2.4.1 Chemical classification

In this classification method, colorants are grouped according to certain common chemical structural features. The most important organic dyes and pigments belong to the azo, carbonyl (including anthraquinones), phthalocyanine, arylcarbonium ion (including triphenylmethanes), sulphur, methine and nitro chemical classes. An outline of the characteristics of each of these classes follows, together with selected examples of typical chemical structures, and the section concludes with a brief review of the most important inorganic pigments, a group of colorants which has no counterpart in dye chemistry. A summary of the principal features of each of the chemical classes is given in Table 2.2.

Azo dyes and pigments

Azo colorants constitute by far the most important chemical class. They account for around 60–70% of the dyes used in traditional textile applications and they occupy a similarly prominent position in the range of commercial organic pigments. Azo colorants, as their name implies, contain as their common structural feature the azo (—N=N—) linkage which is generally attached at either side to two aromatic ring systems. The majority of these products contain a single azo group, as for example in C.I. Disperse Orange 25 (**1**), the structure of which is given in Fig. 2.2. As such they are referred to as monoazo dyes or pigments, but many contain two (disazo), three (trisazo) or more such groups. They are capable of providing a wide range of colours of high intensity and brightness, although they more readily give yellows, oranges and reds than blues and greens. In addition, they are capable of providing reasonable to very good technical properties, although they are often inferior to other chemical types in this respect.

The synthesis of dyes and pigments is for the most part beyond the scope of this book. However, a brief outline of the general method used to prepare azo dyes and pigments is included here to assist the reader to appreciate why these products have become the most important chemical class. The synthesis of azo colorants involves a two-stage sequence of reactions known as diazotisation and azo coupling. Diazotisation involves the treatment of a primary aromatic amine ($ArNH_2$), referred to as the diazo component, under acidic conditions at relatively low temperatures

Table 2.2 A summary of the characteristics of the main chemical classes of dyes and pigments.

Chemical class	Distinctive structural feature	General characteristics	Main application class(es)
Azo	—N=N—	All hues, but yellow to red most important	Dominate most, but not vat dyes
Carbonyl	C=O	All hues, but blue most important	Important in most applications
Phthalocyanine	16-membered heterocyclic ring, metal complex	Blue and green only	Most important in pigments
Triarylcarbonium ion	Positively-charged carbon atom	All hues, but reds and blues most important	Cationic dyes and pigments
Sulphur dyes	Complex polymeric S-containing species	Mostly dull colours, such as blacks and browns	Often considered as an application class itself
Methine dyes	—C=	All hues, but yellows most important	Disperse, cationic
Nitro	$—NO_2$	Mainly yellows	Disperse, hair dyes
Inorganic colorants	Range of inorganic types	All hues, white and metals	Exclusively pigments

with sodium nitrite ($NaNO_2$) to form a relatively unstable material known as a diazonium salt ($ArN_2^+Cl^-$). In the second stage (azo coupling) the diazonium salt thus formed is allowed to react with a compound known as a coupling component, generally a phenol or aromatic amine, to form the azo dye or pigment. Probably the most important reason for the dominant position of azo colorants is that they are relatively inexpensive. This is because the manufacturing processes, using the chemistry outlined above generally give high yields of the products from low cost and readily available starting materials. In addition, water is the solvent in which the reactions are carried out, which gives obvious economic and environmental benefits. The versatility of the chemistry, which involves bringing together two separate organic components, means that an immense number of azo

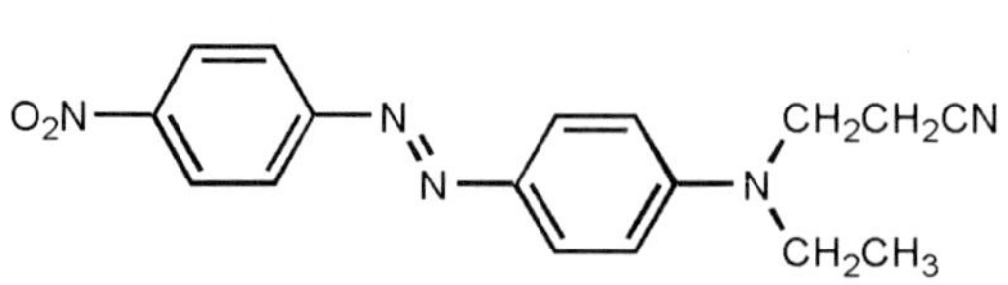

1

Fig. 2.2 C.I. Disperse Orange 25 (**1**), an example of a monoazo dye.

colorants may be prepared and so they have been adapted structurally to meet the requirements of most colour applications. In the future, it is likely that azo dyes are likely to assume even greater importance as some of the other chemical types, notably anthraquinones, become less economic (see section below).

Carbonyl dyes and pigments

Carbonyl dyes and pigments are characterised chemically by the presence the carbonyl (C=O) group. Usually there are two or more such groups linked by a conjugated system, frequently involving aromatic rings. Examples of the chemical structures of typical carbonyl dyes and pigments are given in Fig. 2.3. The most commonly encountered carbonyl dyes are anthraquinones which contain a characteristic system of three linear fused six-membered rings with the carbonyl groups in the central ring, as illustrated by C.I. Disperse Green 5 (**2**). Although anthraquinones are the most important carbonyl colorants, there are a number of other structural types, such as the well-known blue vat dye, indigo (**3**) and pigments such as the quinacridone (**4**). Carbonyl dyes and pigments are capable of providing a wide range of colours and a set of fastness properties which are often superior to those shown by azo colorants. However, largely because of the more elaborate chemistry involved in the manufacture of carbonyl

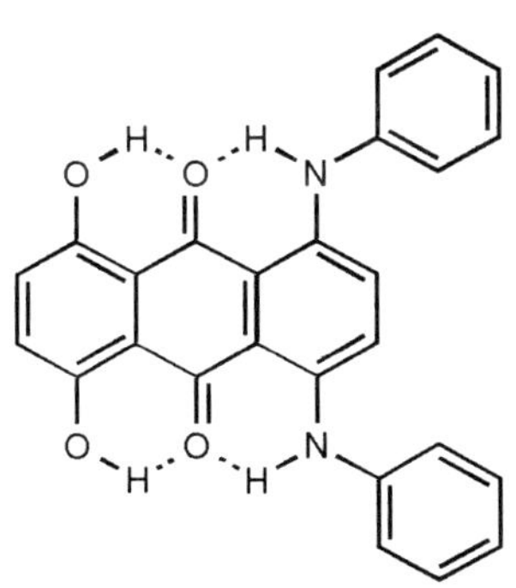

2

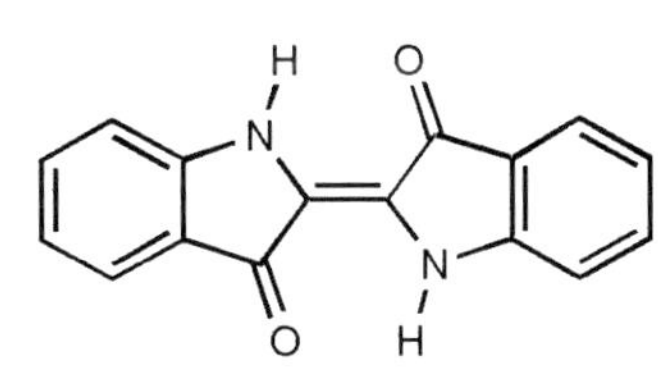

3

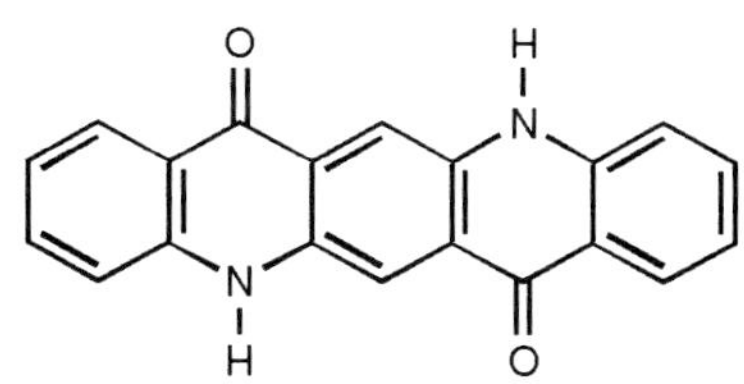

4

Fig. 2.3 Examples of the structures of carbonyl dyes and pigments. C.I. Disperse Green 5 (**2**), an anthraquinone dye; C.I. Vat Blue 1 (**3**), indigo; C.I. Pigment Violet 19 (**4**), quinacridone.

colorants, they tend to be rather more expensive and for this reason anthraquinones in particular are becoming progressively less important commercially.

In most textile dye application classes, carbonyl dyes rank second in significance to azo dyes. They are often of particular importance in the blue and green shade areas thus complementing the colours available from azo dyes (see section above). However, they dominate the vat dye class (section 2.4.2), an important application class for cellulosic textile fibres such as cotton, where the ability of the carbonyl groups to undergo a reversible reduction to a water-soluble form is put to use. Azo dyes are inappropriate as vat dyes because of the instability of the azo group to the reduction process.

Phthalocyanines

The phthalocyanines are complex cyclic structures containing an internal 16-membered ring of alternate carbon and nitrogen atoms and four outer benzene rings. In addition, most contain a central complexed metal atom. The colours of phthalocyanine dyes and pigments are restricted to blues and greens. While textile dyes based on phthalocyanines are of limited importance, copper phthalocyanine (**5**), the structure of which is given in Fig. 2.4, is by far the most important blue pigment for paint, printing ink and plastics applications. It owes this dominance to its intense brilliant blue colour and its excellent technical performance which includes outstanding fastness to light, heat, solvents and chemicals. In addition, it is a relatively low cost product because, in spite of its structural complexity, its manufacture is straightforward giving high yields from inexpensive starting materials. Copper phthalocyanines with the outer benzene rings substituted heavily with chlorines and bromines provide equally high performance green pigments.

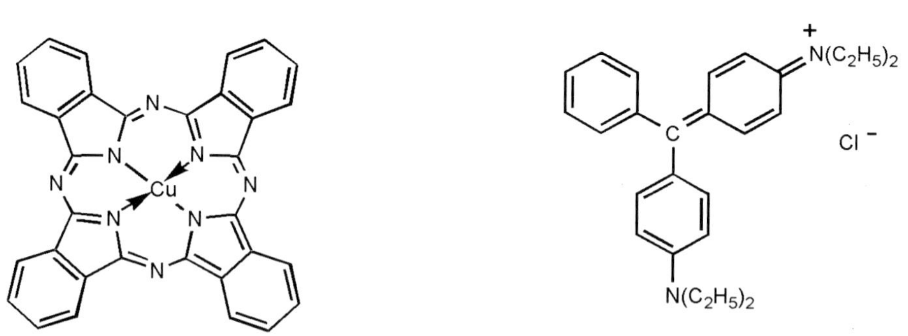

5 **6**

Fig. 2.4 Examples of phthalocyanine and arylcarbonium ion colorants. Copper phthalocyanine, C.I. Pigment Blue 15 (**5**) and Malachite Green, C.I. Basic Green 4 (**6**), an example of a triphenylmethane dye.

Arylcarbonium ion dyes and pigments

Arylcarbonium ion colorants were historically the first group of synthetic dyes developed for textile applications. They are used considerably less today, but many are still of some importance, particularly as basic (cationic) dyes and as pigments. This group of colorants are relatively low cost products and provide extremely intense, bright colours covering virtually the complete shade range, but they are generally inferior in technical properties compared with the azo, carbonyl and phthalocyanine classes. The best-known arylcarbonium dyes are triphenylmethanes, which contain a central carbon atom surrounded by three aromatic rings. Malachite Green (**6**) is illustrated in Fig. 2.4 as an example of a triphenylmethane dye.

Sulphur dyes

Sulphur dyes are a long-established group of dyes for cellulosic fibres which are fairly small in number, although some of the individual products are manufactured in large quantities. They are capable of providing a wide range of shades although they tend to be rather dull colours. C.I. Sulphur Black 1 is by far the most important product in the series. The chemical structures of sulphur dyes are not completely established. They are known to be complex mixtures of polymeric species containing sulphur in the form of sulphide (—S—), disulphide (—S—S—) and polysulphide ($—S_n—$) links and in heterocyclic rings. These sulphur dyes are initially insoluble in water, but are converted to a soluble form when treated with aqueous alkaline sodium sulphide solution. This is thought to be mainly due to the reduction of disulphide (—S—S—) links to alkali-soluble thiol (—SH) groups. After application to the fibre, oxidation regenerates the insoluble polymeric structure of the dye within the fibre. Unfortunately sulphur dyes present some environmental problems, largely associated with sulphide residues which remain in the dyehouse effluent, a factor which restricts their use considerably.

Methine dyes

Methine dyes contain as their essential structural feature one or more (in which case they are referred to as polymethines) methine (—CH=) groups. This group of dyes is capable of providing a wide range of shades. They are used extensively in colour photography, but their use on textiles is restricted largely to a few disperse and basic dyes. An example of a methine dye, illustrated in Fig. 2.5, is C.I. Basic Yellow 11 (**7**).

Nitro dyes

Nitro dyes are a small group of dyes which are of some importance, particularly as disperse dyes for polyester and as semi-permanent hair dyes.

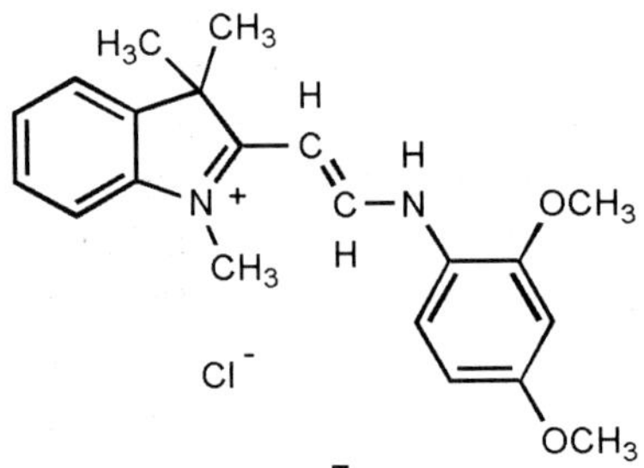

7

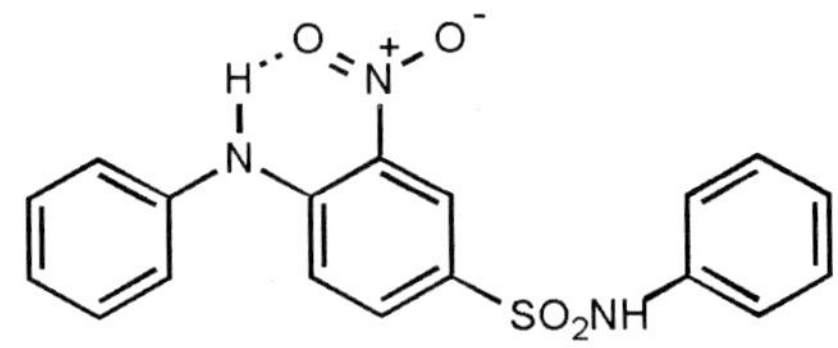

8

Fig. 2.5 Examples of methine, C.I. Basic Yellow 11 (**7**) and nitro, C.I. Disperse Yellow 42 (**8**), dyes.

These dyes, for example C.I. Disperse Yellow 42 (**8**) shown in Fig. 2.5, contain the nitro (NO_2) group as their essential structural feature. They provide bright yellow, orange and red shades, with good lightfastness but the colours tend to be rather weak.

Inorganic colorants

A notable feature of commercial synthetic dyes is that they are exclusively organic compounds. Pigments, in complete contrast, may be either inorganic or organic. Natural inorganic pigments, derived mainly from mineral sources, have been used as colorants from prehistoric times and a few, notably some iron oxides, remain of some significance today. However, these have been largely superseded by a range of technically superior synthetic inorganic pigments. These products include white pigments, by far the most important of which is titanium dioxide (TiO_2), as well as coloured pigments of a number of chemical types. The range of coloured inorganic pigments include oxides (e.g. of iron and chromium), sulphides (principally of cadmium) and chromates (principally of lead). These pigments are essentially ionic in nature with densely-packed crystal lattice structures. This gives the pigments a high refractive index as a consequence of which they are capable of providing the property of high opacity which is especially important for paint applications. The most important blue inorganic pigments are Prussian blue and ultramarine blue. These are long-established products which are structurally complex inorganic materials. Carbon black is by far the most important black pigment for a wide range of applications, and is also usually categorised as an inorganic pigment. A summary of the chemistry of inorganic pigments is shown in Table 2.2 and discussed in more detail in Chapter 6.

2.4.2 Application classification

To the textile dyer whose role it is to apply colour to a particular textile fibre, the classification of dyes according to the method of application is arguably of greater interest than the chemical classification. Dye and pigment molecules are carefully designed and constructed to ensure that they have a set of properties which are appropriate to their particular application. Obvious requirements for both types of colorant are that they must possess the desired colours (in terms of hue, strength and brightness) and an appropriate range of fastness properties. For textile applications, dye molecules are designed to be attracted strongly to the molecules of the fibre to which they are applied. The chemical and physical nature of the different types of textile fibres, both natural and synthetic, require that the dyes used have an appropriate set of chemical features to provide affinity for the particular fibre concerned. In contrast to textile dyes, pigments tend to be versatile colouring materials requiring less tailoring to individual applications. With pigments, therefore, classification according to application is relatively unimportant. Pigments are simply designed to resist dissolving in solvents with which they may come into contact in paint, printing ink or plastics applications.

The section which follows contains selected examples which illustrate how this 'molecular architecture' may be used for some important application classes of textile dyes and for organic pigments. The discussion of the chemical principles involved is illustrated with selected examples from the more important application classes of dyes, namely acid dyes, mordant dyes and premetallised dyes for protein fibres, direct dyes, reactive dyes and vat dyes for cellulosic fibres, disperse dyes for polyester and basic (cationic) dyes for acrylic fibres. For a fuller appreciation of the chemistry of textile dyeing, the reader is directed to the discussion of textile fibres in Chapter 3 and of dye application in Chapter 5. Similarly, Chapter 4 which deals with surface coatings should be read in conjunction with Chapter 6 for an appreciation of the chemical principles underlying pigment applications.

Dyes for protein fibres

Protein fibres, the best-known of which are wool and silk, are a group of natural fibres obtained mainly from animal sources. The polymer which makes up the structure of protein fibres (see Chapter 3) is a complex polypeptide containing a range of side-chains derived from the constituent aminoacids. Protein fibres may be dyed using a number of application classes of dyes, the most important of which are *acid*, *mordant*, *premetallised* and *reactive* dyes. The chemical structures of some selected dyes for protein fibres are illustrated in Fig. 2.6.

Acid dyes, exemplified by the structurally-related monoazo dyes C.I. Acid Red 1, (**9a**) and C.I. Acid Red 138 (**9b**), are commonly used to dye protein fibres (wool, silk) from an acid dyebath. They are also used to a considerable extent to dye polyamide fibres (nylon). A notable feature of acid dyes is the presence of one or more sulphonate (—SO_3^-) groups usually present as sodium (Na^+) salts. These groups have a dual role. Firstly, they provide solubility in water, the medium from which the dyes are applied to

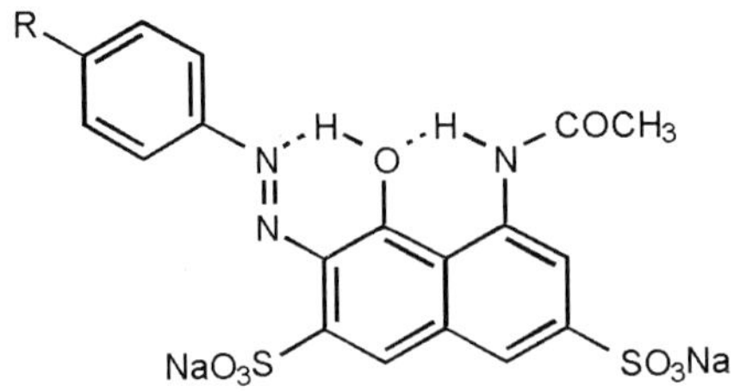

9a: R = H; **9b**: R = $C_{12}H_{25}$

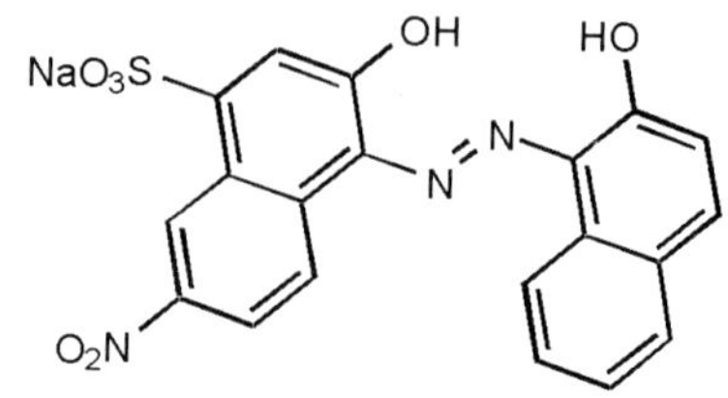

10

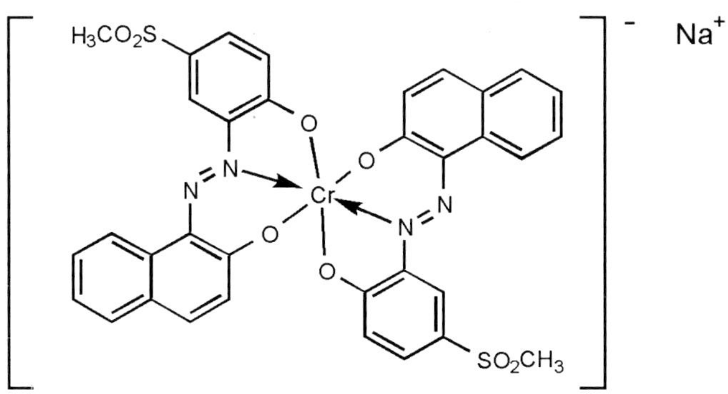

11

Fig. 2.6 Examples of dyes for protein fibres. C.I. Acid Red 1 (**9a**), C.I. Acid Red 138 (**9b**), C.I. Mordant Black 1 (**10**) and C.I. Acid Violet 78 (**11**).

the fibre. Secondly, they ensure that the dyes carry a negative charge (i.e. they are anionic). When strongly acidic conditions are used in the dyeing process, the protein molecules acquire a positive charge and this attracts the acid dye anions by ionic forces. As well as these ionic forces of attraction between dye and fibre molecules, van der Waals forces, dipolar forces and hydrogen bonding may also play a part in the affinity of acid dyes for protein fibres. Most acid dyes, especially yellow, oranges and reds, belong to the azo chemical class while blues and greens are often provided by carbonyl dyes, especially anthraquinones. Acid dyes may be conveniently classified as either acid-levelling or acid-milling types. In terms of size and shape, often an important consideration in the design of dye molecules, compound **9a** may be described as a small to medium-sized planar molecule. This allows the dye to penetrate easily into the fibre and also permits a degree of movement or migration within the fibre as the ionic bonds between the dye and the fibre are capable of breaking and then re-forming. This type of dye is termed *acid-levelling* since, as a result of its ability to migrate, it produces a level or uniform colour. However, as the dye is not very strongly bonded to the fibre, it may show only moderate fastness towards wet-treatments such as washing. *Acid-milling* dyes (such as **9b**) are significantly larger molecules than acid-levelling dyes and they show enhanced affinity for the fibre and hence improved fastness to washing, as a result of the contribution from an extensive set of van der Waals forces, dipolar forces and hydrogen bonding. Dye **9b** shows excellent resistance to washing as a result of the presence of the long alkyl chain substituent ($C_{12}H_{25}$) which is attracted to hydrophobic or non-polar parts of the protein fibre molecules by van der Waals forces. Because of their strong affinity, dyes of this type are known as *acid supermilling* dyes. A notable aspect of the structure of dyes **9a** and **9b** is the strong intramolecular hydrogen bonding which exists contained in 6-membered rings. This feature enhances the stability of the compounds and, in particular, confers good lightfastness properties. For this reason, intramolecular hydrogen bonding is a feature commonly encountered in the structures of a wide range of dyes and pigments. Note, for example that it is also a feature of the structures of disperse dyes **2** and **8**, reactive dyes **14a** and **14b**, and pigments **17** and **18**.

The ability of transition metal ions, and especially chromium (as Cr^{3+}), to form very stable metal complexes or coordination compounds may be used to produce dyeings on protein fibres with superior fastness properties, especially towards washing and light. There are two methods by which this is achicved, mordant dyes and premetallised dyes, differing significantly in application technology but involving similar chemistry.

Mordant dyes generally have the characteristics of acid dyes but with the ability to form in addition a stable complex with chromium. Most commonly, this takes the form of two hydroxy (OH) groups on either side of (*ortho* to) the azo group of a monoazo dye, as illustrated in the case of C.I. Mordant Black 1 (**10**) (Fig. 2.6). The dye is generally applied to the fibre as an acid dye and then treated with a source of chromium, commonly sodium or potassium dichromate. As a result, a chromium complex of the dye is formed within the fibre as illustrated in Fig. 2.7. A dye of this type acts as a tridentate ligand. This term implies that the dye molecule forms three bonds simultaneously with the chromium; in this case the chromium bonds with two oxygen atoms derived from the hydroxy groups and with one nitrogen atom of the azo group. The chromium atom in complexes of this type always bonds to six atoms and the complexes show octahedral geometry. It is not established with certainty how the remaining three valencies of chromium are satisfied in the mordant dyeing of protein fibres, although

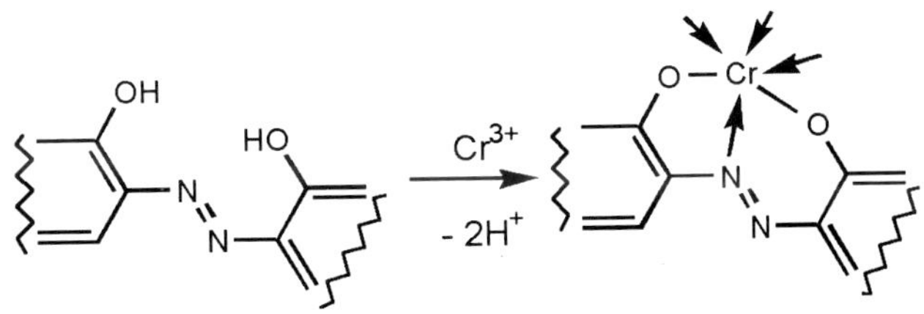

Fig. 2.7 The chrome mordanting process.

the possibilities include bonding with water molecules, or coordinating groups on the fibre or with a second dye molecule. The principal problem currently with chrome mordant dyes is environmental associated mainly with the undesirable presence of chromium residues in dyehouse effluent (see Chapter 12).

Premetallised dyes, as the name implies, are pre-formed metal complex dyes. They are usually six-coordinate complexes of chromium with octahedral geometry, as exhibited for example by C.I. Acid Violet 78 (**11**) in Fig. 2.6. Since in this case there are two azo dye molecules coordinated with one chromium atom, compound **11** is referred to as a 2:1 complex. Less widely used but also known are 1:1 complexes. Premetallised dyes of this type, like traditional acid dyes, are anionic in nature even though, as is the case with compound **11**, they may not contain sulphonate groups. They are applied to protein fibres as acid dyes and, because of the special stability of chromium complexes, provide dyeings with excellent lightfastness. Most chrome mordant and premetallised dyes are azo dyes, with one nitrogen of the azo group playing a part in complexing with the chromium.

Dyes for cellulosic fibres

Cellulosic fibres, the best-known of which is cotton, are natural fibres obtained mainly from plant sources. As the name implies, the principal component of these fibres (see Chapter 3) is cellulose, a high molar mass, linear polymeric sugar or polysaccharide. There are a number of application classes of dyes which may be used to dye cellulosic fibres. Three of the most important, *direct*, *vat* and *reactive* dyes, are considered in this section and selected examples of typical chemical structures are given in Fig. 2.8.

Direct dyes are a long-established group of dyes capable of direct application to cellulosic fibres. In some ways they show structural similarities to acid dyes for protein fibres. For example, they are anionic dyes as a result of the presence of sulphonate ($—SO_3^-$) groups. However, the role of the sulphonate groups in the case of direct dyes is simply to provide water-solubility. In contrast to the acid dyeing of protein fibres, ionic attraction to the fibre is not involved in the direct dyeing of cellulosic fibres. In fact, the anionic nature of the dyes can reduce affinity for the fibres because cellulosic fibres generally carry a small negative charge. For this reason, usually only as many sulphonate groups as are required to give adequate solubility in water are present in direct dyes and, in addition, the groups are distributed evenly throughout the molecules. Arguably the most important feature of direct dye molecules which influences their application properties is their size and shape. They are, in general, large molecules and long, narrow and planar in shape. Direct dyes show affinity for cellulose by a combination of van der Waals, dipolar and hydrogen bonding forces. Individually, these forces are rather weak. The long, thin and flat geometry of the molecules is essential to allow the dye molecules to align with the

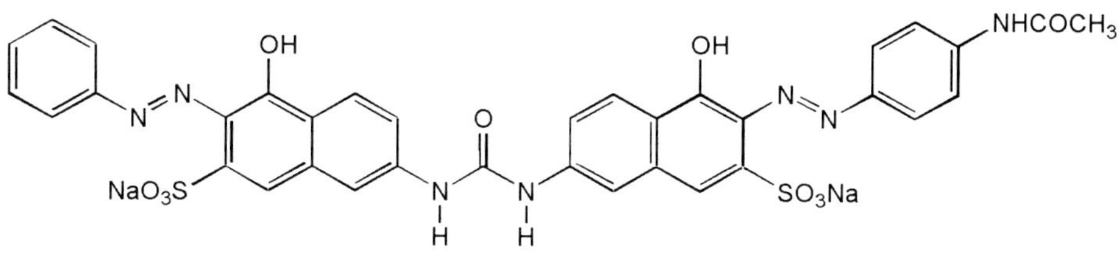

12

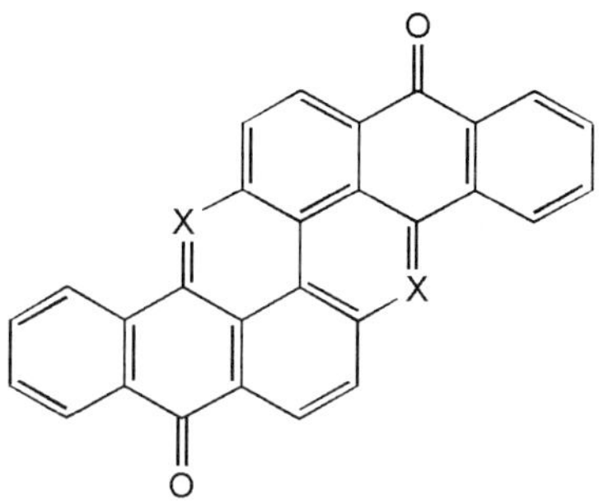

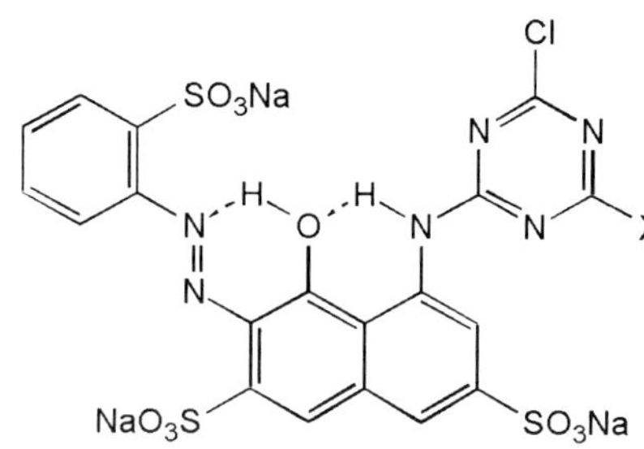

13a: X = CH; **13b**: X = N

14a: X = Cl; **14b**: X = NHPh

Fig. 2.8 Examples of dyes for cellulose fibres. C.I. Direct Orange 25 (**12**), C.I. Vat Orange 9 (pyranthrone) (**13a**), C.I. Vat Yellow 1 (flavanthrone) (**13b**), C.I. Reactive Red 1 (**14a**) and C.I. Reactive Red 3 (**14b**).

long polymeric cellulose fibre molecules and hence to maximise the overall effect of the combined set of intermolecular forces. Chemically, direct dyes, of which C.I. Direct Orange 25 (**12**) is a typical example, are almost invariably azo dyes, commonly containing two or more azo groups. The long, flat linear shape of compound **12** allows in principle groups such as the —OH, —NHCO— (amide), and —N=N— to form hydrogen bonds with hydroxy groups on cellulose as the dye molecule lines up with the polymer molecule. There are only two sulphonate groups in compound **12** and these are well separated from one another. This is sufficient to give adequate water-solubility for their application. It may also be argued that the sulphonate groups are on the opposite side of the molecule from groups which may be participating in hydrogen bonding with the fibre and this means that they will be pointing away from the cellulose molecule, thus minimising any negative charge repulsion effects.

Vat dyes are a group of totally water-insoluble dyes and in this respect they are essentially pigments. In fact, some vat dyes, after conversion to a suitable physical form, are used as pigments. For application to cellulosic fibres, vat dyes are converted to a water-soluble 'leuco' form by an alkaline reduction process. Commonly, this conversion is carried out using a reducing agent such as sodium dithionite ($Na_2S_2O_4$) in the presence of sodium hydroxide. In the 'leuco' form, the dyes are taken up by the cellulosic fibre. Subsequently, they are converted back by an oxidation reaction to the insoluble pigment. Hydrogen peroxide is commonly used as the oxidising agent for this process although atmospheric oxygen can also effect the oxidation under appropriate conditions. The pigment thus produced becomes trapped within the fibre and its insolubility gives rise to the excellent washfastness properties which are characteristic of vat dyes.

The chemistry of the reversible reduction/oxidation process involved in vat dyeing which is illustrated in Fig. 2.9 requires the presence in the vat dye of two carbonyl groups linked via a conjugated system. The carbonyl (C=O) groups are reduced under alkaline conditions to enolate (—O^-Na^+) groups which give the leuco form water-solubility. Vat dyes are thus exclusively of the carbonyl chemical class. Dyes of other chemical classes, including azo dyes, are inappropriate as vat dyes because they undergo a reduction which cannot be reversed. Vat dyes are generally large planar molecules, often containing multiple ring systems, to provide the leuco form of the dyes with some affinity for the cellulosic fibres. There is usually a notable absence of other functional groups in the molecules because such groups can be sensitive to the reduction and oxidation reactions, although halogen substituents (Cl, Br) are encountered on occasions. Pyranthrone (**13a**), C.I. Vat Orange 9, and flavanthrone (**13b**), C.I. Vat Yellow 1, are illustrated in Fig. 2.8 and provide examples respectively of carbocyclic and heterocyclic anthraquinone vat dyes. Indigo (**3**), C.I. Vat Blue 1, (Fig. 2.3) is a further example of a long-established vat dye, used commonly in the dyeing of denim fabric. Indigo was formerly obtained from natural sources, but for many years has been made exclusively by synthetic routes.

Arguably the most significant advance in textile dyeing technology this century has been the development of *reactive dyes*. Initially, reactive dyes were introduced for application to cellulosic fibres and this remains their most important use, but they were later adapted for application to protein

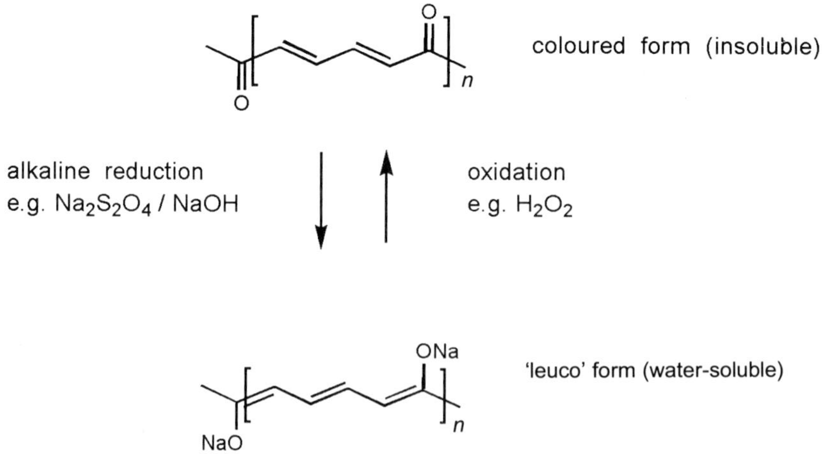

Fig. 2.9 The chemistry of vat dyeing.

fibres. These dyes, after application to the fibre, react chemically (as the name implies) to form a covalent bond between the dye and the fibre. Because of the strength of the covalent bond, reactive dyes show outstanding washfastness properties. Reactive dyes, such as the monoazo dyes (**14a**), C.I. Reactive Red 1, and (**14b**), C.I. Reactive Red 3, show many of the features of acid dyes. For example, they contain one or more sulphonate group to provide the water-solubility which is required for their application. The essential structural feature of reactive dyes is a group which is capable of reacting chemically with the fibre. In the case of dyes (**14a**) and (**14b**), the fibre-reactive group is the heterocyclic triazine ring, a six-membered aromatic ring with alternate carbon and nitrogen atoms, containing at least one chlorine substituent. This ring system is capable of reacting by nucleophilic substitution, a process in which a nucleophile (usually a negatively-charged species) displaces the chlorine (as the chloride anion) from the ring. Reactive dyes of this type (marketed as Procion dyes), are applied to cellulose under alkaline conditions. The presence of alkali causes the hydroxy (—OH) groups on the cellulose molecule to be converted into anions ($—O^-$). These so-called cellulosate anions are the nucleophiles which react with the dyes as illustrated in Fig. 2.10 to form a dye–fibre covalent bond.

Dichlorotriazinyl (Procion M) dyes, such as dye **14a**, are more reactive than the monochlorotriazinyl (Procion H) types, such as **14b**, and are capable therefore of application at lower temperatures. A feature of the chemistry of reactive dyes, also illustrated in Fig. 2.10, is that they undergo, to a certain extent, a hydrolysis reaction, a competing reaction involving reaction of the dye with hydroxide (OH^-) anions present in the aqueous alkaline dyebath. The hydrolysed dye which is formed is no longer capable of reacting with the fibre and so must be washed out of the fibre after dyeing

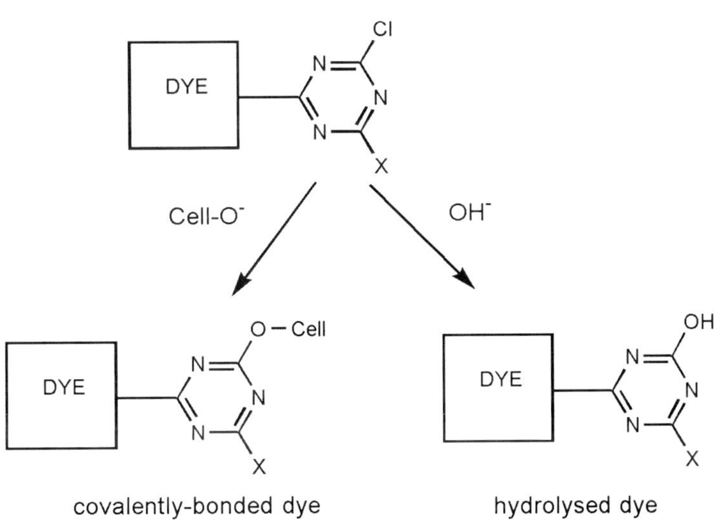

Fig. 2.10 The chemistry of reactive dyeing using chlorotriazinyl dyes.

is complete. The hydrolysis reaction is highly undesirable both on the grounds of cost-effectiveness and for environmental reasons, since hydrolysed dye will be found in the effluent, but as yet, in spite of considerable research effort, no reactive dyeing systems have been developed which eliminate the problem completely.

A range of other fibre-reactive groups are in use, including the vinylsulphones which react with cellulose under alkaline conditions by a nucleophilic addition process, the chemistry of which is illustrated in outline in Fig. 2.11. In this process, hydrolysis is also encountered in competition with the dye–fibre reaction.

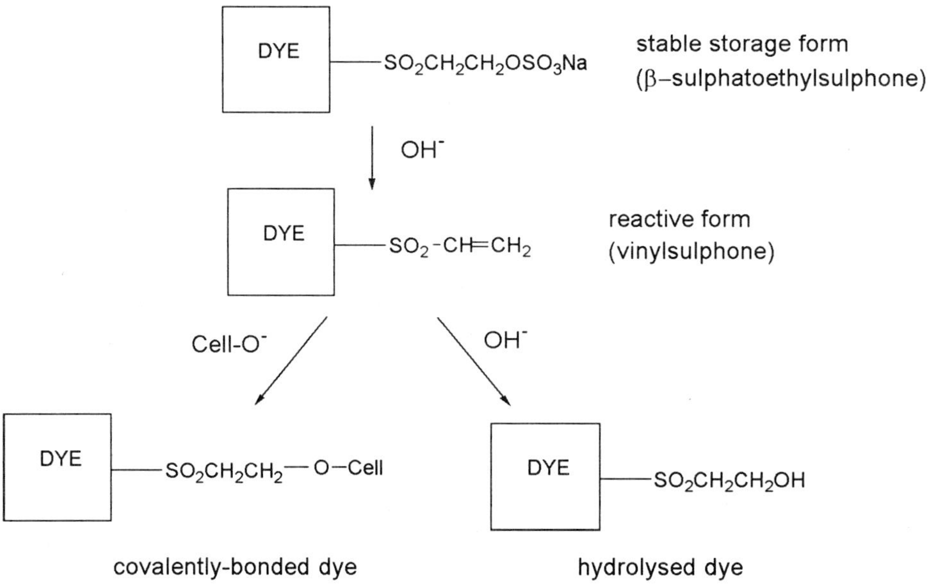

Fig. 2.11 The chemistry of reactive dyeing using vinylsulphone dyes.

Dyes for synthetic fibres

The three most important synthetic fibres are polyester, polyamides (nylon) and acrylic fibres. The structures of the polymers used in each of these fibres is discussed in Chapter 3. Polyester is dyed almost exclusively with the use of disperse dyes. Polyamide fibres may be coloured using acid dyes, the principles of which have been discussed previously in the 'Dyes for protein fibres' section, or with disperse dyes. Acrylic fibres are dyed mainly using basic (cationic) dyes. Examples of disperse and cationic dyes are illustrated in Fig. 2.12.

Disperse dyes are dyes of relatively low water-solubility applied as a fine dispersion in water. They are applied to relatively non-polar (hydrophobic) fibres, principally to polyester, but also to cellulose acetate and to a certain extent to polyamides. The most important disperse dyes are azo dyes, an

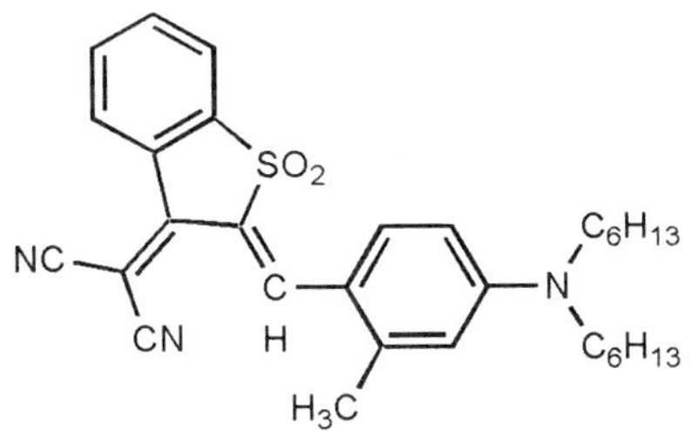

15

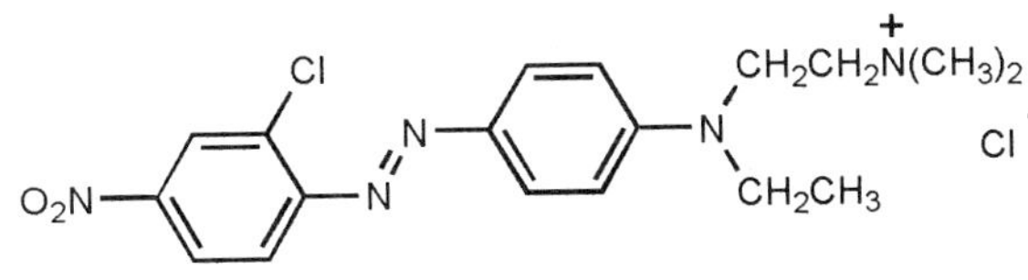

16

Fig. 2.12 Examples of dyes for synthetic fibres. C.I. Disperse Blue 354 (**15**) and C.I. Basic Red 18 (**16**).

example of which is compound **1** (Fig. 2.2), but the range of commercial products also contains representatives of the carbonyl **2** (Fig. 2.3), nitro **8** (Fig. 2.5) and methine **15** (C.I. Disperse Blue 354) chemical types. Polyester is a highly crystalline fibre consisting of tightly-packed, highly ordered polymer molecules. Disperse dyes are required to be relatively small, planar molecules to allow the dyes to penetrate between the polymer chains and into the bulk of the fibre. Polyester is a relatively non-polar or hydrophobic fibre so that, for compatibility, dyes containing ionic water-solubilising groups such as the sulphonate group are inappropriate. Disperse dyes are therefore non-ionic or neutral molecules which effectively dissolve in the polyester. In the solid solutions which form the dye–fibre affinity is generally considered to involve a combination of van der Waals and dipolar forces and hydrogen bonding. These dyes do, however, require the presence of a number of polar groups, such as the sulphone (—SO_2—), amino and cyano groups in dye **15**. One role of these groups is to enhance affinity as a result of dipolar intermolecular forces with the ester groups of the polyester molecule. A second function of the polar groups is to provide a degree of water-solubility at the high temperatures at which the dyes are applied. Disperse dyes can therefore be considered as compromise molecules possessing a balanced degree of polar (hydrophobic) and non-polar (hydrophilic) character, similar to that of the polyester molecule.

A consequence of the fact that disperse dyes are relatively small, non-polar molecules is that they may have a tendency to be volatile, and prone to sublimation out of the fibre at high temperatures. This can lead to a loss of colour and the possibility of staining adjacent fabrics when dyed polyester is subjected to high temperatures, e.g., in heat setting and ironing. Fastness to sublimation may be enhanced by increasing the size and/or the polarity of the dye molecules. However, as a consequence, these larger, more polar dye molecules will require more forcing conditions, such as high temperatures and pressures, to enable the dyes to penetrate into the fibre.

The principal application class used to dye acrylic fibres is cationic dyes, such as the azo dye **16**, C.I. Basic Red 18 (Fig. 2.12), the arylcarbonium ion (triphenylmethane) dye **6** (Fig. 2.5) and the methine derivative **7** (Fig. 2.6).

These dyes are classified by the *Colour Index* as basic dyes, a term which was derived from their use, now largely obsolete, to dye protein fibres from a basic or alkaline dyebath. As the name implies, the dyes are coloured cationic species, generally as a result of the presence of positively-charged quaternary nitrogen atoms ($—NR_3^+$, or $=NR_2^+$). These groups serve two purposes. Firstly, they provide water-solubility due to their ionic character. Secondly, they provide affinity for the acrylic fibres as a result of ionic attraction between the dye cations and the anionic groups ($—SO_3^-$ and $—CO_2^-$) which are present in acrylic fibre molecules. It may be seen that the means of attachment of cationic dyes to acrylic fibres is in a sense the converse of that involved in the acid dyeing of protein fibres in which dye anions are attracted to cationic sites on the fibre.

Pigments

Pigments are colouring materials, used mainly in paints, printing inks and plastics, which are required to be completely insoluble in their application medium. Both organic and inorganic types are used. The technical properties of organic pigments are dependent not only on the detail of their molecular structure but also on the way the molecules pack together in their crystal lattice. Since organic molecules will usually tend to exhibit some tendency to dissolve in organic solvents, organic pigments frequently contain structural features designed to enhance their resistance towards dissolving in solvents. One way of improving solvent-resistance is to increase the molecular size of the pigment. Secondly, the amide (—NHCO—) group is commonly a prominent feature of the chemical structures of organic pigments. Its presence frequently enhances fastness to solvents, light and heat, a property which may be attributed to the ability of the amide group to participate in strong dipolar interactions and in hydrogen bonding, both intramolecular and intermolecular. These features are illustrated in the structures of the red pigments shown in Fig. 2.13.

C.I. Pigment Red 3 (**17**), a monoazo pigment, shows good lightfastness as a result of extensive intramolecular hydrogen bonding in six-membered rings. However, the molecule is rather small and so it tends to dissolve readily in organic solvents. The pigment may therefore only be used in applications, such as in water-based paints, where it does not come into contact with powerful organic solvent. In contrast, C.I. Pigment Red 183 (**18**) shows excellent solvent resistance as a result of its larger molecular size and, in particular, to the presence of the amide groups which provide strong intermolecular forces of attraction in the crystal lattice. One of the clearest illustrations of the effect of strong intermolecular association on the fastness properties of organic pigments is provided by quinacridone (**4**), C.I. Pigment Violet 19 (see Fig. 2.3). Although the quinacridone molecules are fairly small, they are strongly attracted to one another in their crystal lattice by a combination of hydrogen bonding and dipolar forces as illustrated in Fig. 2.14. As a result, quinacridone pigments show outstanding resistance

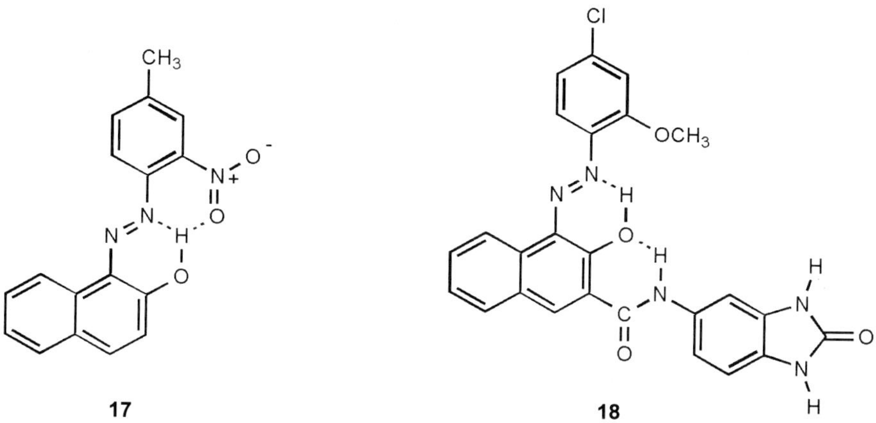

17

18

Fig. 2.13 Examples of organic pigments. C.I. Pigment Red 3 (**17**) and C.I. Pigment Red 183 (**18**).

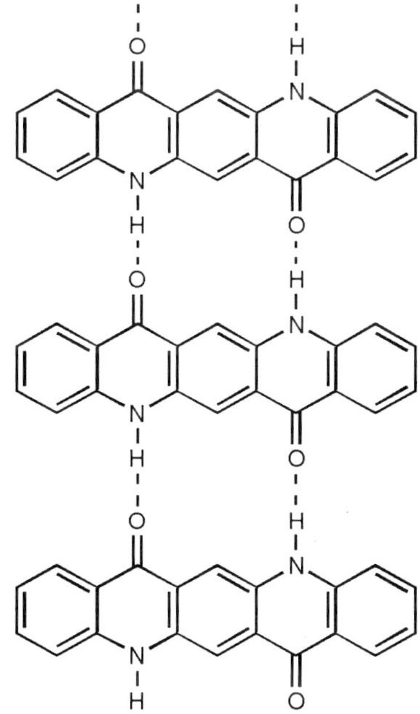

Fig. 2.14 Intermolecular hydrogen bonding in the crystal lattice of quinacridone (**4**).

towards heat, light and solvents. The structures and properties of pigments are considered in further detail in Chapter 6.

Bibliography

Organic Chemistry in Colour, P.F. Gordon and P. Gregory, Springer-Verlag, New York (1983).

Colour Chemistry, R.L.M. Allen, Nelson, London (1971).

Colour Index, Vols 1–9, Society of Dyers and Colourists, Bradford, 3rd revision (1988).

Color Chemistry, H. Zollinger, VCH, Cambridge (1991).

Colorants and Auxiliaries, Vol. 1, J. Shore (ed.), Society of Dyers and Colourists, Bradford (1990).

Dyes and their Intermediates, E.N. Abrahart, Edward Arnold, London (1977).

Fundamentals of the Chemistry and Application of Dyes, P. Rys and H. Zollinger, John Wiley, London (1972).

*The Chemistry and Application of Dye*s, D.R. Waring (ed.), Plenum Press, London (1990).

The Chemistry of Synthetic Dyes, Vols 1–8, K. Venkataraman (ed.), Academic Press, London (1952–1978).

Chapter 3

Introduction to Substrates

3.1 Introduction to polymers

There are generally two ways in which materials are coloured. One method involves the application of colour directly to a substrate. Substrates coloured in this way include most notably textiles, but also a variety of others, such as paper and leather. They are usually coloured for aesthetic reasons or for ease of identification. Alternatively, the colour may be incorporated into a medium (usually liquid), which is then applied as a surface coating, for example, a paint or printing ink. In this case, the coating may additionally have a protective effect and indeed this is often considered to be the principal function of paints.

Despite the diverse nature of these substrates and surface coatings, the majority have one common property, in that they are composed very largely of polymers: these are compounds which consist of giant molecules (macromolecules) containing repeating structural units many, many times over. Most materials to which colour is applied directly consist substantially of polymers. Surface coatings prior to use are liquids containing polymeric resins and are required to set or dry after application.

To have even a cursory appreciation of polymeric materials, one must understand something of the behaviour of their constituent macromolecules. One important feature which sets them aside from smaller molecules is that they cannot be assigned a precise molar mass, for there is a distribution of molecular chain lengths. For this reason, it is customary to characterise a polymer by at least two average molar masses, the types of average molar mass determined depending on the methods used. If in effect the *number* of macromolecules of each size is counted, a number average molar mass, M_n, is determined. By contrast, if the method depends on the *size* of the constituent macromolecules rather than on their number, a weight average molar mass, M_w, is determined. The ratio of the two, M_w/M_n, provides an indication of the breadth of the distribution of molar masses. A typical distribution is illustrated in Fig. 3.1. Another useful parameter is the degree of polymerisation (D.P.), which represents the average number of monomeric units in a polymeric chain.

Although polymeric molecules are formed from repeating monomeric units, they are not necessarily simple chains, giving rise to linear polymers.

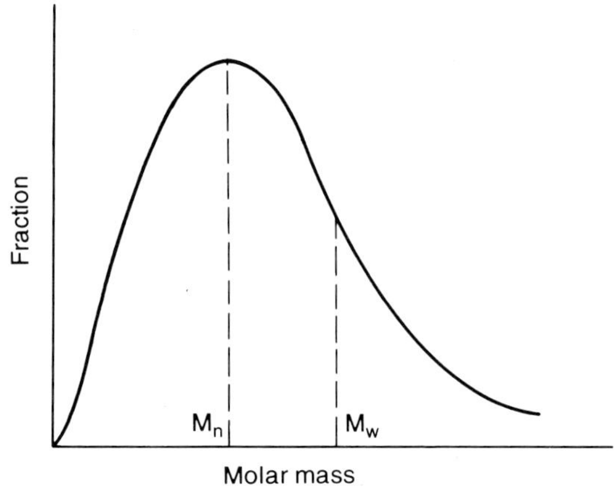

Fig. 3.1 A typical distribution of molar masses for a synthetic polymer.

As Fig. 3.2 illustrates, the polymer chains may be branched or even completely cross-linked. Certainly, many textile fibres, for example, are composed of linear polymers, but cross-linking is often crucial for the efficacy of surface coatings. On the other hand, the keratin chains in wool are substantially cross-linked and the polymer chains in some surface coatings, e.g. nitrocellulose, are linear.

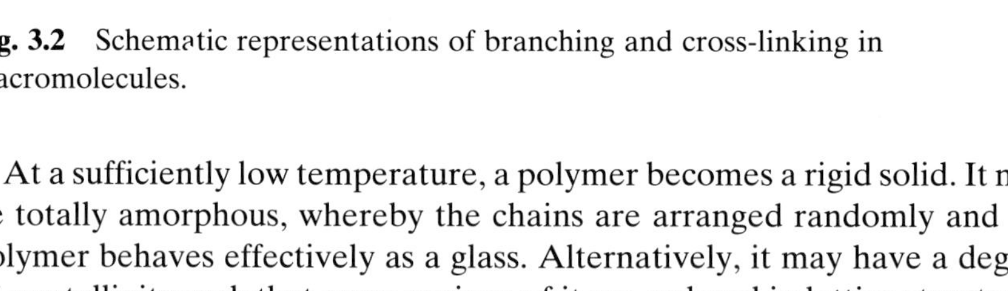

Fig. 3.2 Schematic representations of branching and cross-linking in macromolecules.

At a sufficiently low temperature, a polymer becomes a rigid solid. It may be totally amorphous, whereby the chains are arranged randomly and the polymer behaves effectively as a glass. Alternatively, it may have a degree of crystallinity such that some regions of it are ordered in lattice structures. Perfectly crystalline polymers are not found in practice. As the temperature of a polymer rises, a critical temperature is reached, at which the polymer softens and becomes like a rubber. At this temperature, known as the glass transition temperature, T_g, the chains in the amorphous regions of the polymer begin to acquire sufficient thermal energy that they can move independently of each other. At a higher temperature, T_m, the polymer melts.

Polymeric materials can be grouped into three categories: fibres, plastics and elastomers, although the division between the categories is by no means a sharp one. Plastics and elastomers are discussed in section 3.5.

3.1.1 *Fibres*

Many of the substrates to which colour is applied are fibrous, such as textiles and paper. Fibres are most readily distinguished by their fineness and their very high length-to-breadth ratio (at least 100:1 and often very much greater). The constituent macromolecules are substantially linear. One of the most significant features of a fibre, however, is that the macromolecules within it have a high degree of orientation along its length, even in the amorphous regions. Orientation enhances intermolecular attraction between adjacent chains and so helps to prevent them from sliding over one another. However, some degree of entanglement is also necessary. The extent of entanglement increases with the length of the polymer chain and the optimum molar mass to produce fibres of desirable strength normally lies within 10 000 and 50 000 g mol^{-1}. In wool, slippage of the polymer chains is substantially prevented by covalent links between them, so that effectively the polymeric structure is heavily cross-linked.

In the crystalline regions of a fibre, slippage is prevented by the ordering of weak forces between adjacent chains. However, within these crystalline regions, only parts of each macromolecule are accommodated. Indeed, different sections of a macromolecule may lie in different crystalline regions, as illustrated schematically in Fig. 3.3.

Some fibres also have elastomeric properties (section 3.5.3). The macromolecular chains of many of these elastomeric (Spandex) fibres consist of long segments of a very flexible, amorphous nature, joined together by short non-flexible segments, which are crystalline.

Fig. 3.3 The amorphous and crystalline regions of a fibre.

3.2 Textiles

By far the commonest fibrous substrates to which colour is applied are textiles, especially those used for clothing, carpets and furnishings. Many fibrous materials exist naturally or can be manufactured, but only a few of these have the necessary combination of properties to be suitable as textiles. One essential feature is sufficiently high tenacity, to provide both strength during mechanical processing and durability during use. Tenacity refers to the tensile stress of a fibre, expressed as force per unit linear density of the unstretched fibre. Textile fibres should often also be flexible, so that they can be bent without breaking, and should possess some degree of softness and elasticity too. Many textiles, particularly from natural sources, have a significant moisture regain, which promotes comfort in wear. Moisture regain is the moisture content, expressed as a percentage of the mass of dried material. Table 3.1 lists some values, determined under the conventional conditions of 65% relative humidity and 20°C.

Table 3.1 Moisture regains of fibres at 65% relative humidity and 20°C.

Fibre	Moisture region (%)
Cotton	7–8
Viscose rayon	12–14
Silk	10
Wool	15–18
Acrylic	1–2
Nylon 6, 6	4.1
Polyester	0.4

The extent of crystallinity within textile fibres has important implications. The crystalline regions confer strength and rigidity and the amorphous regions confer flexibility. Moreover, the balance between crystallinity and amorphism has a strong influence on the application of dyes. Molecules from a dyebath may enter into the body of each fibre through the relatively open amorphous regions, but the closely packed crystalline regions do not allow ready access. In an analogous manner, accessibility is much greater above the glass transition temperature, T_g, of a textile, when the macromolecular chains in the fibres begin to acquire independent motion, so that there is greater space between the chains. This property is exploited in dyeing many synthetic textile fibres.

Textile fibres are often classified in terms of their origin, and an outline of the classification is presented in Fig. 3.4. For the application of dyes, a simpler classification into three broad categories often suffices:

(a) *cellulosic fibres* e.g. cotton, viscose rayon and flax (linen);
(b) *protein fibres* e.g. wool and silk;
(c) *synthetic and cellulose acetate fibres* e.g. polyester, polyamide, acrylic and cellulose triacetate.

The nature of the macromolecules comprising some of these fibres is illustrated in Fig. 3.5.

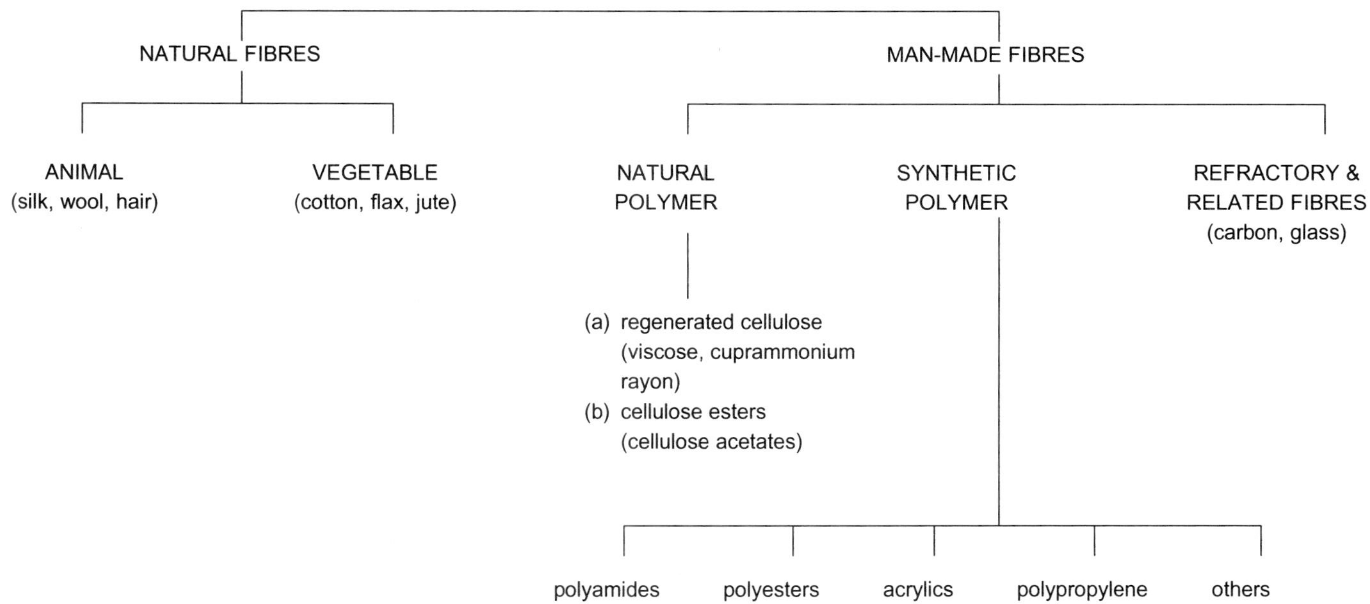

Fig. 3.4 Classification of textile fibres.

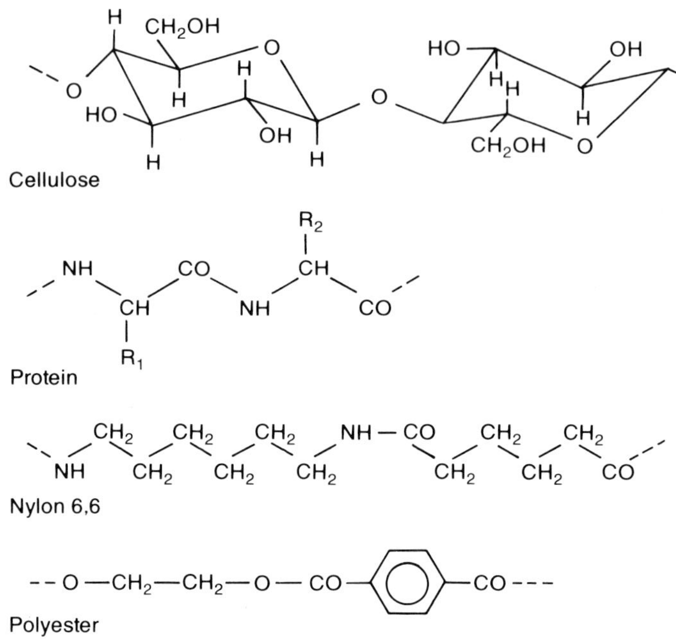

Fig. 3.5 Structures of macromolecules in a variety of fibres.

3.2.1 *Cellulosic fibres*

Cellulosic fibres are all chiefly composed of the polymeric carbohydrate, cellulose. The polymer is formed from molecules of the monosaccharide, β-D-glucose (**1**) linked through carbon atoms in the 1- and 4-positions (Fig. 3.5). The presence of so many hydroxyl groups in cellulose makes cellulosic fibres highly hydrophilic: they are readily wetted by water. Moreover, the presence of these hydroxyl groups enables the polymeric chains to be linked by hydrogen bonds; and if the chains are sufficiently long, they may consequently form highly ordered, crystalline structures.

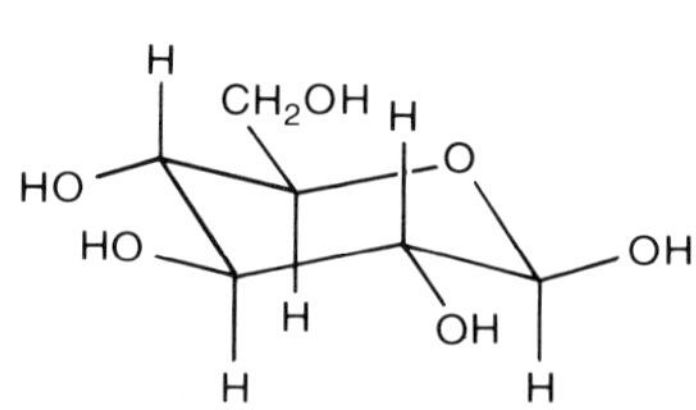

1

The degree of polymerisation varies depending on the type of fibre. In cotton, the number of glucose units in each molecule of cellulose is about 10 000 and in flax as high as 36 000, but in viscose rayon it is only about 400.

Cellulosic fibres, such as those constituting viscose, which are regenerated from natural sources, generally possess shorter macromolecular chains and lower crystallinity than do natural fibres.

In addition, cellulose has a number of different crystalline forms. Cellulose I is the form which occurs in cotton, flax and other natural cellulosic fibres; Cellulose II is the form occurring in viscose and other types of regenerated cellulosic fibre. Cellulose II is also the form present in mercerised cotton. This is cotton which has been swollen under tension in strong solution of sodium hydroxide, to develop a permanent lustre.

The physical structures of cellulosic fibres vary considerably. Viscose fibres are generally made from wood pulp and being regenerated have a relatively simple structure. Each fibre consists of two distinct parts, a skin and a core. The polymeric structure in the skin is more highly ordered than in the core. By contrast, cotton fibres, like many natural fibres, have a complex structure, as illustrated in Fig. 3.6. Each fibre has a ribbon-like structure, comprising several concentric layers and a central channel, called the lumen. The external layer is a thin, hydrophobic film, consisting primarily of waxes, and is removed during processing of the fibres. Immediately beneath this layer is a thin primary wall of cellulosic fibrils, arranged in a criss-cross manner. The bulk of the fibre (about 90%) is formed by the secondary wall, which consists essentially of three layers. The first two, S1 and S2, consist of helical cellulosic fibrils of different pitch. In some cotton fibres, there is also a thin third layer, S3, next to the lumen; this layer consists of mineral salts and proteins.

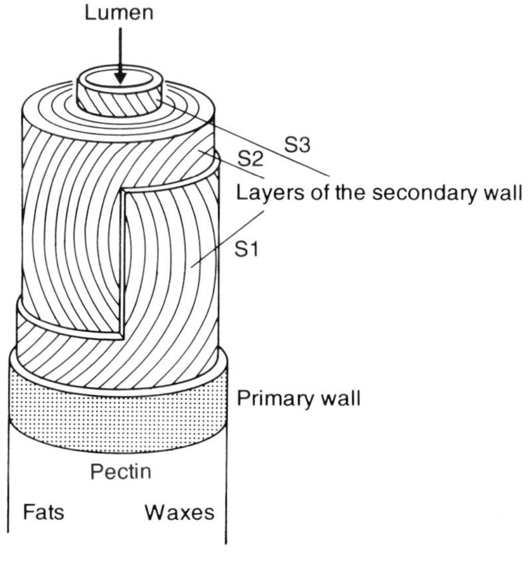

Fig. 3.6 Schematic representation of the physical structure of a cotton fibre.

3.2.2 *Protein fibres*

The most prominent protein textile fibres are wool and silk. Protein molecules are polypeptides, derived from different naturally occurring α-aminoacids, most of which can be represented by the general formula (**2**).

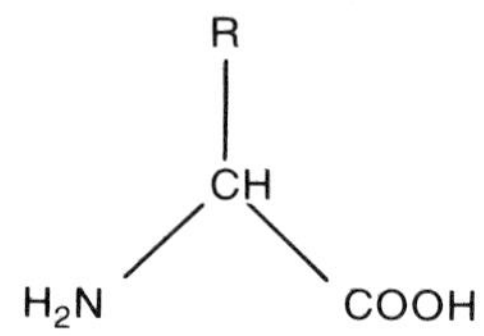

2

Amino-acids are, therefore, distinguished from each other by the nature of the R-group, exceptions being proline and cystine (Fig. 3.7).

Cystine plays a key role in the properties of wool but is absent from silk. Such linking of the aminoacids gives the polypeptide structure illustrated in Fig. 3.5. Thus, each protein chain contains a primary amine —(NH_2) group at one end of a protein chain and a carboxylic acid (—COOH) group at the other end.

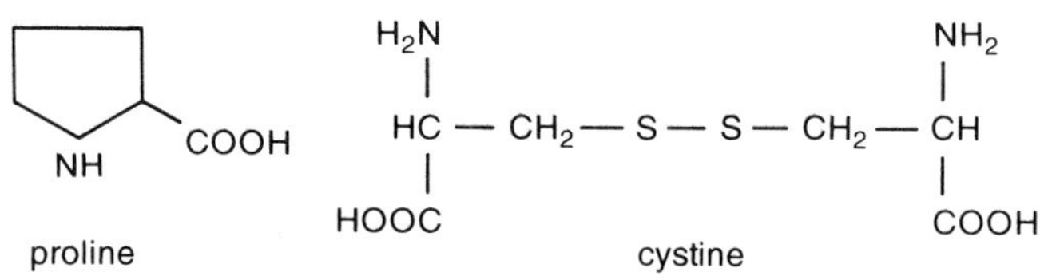

Fig. 3.7 Chemical formulae of aminoacids, proline and cystine.

Wool has a particularly complex physical structure (Fig. 3.8) and, although its predominant source is sheep, wool from other animals such as goats and llamas is also marketed. The external part of the wool fibre is the scaly cuticle, which is particularly important for the application of dyes. Its outer layer, the epicuticle, is hydrophobic (not readily wetted by water), but being extremely thin ($< 100\,\mu m$) it is damaged during wool processing. The extent of damage can have a large influence on the ease of dyeing.

The cortex forms the bulk of the fibre (*c.* 90%). Cortical cells are spindle-shaped and are composed of ordered rod-like microfibrils, stacked in bundles. These bundles are packed in an amorphous matrix, which forms the rest of the cortex.

The proteins comprising wool are keratins. Some of the constituent R-groups (see Fig. 3.5) themselves terminate in primary amine groups (*c.* 10%) and others in carboxylic acid groups (*c.* 18%). Thus, these groups are situated along each protein chain as well as at the chain ends. Both are capable of some degree of ionisation depending on the pH of the envir-

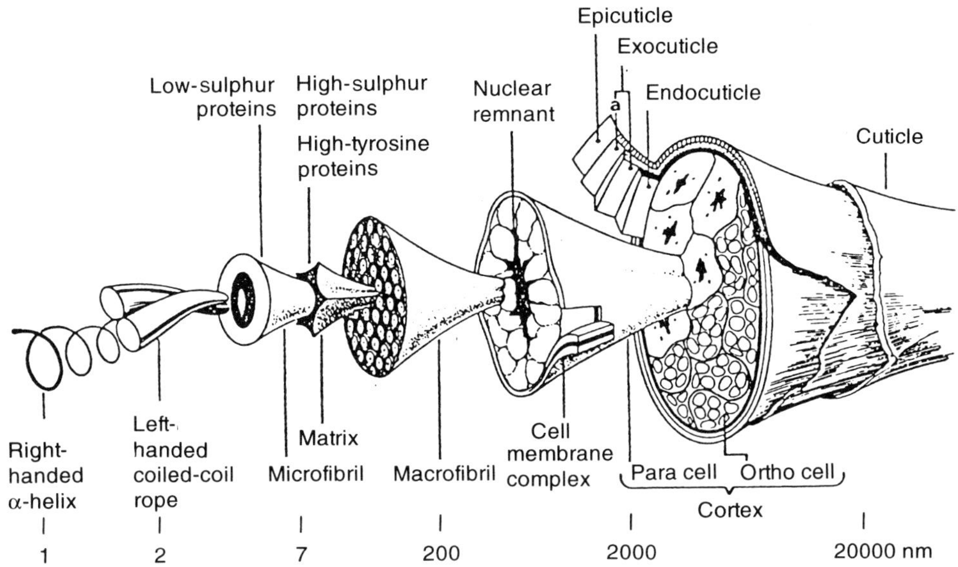

Fig. 3.8 Schematic representation of the physical structure of a wool fibre

onment of the wool. At low pH, the carboxyl groups are hardly ionised at all, but the amino groups are almost completely protonated to form —NH_3^+. Thus, the keratin chains have an overall positive charge at low pH. As the pH is raised, the carboxyl groups become progressively more ionised to —COO^-, whilst the amino groups lose their ionic character. At neutral pH, for example, the keratin chains have an overall negative character. At higher pHs, however, wool tends to be degraded. The pH at which the change from net positive to net negative character occurs (called the iso-electric point) is 4.5–5.0. The overall charge on wool fibres has important implications for dyeing.

The amino and carboxyl groups on some of the side-chains are also the basis for salt-links (**3**) between adjacent keratin chains. However, the most important cross-chain link is provided by the aminoacid, cystine (**4**), for the link consists of covalent bonds:

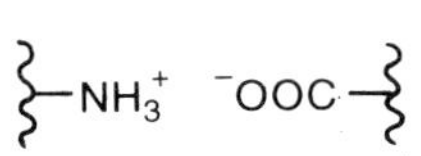

3

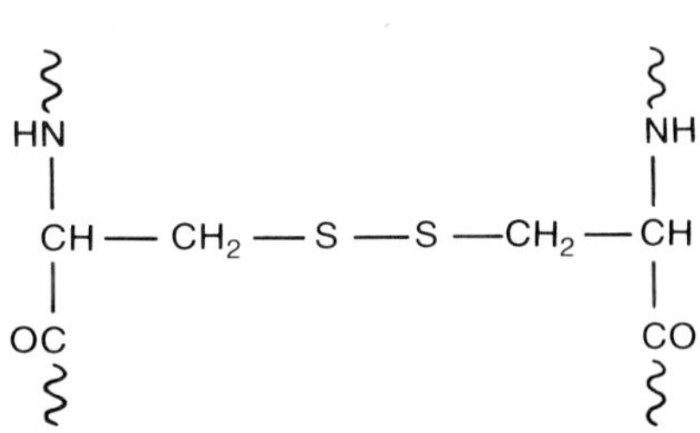

4

Hydrogen bonds can also form links between adjacent chains but are particularly important in maintaining the coiled structure of the keratin molecules in the microfibrils.

The highly heterogeneous nature of wool fibres also affects the application of dyes. The variation in the nature of the scales which surround each fibre can be particularly important. A skittery dyeing may result; some fibres will be far more readily dyed than others, and the dyed material will have a speckled appearance. In addition, the exposed tips of the wool fibres on a sheep are more susceptible than the roots to degradation through weathering and exposure to light. Consequently, the tips are more readily dyed than the roots, and the effect is known as tippy dyeing. When a mixture of dyes is applied, the colours of the tips and the roots may be different.

Raw silk is a secretion of the silk-worm, *Bombyx mori*. It consists of pairs of fibroin filaments joined together by sericin gum. Both fibroin and sericin are proteins. The gum is removed by treatment at *c*. 95°C with soap solution, often with added alkali. Thus, silk appears more resistant than wool to alkaline treatments. When the gum has been removed, the individual fibres have the appearance of structureless threads, although they are composed of a large number of very fine fibrils. The process of degumming brings out the soft qualities of silk.

Fibroin contains *c*. 16 aminoacids. It differs from wool in containing no cystine residues to link adjacent protein chains. However, it contains few bulky R-groups, so that the protein chains can approach one another sufficiently closely to form hydrogen bonds.

3.2.3 Cellulose acetate fibres

Cellulose acetate fibres contain acetylated derivatives of natural cellulose. They can be grouped into two categories: triacetate, which is completely acetylated, and secondary acetate, which is about 80% acetylated. Secondary acetate is obtained by partial hydrolysis of triacetate so is structurally a copolymer of glucose units each containing two or three acetate groups. Both types of acetate polymer are linear, but the bulky nature of the acetate side groups prevents close approach of the polymeric chains. Consequently, the crystallinity of both types of acetate fibre is low. The heterogeneous nature of the chains in secondary acetate reduces crystallinity even further. In addition, the fibres are hydrophobic, for the acetate groups are less polar than hydroxyl groups.

3.2.4 Synthetic fibres

Synthetic fibres comprise a wide variety of chemical types. The commonest include polyamides, polyesters, acrylics and polypropylene, but high-performance fibres, developed more recently, are assuming increasing commercial importance. However, despite their chemical variety, synthetic fibres are all produced in a highly crystalline form, arising from a high degree of alignment between adjacent macromolecular chains. Such high crystallinity is obtained as a result of drawing the freshly spun fibres. Drawing involves hot-stretching of the yarn to as much as five times its original length, with a concomitant reduction in cross-section. The process confers higher tenacities than those possessed by natural fibres, but reduces accessibility to dye.

Polyamides are distinguished by the presence of amide, —CO—NH—, groups along the macromolecular chains. They include all the synthetic nylons (and, strictly, proteins as well). Probably the best known is Nylon 6,6 (Fig. 3.5), a condensate of adipic acid, HOOC—$(CH_2)_4$—COOH, with hexamethylene diamine, H_2N—$(CH_2)_6$—NH_2. The notation 6,6 denotes the number of carbon atoms lying between adjacent amide groups in a chain. Nylon 6,6 is said to be dyadic, in that it contains two moieties, derived from a dibasic acid and a diamine, respectively. Nylon 6, however, is monadic, in that it is derived from one monomer, caprolactam (**5**):

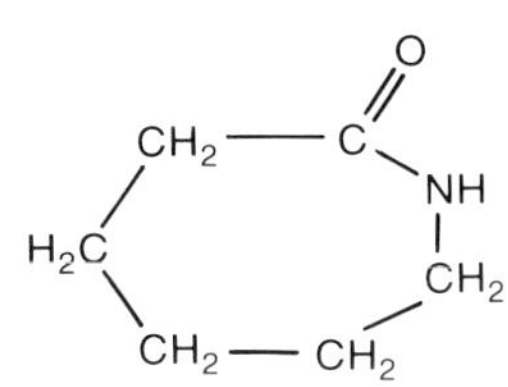

5

The commonest polyester is polyethylene terephthalate (Fig. 3.5) and is indeed often simply called, polyester. It is produced from the reaction of ethylene glycol, HO—CH_2—CH_2—OH, with either terephthalic acid (**6**):

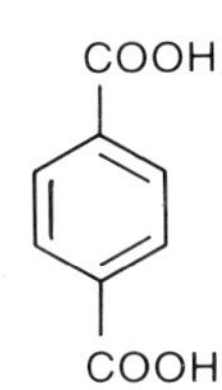

6

or its dimethyl ester. Even amongst synthetic fibres it is noted for its very high crystallinity and tenacity. In addition, it possesses a particularly low moisture regain. Some forms of polyester contain ionic comonomers, designed to attract ionic dyes of the opposite charge.

Acrylic fibres are produced by the polymerisation of acrylonitrile, CH_2=CH—CN, together with a comonomer. Pure polyacrylonitrile has a number of undesirable properties, including a high degree of crystallinity and a high T_g, which render dyeing difficult. Thus, commercial samples are copolymers. They are classified as acrylic, containing not more than 15% comonomer, and modacrylic, containing between 15% and 65% comonomer. The comonomers modify the physical properties of the polymer, by disrupting its crystallinity and so lowering its T_g. The application of dyes is consequently facilitated. Many comonomers are neutral, such as vinyl acetate, but acidic comonomers are also extensively used to provide anionic sites, to which cationic dyes are attracted. In addition, acidic groups, such as sulphates and sulphonates, are present at the polymer chain-ends, by virtue of the redox initiators used in the original polymerisation process. Basic comonomers have been employed as well, to attract anionic dyes, but nowadays these comonomers are seldom used.

Polypropylene is a polymeric hydrocarbon. To be suitable for textile fibres, the isotactic form (**7**) is required, in which the methyl substituents are situated on the same side of each polymer chain. It is both highly crystalline and highly hydrophobic and is consequently very difficult to dye. Even more difficult to dye are the more modern high-performance fibres because of their extreme crystallinity.

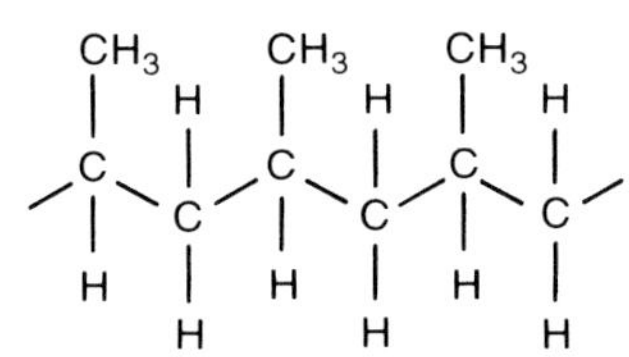

7

3.3 Paper

Paper is primarily a cellulosic material. It consists of small fibres formed into a sheet from a dilute aqueous suspension flowing onto a fine screen. Paper comes in a wide variety of types, such as writing and printing papers, art papers, tissues and packaging papers. Most of the content of paper is normally derived from wood. The rest consists primarily of fillers, such as china clay, chalk or titanium dioxide, to promote opacity.

Conversion of wood to pulp is achieved by a mechanical or chemical process or a combination of both. In a mechanical process, fibres are separated from the wood matrix by solely mechanical means. The resulting pulp contains, in addition to cellulosic material, significant contents of lignin and hemicellulose. Lignins from coniferous trees are complex polymers derived from coniferyl alcohol. Lignins from deciduous trees are polymers derived from both coniferyl alcohol and sinapyl alcohol (Fig. 3.9).

Both types of lignin can impair the strength of a paper and its stability to light. Hemicelluloses are complex polymers of sugars and related compounds. Pulps produced by chemical processes contain far less lignin and

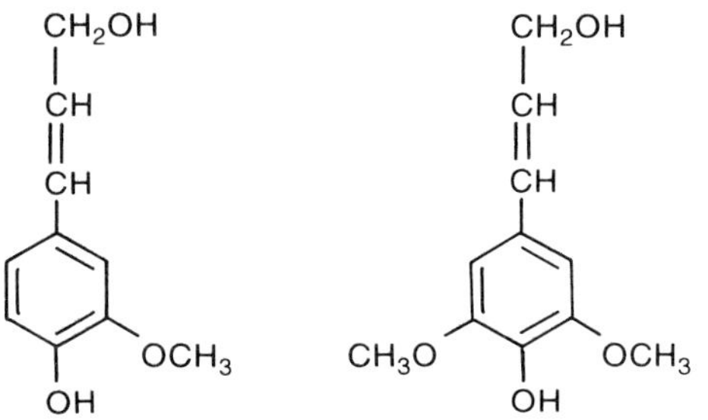

Coniferyl alcohol

Sinapyl alcohol

Fig. 3.9 The monomeric units in lignin.

hemicellulose and give rise to papers of superior quality. Papers produced by both types of process are generally subsequently bleached.

A number of additives are often present in paper, to achieve a range of desired properties. In addition to fillers, these additives include colorants and fluorescent brightening agents to enhance appearance, resins to promote wet strength and special coatings to provide a smooth surface for printing.

3.4 Leather

Leather is obtained from animal hides and skins and has many different applications, such as for shoes, clothing, upholstery and even book-binding. An animal skin consists essentially of three layers:

- epidermis (outer layer)
- corium (from which leather is produced)
- subcutis

The corium, in turn, is divided into two layers: a papillary layer and a bulkier reticular layer. The primary constituent of the corium is the protein, collagen, a material made up of elementary fibres, consisting of up to 1000 fibrils of *c.* 100 μm in cross-section. After the skin has been removed from the carcass, it has to be preserved to avoid putrefaction. It is then transported to the tannery, where a variety of preparative processes are necessary before tanning agents can be applied.

Tanning is a treatment designed to give leather a much greater stability during processing and use. The collagen chains are stabilised by cross-linking. By far the commonest treatment, chrome tanning, is based on the combination of chromium (III) complexes with carboxylic acid groups in adjacent collagen chains, and the cross-links formed make the leather particularly stable. The complexes are derived from basic chromium (III) sulphate, $Cr_2(OH)_2(SO_4)_2$. The activity of a chromium complex is governed by the number of hydroxyl groups it contains, often called the degree of basicity. At the start of tanning, chromium complexes with lower basicity are preferred, since they can penetrate the hide more readily. Later, the basicity is increased by addition of alkali, in order to provide better cross-linking of the collagen chains and so improve the tanning effect. Another important control is to modify the chromium complexes by means of carboxylic acid salts, such as formates, citrates or phthalates. This masking effect, as it is called, renders tanning agents less active.

Vegetable tanning, which is far less widely utilised than chrome tanning, uses natural tannins consisting of polymeric chains containing phenolic groups. These groups cross-link collagen chains, through formation of hydrogen bonds with the peptide linkages. There is also some interaction between amino groups in the collagen chains and carboxylic acid groups in the tannin. Many synthetic tannins function by the same mechanism.

Nowadays, it is common in a tannery to adopt the same tanning procedures initially for all types of leather. The different properties required of different leather products can then be obtained by subsequent retannage. Indeed, retanning is one of the most effective means of governing the properties of the final leather product. Tanning or retanning is followed by fat-liquoring, a process which imparts softness and suppleness to the leather by enveloping each collagen fibre with a thin layer of fat.

3.5 Plastics and rubber

Plastics have become probably the most commonly encountered of all the synthetic materials in use today. Plastics are used in the manufacture of an immense range of products including film, tape, pipes, vehicle body parts, toys, household goods and countless other objects. A plastic material may be defined as a high molar mass polymer which is solid in its final form but which is formed by flow directly into a desired shape at some stage of the manufacturing process. Generally this is achieved by subjecting the material to high temperatures. Plastics are usefully classified into two groups, thermoplastics or thermosets, according to the way the materials are affected by the application of heat. However, this method of classification may be extended to include two further related polymeric materials: elastomers and fibres. Each of these groups of plastic materials is discussed separately in the remainder of this section.

3.5.1 Thermoplastics

Thermoplastics are the polymers most of us associate with the term 'plastic'. They are materials which are hard solids at room temperature, but on heating they soften or melt and as a result they acquire the ability to flow. On cooling they revert to their original solid form. This overall process is reversible so that the heating/moulding/cooling cycle may in principle be carried out on a thermoplastic material many times. Plastic articles are manufactured from a thermoplastic polymer, by shaping the material in its softened or liquefied state at elevated temperatures. There are a number of manufacturing methods used to process the thermoplastic into the appropriate shape. Two of the more important of these production methods are injection moulding, in which the material is forced into an appropriately-shaped cavity under pressure, and extrusion, in which the material is forced through an orifice under controlled conditions.

Around 90% of plastics in use are thermoplastics. The chemical structures of some common thermoplastics are given in Fig. 3.10. These materials are all essentially made up of linear polymer molecules, with little or no cross-linking between the polymer chains. This feature of the molecules allows the individual polymer chains to slip past one another on heating thus giving the material the ability to flow. Thermoplastic polymers may be separated into two groups according to the chemistry of the polymerisation process used to synthesise the polymer and the nature of the monomer from which they are derived. Polymers (**8a**)–(**8d**) and (**9**) are polyalkenes, and are referred to as addition polymers because they are manufactured from alkene monomers by a free radical addition process. Polyester (**10**) and nylon (**11**) are examples of condensation polymers, deriving this name from the fact that they are prepared from the appropriate monomers by condensation reactions.

Polyethylene (**8a**) is the thermoplastic material which has the simplest molecular structure and which is produced in the greatest quantity. As the name implies it is the product obtained by polymerisation of ethylene

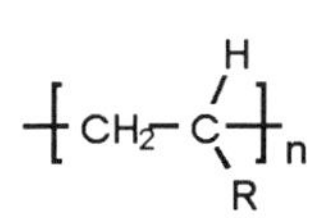

(8)

(8a) R = H polyethylene
(8b) R = CH_3 polypropylene
(8c) R = Ph polystyrene
(8d) R = Cl polyvinyl chloride

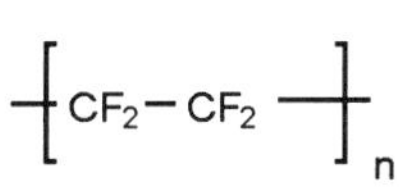

(9) polytetrafluoroethylene (PTFE)

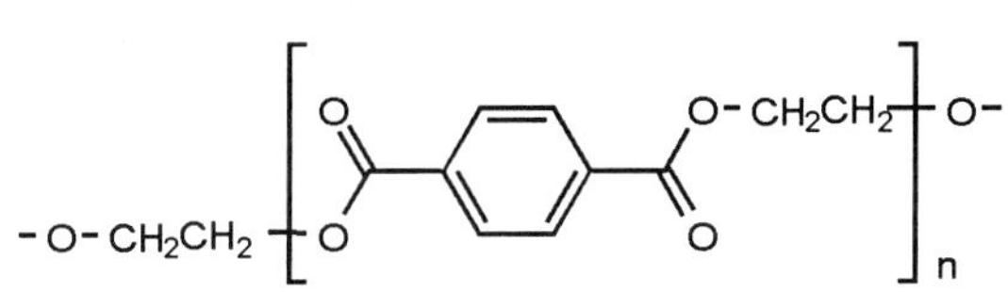

(10) polyethylene terephthalate (polyester)

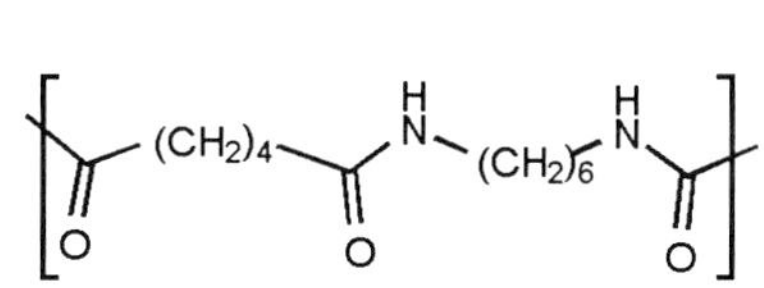

(11) nylon 6,6

Fig. 3.10 The chemical structures of some common thermoplastics.

(ethene, $CH_2{=}CH_2$), and is thus a hydrocarbon polymer. It is a waxy solid which is relatively low cost, easily processed and shows good chemical resistance. Polyethylene is manufactured in two grades, low density (LDPE) and high density (HDPE). It is a relatively low melting thermoplastic, LDPE being processed normally in the range 150–180°C and HDPE in the range 180–200°C. Polyethylene is used in packaging films, containers, toys and a range of household goods. Polypropylene (**8b**) is, like polyethylene, a hydrocarbon polymer and in many ways is similar in properties to polyethylene. It does, however, have a lower density and it is higher melting than polyethylene, the normal processing temperatures being in the range 220–250°C. It is mechanically stronger than polyethylene, for example being less susceptible to environmental stress cracking, but it is more susceptible to oxidation. Polypropylene is used, for example, in packaging, bags, ropes, netting, carpets and upholstery. Polyvinyl chloride (PVC) (**8c**) is a chlorinated hydrocarbon polymer manufactured in both rigid and plasticised forms. It is a widely used plastic material, finding application in the manufacture, for example, of pipes, floortiles, raincoats and electrical cable insulation. PVC is a relatively unstable material undergoing degradation over a period of time when exposed to light and heat, as a result of loss of HCl from the polymer backbone. This causes discoloration of the polymer and the material becomes increasingly brittle. Stabilisers of a variety of chemical types are incorporated into PVC to minimise these problems. Polystyrene (**8d**) is a widely used plastic material which is inexpensive, transparent and rigid. It is used in refrigerators, food containers and in a variety of plastic foam materials. Polytetrafluoroethylene (PTFE) (**9**) or Teflon™ is the completely fluorine-substituted analogue of polyethylene. PTFE is a tough, flexible, very high melting thermoplastic material with excellent mechanical properties and chemical resistance. PTFE is used in bearings and artificial joints and, because of its low surface energy and coefficient of friction, as a 'non-stick' lining on cooking utensils. Of the condensation polymers whose structures are illustrated in Fig. 3.10, polyamides, such as Nylon 6,6 (**11**) show high impact strength, toughness, flexibility and abrasion resistance and find use, for example, in brushes, sports equipment, bearings, gears, surgical sutures, tyre cord and door latches, while polyester (**10**) is used in soft-drink bottles and recording tape. Both nylon and polyester are also used extensively as fibres (see section 3.2).

To become useful materials, a number of additives are frequently incorporated into thermoplastic polymers. Pigments are used to give a plastic article its colour and, if desired, to make it opaque. Plasticisers are small organic molecules which act as lubricants and are incorporated into plastics to prevent them becoming brittle at room temperatures. Fillers are low cost materials which are incorporated into thermoplastics mainly to increase the bulk of the materials and hence to reduce the cost of the products. Fillers may be white powders, which are sometimes referred to as non-hiding white pigments, such as calcium carbonate, or gases which are used to produce plastic foams.

3.5.2 Thermosets

Thermosets are polymeric materials which may be liquefied and moulded to a desired shape on heating and at the same time undergo a chemical change to become rigid materials. In contrast to thermoplastics, because a chemical change is involved, the process is irreversible and the materials therefore cannot be re-processed. In their final form, thermosets possess a very highly cross-linked polymeric structure as a result of the polymerisation reactions which take place on heating. Examples of thermosetting materials include phenolic resins derived from reaction of phenols with aldehydes, amino resins derived for example from condensation of formaldehyde with urea or melamine, epoxy resins and polyurethane resins. Thermosets have numerous uses including electrical parts, heat resistant components and counter tops.

3.5.3 Elastomers

Elastomers are a group of polymeric materials which may be deformed under an applied force but which return to their original dimensions when the force is removed. Such behaviour is the essence of rubber-like elasticity. The chemical structure of natural rubber, the best-known elastomer, is shown in Fig. 3.11. It consists of long folded chains, more or less randomly coiled, which become lined up when the material is stretched. When the stress is removed the polymer molecules return to their original configuration. With natural rubber, the use of a process known as vulcanisation is essential to give the material its elastic properties. This process involves the treatment of the polymer with sulphur at elevated temperatures, which provides a light degree of cross-linking as a result of the introduction of sulphide (—S—) and disulphide (—S—S—) links. A variety of synthetic elastomers are also in use, including polybutadiene, used in tyre treads, SBR (styrene–butadiene rubber), ABS (acrylonitrile–butadiene–styrene rubber), isoprene rubber and silicone elastomers, which are used in flexible seals.

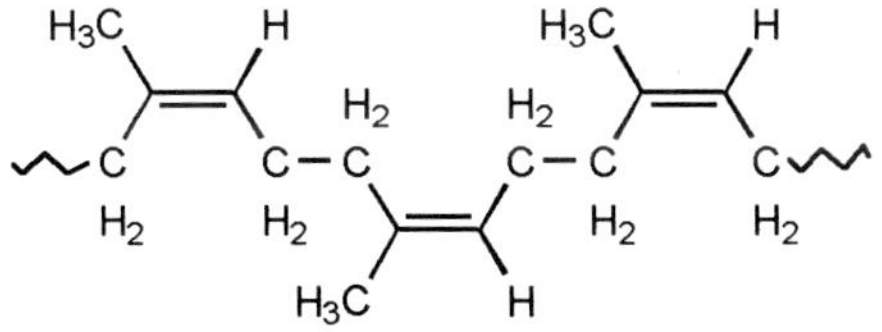

Fig. 3.11 The chemical structure of natural rubber.

3.5.4 Fibres

Fibres are polymeric materials, either natural or synthetic, manufactured in a long, thin, thread-like physical form. In the case of synthetic fibres, they are frequently chemically identical to some of the thermoplastics discussed in section 3.5.2. For example, polyester (**3**) and polyamides such as Nylon 6,6 (**4**) in fibre form are two of the most important synthetic fibres used in textile fabrics. The structure and properties of the most important fibres and their use in textiles are dealt with extensively in section 3.2 of this chapter and so they are not considered further here.

3.5.5 The coloration of plastics and rubber

Pigments and, to a lesser extent, dyes are incorporated into a product manufactured from a plastic material mainly to impart colour in order to enhance its market appeal. Pigments may be introduced into plastics by a variety of methods. Direct dry colouring is frequently used especially with thermoplastics. This process involves incorporation of the coloured powder into the molten polymer at the concentration required to give the desired colour, often along with other additives, using appropriate high shear dispersing equipment such as the two roll mill or the Banbury mixer (see Chapter 10). However, many manufacturers of plastic articles find it more convenient to make use of pre-dispersed concentrates or masterbatches of colorant in a liquid additive such as a plasticiser or in a compatible resin. Such concentrates are then easily incorporated by a fairly simple mixing procedure into the final polymer composition at an appropriate stage of the processing sequence.

Because the pigment is incorporated by dispersion into the polymer at high temperatures and because the coloured polymer may well be processed subsequently at high temperatures in the manufacture of the plastic article, the pigments used must be stable to heat. The degree of heat stability required is dependent on the temperatures and times of processing to which the pigment will be exposed and these vary with the nature of the polymer. In the case of thermoplastics, processing temperatures will be dependent on the melting point of the polymer which may be as low as 120°C in the case of low density polyethylene, but above 300°C with some engineering plastics such as nylon and polytetrafluoroethylene (PTFE). In the case of thermosets and elastomers, processing temperatures and times will be dictated by the requirements of the curing processes used although these are generally less severe than is the case with thermoplastics.

3.6 Ceramics, cement, concrete and glass

These four types of materials are conveniently grouped together in this section because they share the common feature that they are inorganic materials, which are for the most part based structurally on silicate chemistry. This contrasts with the polymers used in textile fibres, leather, paper and plastics which are almost exclusively organic in nature.

3.6.1 *Ceramics*

Ceramics are inorganic engineering materials and products, generally based on silicates, which are manufactured using very high temperature processing. Ceramics are perhaps best known in applications such as tableware, kitchenware, plumbing fixtures, floor and wall tiles, but they encompass a much wider range of less familiar applications ranging from high efficiency turbine parts to lasers. The main raw materials for ceramics are inorganic minerals such as clay (a hydrated aluminosilicate), silica and feldspar. The manufacture of a ceramic product involves the three stages of forming, drying and firing. The first of these three stages, as the name implies, involves forming of the ingredients into the desired shape. This stage of the process, which may be carried out at normal temperatures, usually gives the product its permanent shape although some shrinkage can occur during the subsequent drying and firing processes. The thermal treatment of the formed material is an essential feature of the manufacture of a ceramic product in order to develop the mechanical strength of the material. The temperatures which are ultimately required range from 700°C to 1650°C, or even higher for certain speciality products. Between 0 and 400°C the material loses bound water (drying) and any organic binders used in the forming process. Much higher temperatures are then required (firing) to convert the product into a structural arrangement which provides substantially improved mechanical strength which is essential in the final product. Because of the very high temperatures used, ceramics are coloured using very heat-stable inorganic pigments. No organic dyes or pigments would be capable of withstanding these temperatures.

3.6.2 *Cement and concrete*

The term 'cement' may be used to describe many different kinds of binders or adhesives, but it is most commonly applied to the familiar inorganic materials used for the purpose of binding bricks and other stonework together in building and construction. The essential feature of these cements is their ability to react with water to form aggregates of considerable strength and dimensional stability. Most cements consist of a finely powdered mixture of the principal ingredients calcium oxide (CaO) and silica (SiO_2) together with other materials, such as alumina (Al_2O_3), in varying quantities. Cement may be manufactured by heating an appropriate mixture of limestone, sand and clay to very high temperatures in a rotary kiln. When cement is mixed with water at normal temperatures, reaction takes place over a period of time between the solid ingredients which make up the cement and water, and the result is a conversion to a hard material which is essentially a hydrated calcium silicate. Concrete is a familiar building material which is manufactured by treating cement and sand mixtures with water. Cement and concrete are generally coloured using inexpensive inorganic pigments, such as iron oxides, which are stable to light and to the strongly alkaline conditions present in the materials.

3.6.3 Glass

The term 'glass' may be used to describe a wide range of solid, transparent, brittle materials. However, most commonly it is applied to the familiar inorganic material used in windows, kitchen utensils and ornaments of a variety of types. Glass is formed from the hot liquid phase of the material by cooling in such a way that the viscosity of the material becomes so high that it is effectively rigid. The material adopts a highly disordered or amorphous structure closely related to that of the liquid phase. The basic structural unit of most glasses is the silicon–oxygen (SiO_4) tetrahedron in which a silicon atom is bonded to four surrounding oxygen atoms. In silica glass, virtually all oxygen atoms are bridging in that they link to other SiO_4 tetrahedra in a three-dimensional network. Silica glass has excellent properties, but its production requires very high temperatures so that it is rather expensive. Most glasses in common use fall into the category of silicates containing modifiers, which are generally oxides of a variety of appropriate metals or non-metals. For example, the use of sodium oxide (Na_2O) as a modifier (soda glass) alters the structure by cleaving Si—O—Si bonds to form Si—O^-Na^+ ionic links. Other modifiers used commonly in other types of glass include oxides of potassium, aluminium, calcium, lead, magnesium and boron. Heat-stable inorganic pigments are used in the coloration of glass.

Bibliography

Plastics and Rubber, E.W. Duck, Butterworth, London (1971).
Plastics Materials, J.A. Brydson, Newnes-Butterworth, London (1975).
Polymers: Chemistry and Physics of Modern Materials, 2nd edn., J.M.G. Cowie, Blackie, Glasgow & London (1991).
Encyclopaedia of Polymer Science and Engineering, Vols 1–17, 2nd edn., Wiley, New York (1990).
The Chemistry of Polymers, J.W. Nicholson, The Royal Society of Chemistry, Cambridge (1991).
Textile Science, K.L. Hatch, West Publishing, Minneapolis/St. Paul (1993).
Handbook of Textile Fibres: Natural Fibres, J.G. Cook, Merrow, Durham (1984).
Handbook of Textile Fibres: Man-made Fibres, J.G. Cook, Merrow, Durham (1984).
The Chemistry of Paper, J.C. Roberts, The Royal Society of Chemistry, Cambridge (1996).
Kirk–Othmer Encyclopaedia of Chemical Technology, 3rd edn., John Wiley, New York (1984).

Chapter 4

Surface Coatings: Paints and Printing Inks

Paints have a very ancient history. Prehistoric man used a combination of minerals as pigments together with animal fats as binders to produce rudimentary paints with which he decorated his cave-dwellings with drawings of animals and men. From these early beginnings, the modern range of surface coatings, which are very sophisticated products, has evolved. A large proportion of these materials are the familiar decorative and architectural paints which are used to protect and enhance the appearance of our homes, offices and other buildings. A similarly large proportion are industrial coatings which are formulated to meet the special conditions encountered in numerous industrial situations. The products to which industrial paints are applied include motor vehicles, agricultural equipment, ships, aircraft and a wide range of durable consumer goods such as kitchen appliances, furniture, toys and containers. To many the terms 'paint' and 'surface coating' would appear to be essentially synonymous. The distinction between *paints* and *coatings* which has emerged historically is that the former implies a decorative and aesthetic function while the latter are considered to serve a more serious purpose. For the purposes of this introductory text, no such rigorous distinction is made. A specific reason for this is that, to this author, it is of relevance to consider printing inks as surface coatings also. It is, of course, recognised that printing inks serve a quite different purpose compared with paints. Their purpose is to decorate, illustrate and inform, in newspapers, magazines, books and on a wide range of packaging materials, and they have no protective role. Nevertheless, paints and printing inks as materials frequently show distinct similarities in the way that they are formulated and used and especially in the way they make use of colour, which is the essential theme of this book. Most commonly, colour is introduced into a paint or printing ink by dispersion into a liquid medium and pigments, rather than dyes, are the colorants of choice for this purpose. After application as a fluid suspension by some means to the surface, the paint or ink subsequently dries to a solid resistant film. There are some important exceptions to this description, for example powder coatings which, as the name implies, are applied to a substrate as a solid powder, and which subsequently are converted to continuous films by exposure to high temperatures. The following sections of this chapter provide an introduction to some of the fundamental principles of paint and ink technology. Also included is a discussion of the nature of printing processes which is essential for an understanding of the basis of printing ink formulation. Chapter 10 of this book contains a description of some of the manufacturing technology used in the production of surface coatings, with particular reference to the dispersion equipment used to incorporate pigments into the paint or ink medium.

4.1 Paint

4.1.1 The nature of paint

A paint is a surface coating material which is usually applied to a surface as a fluid suspension by an appropriate method of application, such as brushing, roller, spraying or by electrodeposition. Afterwards, the paint dries, or more correctly converts from a liquid to a solid film, and this film completely covers the surface. Paint films are commonly (but not always) applied at thicknesses of the order of a few tens of micrometers (μm). Paints are applied to surfaces to serve two basic purposes: protection and decoration.

A paint may be used to protect the surface in a variety of ways. The coating may provide protection against mechanical 'wear and tear' by improving resistance to impact and abrasion. It may also protect against the effects of weathering and corrosion in a hostile environment, either actively by the inclusion of ingredients such as anticorrosive pigments, or passively by providing an adhesive and impermeable barrier. Paints can also confer resistance to contamination or hygienic properties, such as fungal, bacterial or fouling resistance. The economic benefits of using coatings for protective purposes are immense. This 'added value' comes from the fact that they increase the useful lifetime of products to whose surfaces they are applied, they reduce down-time for maintenance and repair, and they can produce energy savings. A further feature is that they can allow cost-effective materials substitution. For example, appropriately-coated mild steel performs many of the functions which might otherwise require much more expensive materials such as copper, bronze and stainless steel. It may be argued that these economic benefits are accompanied by environmental benefits, for example in that they play a role in the preservation of natural resources. They ensure that less metal is used in construction, that it lasts longer and that such metal as is used is more easily recycled.

The second role of paints is to decorate the surface and generally enhance its appearance. They do so in a number of ways, the most obvious of which is providing the desired attractive surface colour. In addition, however, paints produce the degree of opacity which is required to hide the surface to which they are applied, and they also provide a smooth finish which can obscure defects in the surface.

There are a number of other features which a paint must demonstrate to be a useful product. Control of the flow properties (rheology) of wet paints is of critical importance not only in relation to their method of application, but also with respect to the film formation process. Another example is that the dry film must adhere to the surface to which it is applied. Coatings are applied to a wide variety of substrates, including brick, wood, metals and plastics, and provision of the appropriate adhesion properties requires careful consideration of the widely differing chemical and physical characteristics of the surfaces. However, it is probably true to say that the most common substrate to which paint is applied, and thus is required to adhere to, is paint itself, since most coatings are applied as several layers. Finally, a coating must be sufficiently durable to withstand the effect of its environment in order to retain its protective and decorative properties for an appropriate period of time. All of these factors require careful consideration in the formulation of paints for specific purposes.

The term 'paint' encompasses an immense variety of products. In order to gain an appreciation of the scope of paint technology, it is useful to consider the ways in which paints may be classified. Three different methods of classification are discussed here, based on, respectively, the order of application, the end-use and the 'drying' mechanism.

The first general method of classification considered here is based on the order in which the coatings are applied to the surface. Three important types of coating which may be identified using this method of classification are *primers*, *undercoats* and *finishing coats*. The first layer of paint applied directly to the surface is generally referred to as a primer. Primers are formulated to meet the requirements of the surface to which they are applied, which may be, for example, metal or wood. For example, they will be required to provide strong adhesion both to this surface and to another coat of paint which is applied subsequently. They may often also be used to fill minor imperfections in the surface. Primers generally require good elasticity in order to accommodate expansion and contraction of the surface. In some cases, especially for direct application to metal surfaces, primers are formulated to provide protection against corrosion. Undercoats are paints which are applied, sometimes over a primer, to provide opacity so as to obscure the surface and to build up film thickness. Undercoats therefore also require good adhesion properties. Finishing paints, or topcoats, are applied last and are largely responsible for the final desired decorative effect. The decorative effect provided by the finishing coat includes not only the colour but also the degree of gloss observed. Depending on the desired visual effect, finishing paints may be designed to dry to films which display either high gloss (i.e. possessing a shiny mirror-like appearance), semi-gloss or matt (i.e. possessing a dull, rather flat appearance) surface appearance. However, since finishing coats exist at the interface with their environment, they may also be required to play an important protective role, such as resistance to impact, to light, oxidation and the action of aggressive fluids with which they may come into contact.

A second way of classifying paints is according to their end-use. In this context, paints are commonly referred to as either *decorative* or *industrial*, although each of these groups contains a diverse range of products which can be categorised further. We are all probably most familiar with the various types of decorative paints used in and around the home. These paints, as the name implies, are used mainly for the decorative effect which they give. They are generally slow drying (i.e. taking up to a few hours) at normal temperatures. The durability requirements of decorative paints for indoor use are not particularly high, but they are rather more demanding when exterior use is envisaged. Currently, both water-based and organic solvent-based decorative paints are in use, although it is probably only a matter of time before health, safety and environmental considerations will ensure that solvent-based decorative paints are phased out completely.

There are also a wide range of paint products designed for specific industrial uses. These so-called industrial paints include, for example, coatings for application to household appliances, structural steel, automotive finishes and re-finishes, and for marine applications such as ships, harbour installations and oil-rigs. Each particular product places different demands on the formulation of the paint in terms both of the degree of durability required and the desired decorative effect, but frequently with industrial paints protection is the priority. The coatings will usually be required to give very good mechanical properties, and may be required to be resistant to heat, solvents and other chemicals with which they may come into contact. Industrial paints which are applied to factory-built articles such as domestic appliances frequently are dried rapidly, as is required by the need to satisfy production schedules in continuous 'conveyer-belt' processes, in ovens by an elevated temperature curing process. Such paints are frequently referred to as 'stoving' paints. The high temperatures induce polymerisation reactions within the coating and lead to highly-resistant, durable cross-linked polymeric films. However, for many applications of industrial paints especially when they are required to be applied to very large objects, this is impractical due to the scale of operation, and rather slower, normal temperature drying processes are involved. In these cases, the chemistry and technology of film-formation has been developed to ensure that cross-linked polymeric films with excellent durability characteristics are also obtained. Buildings of one sort or another account for about half of the coatings materials used worldwide. The substrates which are coated include concrete, plaster, brick, wood, steel, galvanised steel and plastics. Concrete and plaster present porous, alkaline substrates and consideration has to be given to the sensitivity of some coatings components to alkali and to the movement of water and inorganic salts. Wood is also porous and is inclined to expand and contract with variations in humidity. Of importance in wood coatings therefore is flexibility, as is biocidal pretreatment of the wood. Land-based steel structures such as bridges, pipelines, tanks and chemical plant provide examples of surfaces which require coatings which must be applied at ambient temperatures to provide durability and corrosion protection. Since large amounts of these paints may be consumed in one project, the emphasis is commonly also on cost-effectiveness. Marine paints, such as those used on ships, harbour installations and oil drilling platforms, have similar general requirements to their land-based counterparts, but with a vastly increased potential for corrosion by sea-water. Coatings used on ships need excellent exterior durability and an important specific requirement is the need to provide anti-fouling properties, for example to inhibit the growth of marine organisms such as seaweed and barnacles. The adhesion of these marine growths to the hull of a ship gives rise to a very rough surface which can seriously reduce the speed of the ship and increase fuel consumption. Compounds containing copper and tin which are marine toxins are used as anti-fouling agents, although the tin-based compounds especially are being called into question increasingly on environmental grounds and are likely to be phased out when suitable alternatives are available. Aircraft exterior coatings are amongst the most severely-tested in service. Flying at high altitudes, they are subjected to intense ultraviolet radiation, mechanical stress and rapid extreme temperature changes. Efforts are being made to develop coatings which provide all of the required attributes and which are, at the same time, easily strippable to allow periodic inspection of the airframe, before re-application. Car plants are large-scale production line operations in which the car bodies are manufactured continuously and undergo coatings processes, with rapid high-temperature drying, before being built up into the finished vehicles. Car owners demand a very high standard of appearance and resistance to mechanical impact and corrosion. An automotive paint, such as that used on a car body, will therefore be required to supply an attractive coloured glossy finish, which retains its colour for many years, yet at the same time to provide a considerable degree of protection to the car body. Car-refinishing will be judged by the owner of the car in a similar way, in that he will expect a match to the original finish in all respects, even though the method of application by spraying with low temperature drying is significantly different. Food and beverage cans require to be coated to prevent attack by their contents. The coatings are required to withstand considerable deformation without failure, and are permitted to contain only ingredients which are acceptable for food contact. Similar constraints apply to inks which are printed on to the exteriors of the cans.

A third method of classifying paints is according to the mechanism by which the paint dries, or more correctly converts from a fluid to a solid form. This aspect of paints is discussed as a separate topic in section 4.1.5.

4.1.2 The composition of paints

The wet paint, before application to the surface, contains three major ingredients: the pigment, the film-former and the liquid, together commonly with other additives or minor ingredients which are incorporated into the formulation for specific purposes. The following section presents an outline of the functions of each of these major ingredients in the formulation of the paint.

A pigment is a finely divided solid colouring material which is insoluble in the liquid used in the formulation of the paint and is incorporated into the paint as an intimate dispersion. The reader is directed to Chapter 2 of this book for a discussion of the distinction between dyes and pigments as colouring materials and to Chapter 6 for a more detailed discussion of the structures and properties of the pigments commonly used in paints and printing inks. When the paint dries, or solidifies, the pigment is held, trapped mechanically as individual solid particles in the polymeric matrix provided by the film-former. The primary function of the pigment in a paint is to produce the required visual effect. The pigment obviously is primarily responsible for the colour of the paint film. However, this is not its only optical function and in many surface coatings applications may not be its most important function. It is the pigment which provides the degree of opacity necessary for the paint to obscure the surface to which the paint is applied. Of special importance in this respect is titanium dioxide, a high refractive index, highly scattering white pigment, which is commonly used in paints to provide the opacity required. Titanium dioxide may be used together with coloured pigments to opacify paints when pale shades are required. Paints of deeper shades require the use of high refractive index coloured pigments, mostly inorganic, to give opacity. Of course, some coatings, for example wood stains, are designed to be transparent and so will be coloured using transparent grades of pigments, while in others, for example clear varnishes, the pigment may be completely absent. A third optical property associated with paints which may also be influenced by the nature of the pigment is the gloss of the dried paint film. To show high gloss, which is essentially a mirror-like reflection, the painted surface must be extremely smooth. To ensure that the surface shows this high gloss, pigments with fine particle sizes are incorporated into the paint to a very high degree of dispersion. This minimises the extent to which the particles protrude through the surface of the paint film, ensuring that the surface retains its smoothness.

In addition, pigments may have secondary roles in paints which extend beyond their optical functions. For example, they may play a part in enhancing the mechanical properties and durability of the film. They may also assist in protecting the surface to which the paint is applied against corrosion, for example by presenting a barrier to water penetration. Pigments also influence the flow properties of wet paints. The way in which paint flows is of critical importance in determining the ease of application of the paints to the surface and in ensuring that the wet film is distributed evenly on the surface allowing it to form a smooth uniform dry film (see section 4.1.3).

The second major ingredient in paints is the *film-former*, or binder. The film-former serves the purposes of providing the solid polymeric matrix which binds the pigment particles together and ensuring that the paint adheres to the surface to which the film is applied. However, arguably the most important function of the film-former is that it is the ingredient which is primarily responsible for the mechanical properties and the durability of the dry paint film, ensuring that the paint provides the required degree of protection of the surface to which it is applied. The chemical nature of some of the most commonly used film-formers, which are a very diverse group of materials, and the physical and chemical processes which occur when the paints dry to a resistant film are discussed in section 4.1.5.

The third major ingredient of most paints is the *liquid*, or diluent. The liquid serves as the carrier for the other ingredients of the wet paint making the paint sufficiently fluid for application as a uniform film. After application of the paint to the surface, the liquid evaporates into the atmosphere and plays no further part. The term 'solvent' is frequently used in the context of paints because commonly it is required to dissolve the film-former and other ingredients in the paint formulation. However, this term may be rather misleading as the pigments and in some instances, for example in the case of water-based latex (emulsion) paints, the film-formers do not dissolve in the liquid but rather are incorporated as a fine dispersion of solid particles. Important considerations in the selection of liquids for paint formulations are solvent power, rate of evaporation, viscosity, flammability and toxicity.

If the coatings liquid is a true solvent, it should be capable of dissolving the film-former completely over a wide range of concentrations to give a clear solution. For this to happen, the intermolecular forces between the solvent molecules and the film-former molecules must be sufficiently large to overcome the intermolecular forces between film-former molecules themselves. The intermolecular forces which affect solvency are usually dipolar and van der Waals forces and hydrogen bonding (see Chapter 2 for further discussion of the principles of the various types of intermolecular force). A commonly stated general rule which is often used to describe solvency is that '*like dissolves like*'. As an example, a polar solvent will dissolve a film-former of high polarity, since it will provide the dipolar forces, together with some contribution from van der Waals forces and, possibly, hydrogen bonding, which are required to disrupt the intermolecular forces within the film-former. The solubility requirements of most of the commonly used organic film-formers means that a wide variety of organic solvents are used in paints. Paint solvents in common use include a range of aliphatic hydrocarbons (hexane, cyclohexane and 'white spirit', a term which encompasses petroleum hydrocarbons of certain distillation ranges), aromatic hydrocarbons (toluene, xylenes and trimethylbenzenes), ketones (acetone, methyl ethyl ketone, methyl isobutyl ketone, diisobutyl ketone, cyclohexanone), esters (ethyl acetate, butyl acetate), alcohols (isomers of propanol and butanol) and ether alcohols (ethoxyethanol, butoxyethanol, methoxypropanol). In practice, organic solvents are often used as mixtures rather than as individual components to provide the appropriate solvency, to control evaporation rates, to lower cost and to reduce health and safety risks.

Water is also used to a considerable extent as the liquid medium in paint formulations. As well as the obvious economic advantages of water as a paint solvent, it has numerous environmental advantages. These include health and safety features such as its non-toxicity, lack of odour, and non-flammability, and its free and essentially unlimited availability. It also provides the advantage of ease of cleaning of brushes and other application equipment. In contrast, most organic solvents must be treated as potentially toxic if they enter the body by inhalation, skin contact or ingestion, and legislation is in force in industialised countries to control their use. There is also a growing concern over the use of organic solvents in paints from wider environmental considerations. There is a convincing argument that water is the only paint solvent which is recycled, by natural means, while the others must be considered as potential atmospheric pollutants. There is considerable concern globally about the emission of volatile organic compounds (VOCs) into the atmosphere and the adverse effects which can result. Probably the most familiar effect of this type of pollution is the depletion of the ozone layer and the potential consequences, such as the reduced filtration by the ozone layer of harmful ultraviolet radiation and the possibility that this might cause increased levels of skin cancer and eye cataracts. It is generally accepted that the principal cause of this effect is a group of compounds known as the chlorofluorocarbons (CFCs) which have been extensively used as coolant liquids in refrigeration, and in aerosols. However, this particular group of solvents has never been used extensively in coatings formulations. VOCs are also cited as a cause of smog in urban and industrial areas, in the formation of toxic low level ozone and as a possible contributor to the 'greenhouse effect' and its influence on global warming. There is therefore continuous and growing pressure, reinforced by legislation, on coatings manufacturers to reduce steadily the levels of VOC emission from their products by reducing the amount of organic solvents which are incorporated into paints and printing inks. A counter-argument which is often presented by the coatings manufacturers is that the amount of VOC emissions directly attributable to coatings is very low in comparison with other sources of emission, both natural (from plants and animals) and synthetic, for example from car exhaust fumes. Nevertheless, the reaction of the paint industry to these pressures has been to reduce the amount of organic solvents used in a number of key ways, for example by developing high solids content coatings, solvent-free systems such as powder coatings and radiation-cured coatings (see section 4.2.2), and a wider range of water-based products. However, water-based coatings which are capable of matching the technical and cost performance of coatings based on organic solvents have not yet been developed for many industrial applications.

In addition to the three main ingredients, a paint may contain small quantities of other components. Examples of such additives include agents which help to stabilise the dispersions of pigments and sometimes of film-formers. Compounds which assist in the control of the flow properties of wet paints are also used frequently (see section 4.1.3). For example, in water-based decorative paints, water-soluble polymers may be used to provide 'non-drip' properties. Other additives used in paint formulation include biocides which protect against bacterial and fungal attack in water-based formulations, and compounds which may act as catalysts for the chemical reactions taking place in the film formation process.

4.1.3 The rheology of paints

If a paint film is to function effectively as a barrier it must be continuous and free from defects. Conventional paints exist in the liquid state at the point of application. An obvious reason for this is that the paint must be easily transferred from its container to the surface and applied at an appropriate even film thickness, and this is most easily achieved as a liquid. In the case of paints applied by spraying, the droplets of paint must coalesce by liquid flow to a form a continuous film. After application to the surface (which may be vertical), the paint must 'hold up' without running or dripping as it dries, or rather transforms from a liquid to a solid, and at the same time surface imperfections, such as brushmarks, should 'flow out' leaving a smooth film of the required thickness. These two features are at first sight contradictory, since the former involves the avoidance of fluid flow while the latter requires flow. The science of the way in which fluids flow or deform is referred to as *rheology*.

In the kitchen, we might commonly describe the rheological behaviour of materials used in food preparation using the term 'consistency' using adjectives such as thick or thin. A term commonly used to provide a quantitative impression of the ease of flow of materials is *viscosity*. In order to appreciate the principles of the rheology of paints in general, and viscosity in particular, it is appropriate to consider the physical principles of shear flow.

When a force is applied to the top surface area of a cube of fluid material, a deformation takes place which diminishes with depth into the material, as illustrated in Fig. 4.1. This force, F, per unit area, A, is termed the *shear stress*, σ. The shear stress creates a deformation characterised by the angle γ, which is known as *shear*. If the shear stress is maintained on the surface of the fluid, the angle γ will change with time but the rate of change of γ will remain constant. This rate of change is referred to as the *shear rate*, τ. Thus shear rate is a measure of how rapidly a material is deformed or caused to flow. The *viscosity*, η, is defined as the ratio of the applied force per unit area to the shear rate. Thus:

$$\eta = \sigma/\tau$$

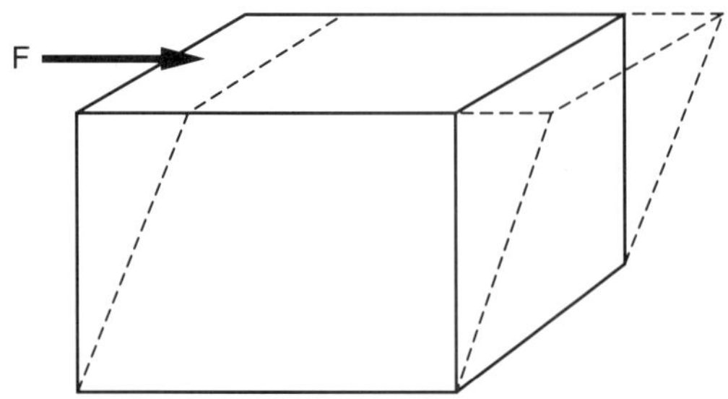

Fig. 4.1 Schematic diagram of shear flow.

Given these defined parameters for shear flow, a number of different types of fluid flow behaviour may be described. Commonly the principles of these various types of behaviour may be described using their characteristic *shear stress* vs *shear rate* relationships, as illustrated in Fig. 4.2.

The simplest rheological behaviour occurs when the applied stress is directly proportional to the shear rate. This type of flow, which is given by most low molar mass liquids such as water and simple organic solvents, is referred to as *Newtonian* fluid behaviour, since the material is by definition obeying Newton's law of viscosity. This type of fluid behaviour is illustrated in Fig. 4.2(a) as a line of constant slope, η, which passes through the origin and in Figure 4.2(b) as a horizontal line indicating constant viscosity. It should be noted in passing that the viscosity, η, changes with temperature, hot fluids being generally less viscous than cold fluids. With solutions of polymers and dispersions of insoluble solid particles (such as pigments and polymers) in liquids, it is commonly found that the viscosity is no longer

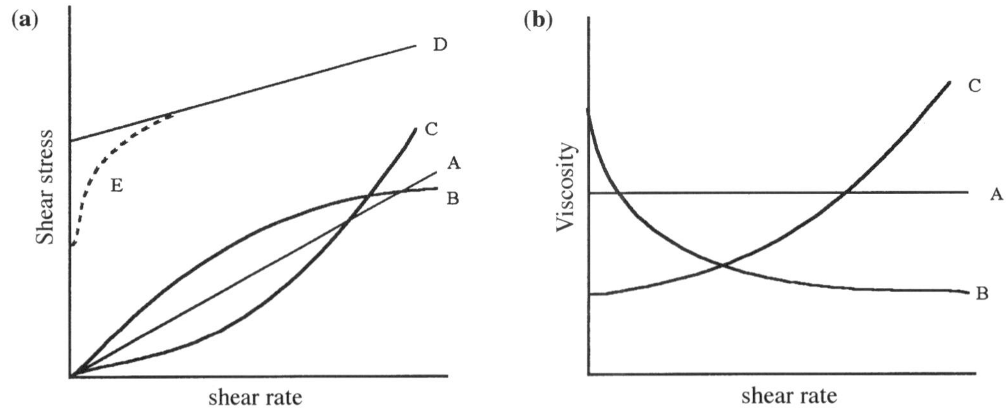

Fig. 4.2 (a) Shear stress-shear rate behaviour of fluids. (b) Viscosity-shear rate behaviour of fluids. (A) Newtonian; (B) shear-thinning; (C) shear thickening; (D) plastic; (E) pseudoplastic.

independent either of the shear rate or shear stress. Such behaviour is termed, for obvious reasons, *non-Newtonian* flow. The various types of non-Newtonian flow behaviour are illustrated in Fig. 4.2. The most common of these is *shear-thinning* in which the viscosity decreases as the shear rate increases. Shear-thinning behaviour is the most commonly-encountered type of non-Newtonian flow behaviour shown by surface coatings in their liquid state, associated with the fact that these materials commonly contain polymers in solution and pigments as solid, insoluble, finely-dispersed particles. The converse of shear-thinning is shear-thickening, in which the viscosity increases as the shear rate is increased. This type of behaviour may be observed in materials containing high concentrations of pigments. Another class of non-Newtonian behaviour is that of *plastic flow*. This flow behaviour requires that a finite shear stress, referred to as the *yield stress*, is required to be applied to the materials before flow takes place. With *pseudoplastic* flow, the material shows behaviour similar to plastic flow at high shear rates, although at successively lower shear rates, the curve tends towards the origin, and the overall behaviour is closely related to shear-thinning.

A further variable which can have a profound influence on the rheology of a fluid material in its shear flow behaviour is time. Fig. 4.3(a) illustrates the behaviour of a polymer solution when it is subjected to constant shear rate for a time, t, after which the shear is removed and the material is allowed to 'recover'. The diagram illustrates that viscosity decreases with time until a steady state value is reached and then, when the shear is removed, the viscosity increases over time to its original value. This behaviour is referred to as *thixotropy*. Alternatively, this phenomen may be demonstrated by observing the viscosity variation with a cycle of first increasing and then decreasing shear rates, resulting in a loop as shown in Fig. 4.3(b). True thixotropy would be confirmed if after a period of rest, the loop was repeated exactly.

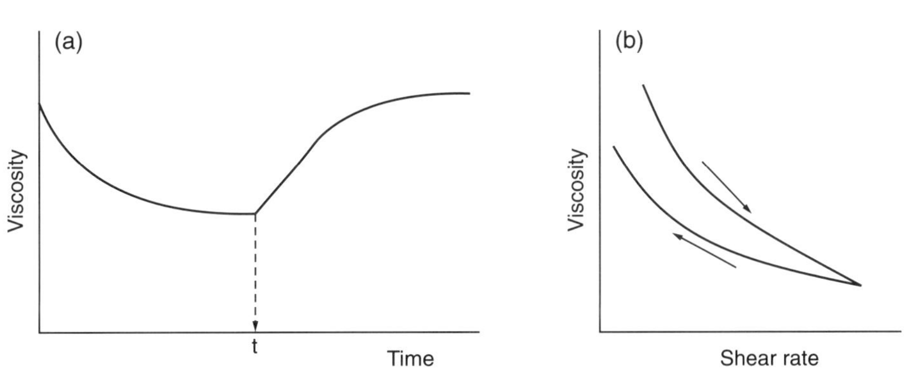

Fig. 4.3 Time dependence of viscosity in thixotropic behaviour. (a) Viscosity-time behaviour; (b) viscosity-shear rate behaviour.

Thixotropic behaviour may be explained from a consideration of the interactions between polymer molecules, especially linear polymers, in solution as illustrated diagrammatically in Fig. 4.4. At rest the polymer molecules are to a certain extent aligned with one another forming a loose three-dimensional structure which gives rise to a high viscosity. The application of shear causes a breakdown of the structure and the molecules become essentially randomly aligned, so that there is a reduction in viscosity and the fluid behaviour becomes more or less Newtonian. When the shear is removed this structure, and hence the viscosity of the solution, 'rebuilds' over a period of time. The converse behaviour to thixotropy in which the viscosity of a fluid increases with time during shear and is then lost when the shear is removed is termed *anti-thixotropy* or *rheopexy*. This is a much less commonly-encountered phenomenon.

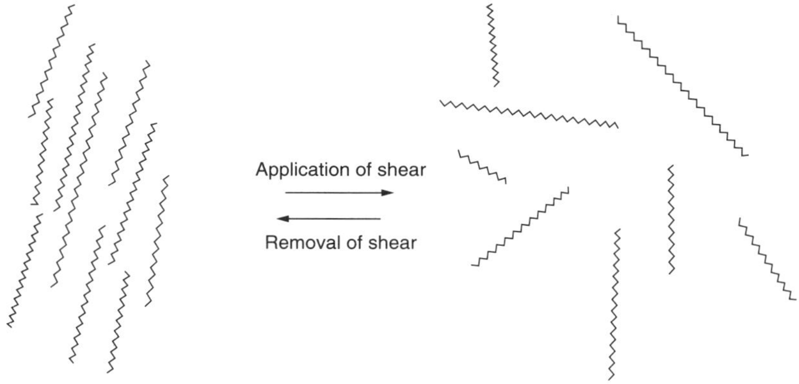

Fig. 4.4 Disruption and recovery of structure in thixotropic polymer solutions.

Control of the flow behaviour is a vital feature in the formulation of surface coatings for particular applications for a wide variety of reasons. Rheology plays a fundamental part in the manufacture of paints, influences storage stability and is also critical in the performance of a paint both during and after application to the surface. Arguably the most critical phase in the manufacture of a paint or printing ink is the dispersion process, the purpose of which is to distribute the insoluble pigment particles through the liquid medium and to reduce the pigment to as finely-divided a form as possible. The dispersion of a pigment into a liquid medium may be separated into three stages: (i) wetting of the pigment particles by the liquid medium, (ii) mechanical breakdown of the clusters of pigment particles to a much more finely-divided form and (iii) dispersion stabilisation so that the particles do not show a tendency to come together again, or flocculate. Control of the rheology during the milling or dispersion process is of fundamental importance in ensuring the optimum efficiency of the dispersion and the quality of the final product. The reader will find further discussion of this topic in Chapter 10 which contains a more detailed description of the manufacturing technology used in the production of surface coatings.

Most paints exhibit either highly shear-thinning or thixotropic behaviour. The high viscosity of certain paints at rest (i.e. low shear conditions) are useful in providing the 'non-drip' properties which are commonly regarded as desirable especially in decorative paints designed for 'DIY' purposes. The high shear rates which are generated during the application of paints, whether by brushing, roller or spraying, lower their viscosity by breaking down the inherent structure within the medium, and thus facilitate the ease with which they flow during application. This effect not only aids the application process but also assists control of the thickness and uniformity of the film. Droplets of paint applied by spraying require to coalesce soon after application by flow to give a continuous film, and this is facilitated by the reduced viscosity. Furthermore, wet films, irrespective of the method of application, require to flow and level in order to produce a smooth finish which is free from defects. However, once the paint is at rest on the surface of the substrate a high viscosity becomes desirable from the point of view of preventing running or sagging. This may be achieved by thixotropic recovery of the structure within the paint, assisted in some cases also by solvent evaporation and crosslinking polymerisation reactions of the film-forming materials (see section 4.1.4). The thixotropy is controlled within a time frame which allows coalescence, flow out , and levelling to occur to a significant extent before appreciable running or sagging takes place. This rheogical behaviour is determined to a large extent by polymers in solution in the paint liquid phase. In many paints, these are also the polymers which act as film-formers or film-forming precursors. In water-based latex paints, in which the film-formers are discrete solid particles of polymer, water-soluble polymers, commonly cellulose derivatives, are incorporated into the formulation in order to control the rheology. These materials also play a part in stabilising the dispersions of polymer and pigment particles.

4.1.4 The physical and chemical principles of film formation

An extremely important step in any coating process is film formation. This stage of the process refers to the conversion of the coating from a liquid to a solid after application, and is commonly referred to, although arguably incorrectly, as 'drying'. Two distinct phases of the film formation process may be identified. The first phase is complete when the paint may be 'touched' without disturbance (touch dry). However, the process of film formation must continue until the coating is 'hard dry' or 'block resistant', when it behaves more like a solid material as required for its ultimate purpose. We have all probably experienced the effect of closing a window with a newly painted frame before the second stage of film formation is complete. Completion of the process in such a situation causes the window to become stuck. These two phases of film formation with paints are closely related to the 'setting' and 'hardening' of inks during printing processes (see section 4.2.2).

An extremely important parameter in explaining film formation behaviour is the glass transition temperature (T_g). This refers to the temperature below which the polymer chains of the film-former have significant mobility so that the polymer has rubber-like properties such as flexibility, and above which this mobility effectively ceases and the material becomes a glass, with the properties of rigidity and brittleness (see Chapter 3). In general, in the early stages of film formation, the T_g of the film is required to be below ambient temperature so that the polymer molecules are capable to a certain extent of the degree of mobility required to form an even film. During the drying process, the T_g rises in a controlled manner until ultimately it is above ambient temperature and the film acquires the solid mechanical properties which are appropriate for its end-use.

Surface coatings are conveniently classified according to the mechanism of film formation into two categories: those drying without chemical reaction (frequently referred to as *non-convertible* coatings) and those products where the drying involves a chemical change (*convertible* coatings). The next two sections contain an overview of these two groups of surface coatings, and outline mechanisms of the film formation process are described for selected chemical types of film-former.

Surface coatings which dry with no chemical reaction

In non-convertible coatings, the film-former is generally a high molecular weight linear thermoplastic polymer which is either dissolved or dispersed as fine particles in the liquid. The film formation process involves evaporation of the liquid to leave a solid film of the chemically-unchanged polymer. The structures of some of the most commonly used film-formers in non-convertible coatings are shown in Fig. 4.5.

Non-convertible coatings in which the film-former is completely dissolved in the liquid are frequently referred to as thermoplastic coatings or *lacquers*. This approach is used in only a few coatings applications nowadays, having been largely superseded by the various types of convertible coatings. Thermoplastic coatings, which may be either clear or pigmented, find applications in, for example, wood furniture finishes and car refinishes. The explanation for this reduction in their importance is that the necessary mechanical properties of the dried coating require the use of very high molar mass linear polymers, large enough for a significant amount of chain entanglement to take place in the final solidified form of the film. The viscosity of solutions of such polymers is high, and consequently their concentration in the wet paint must be low for ease of application. The use of large quantities of organic solvents for this purpose, has become increasingly unacceptable over the years on the grounds of environmental considerations. Commonly thermoplastic coatings use either synthetic

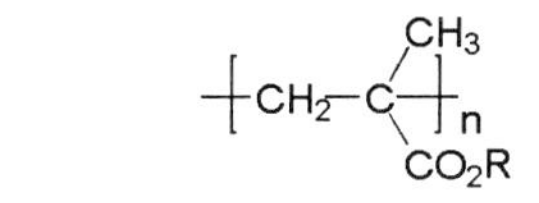

(1) Acrylic polymers

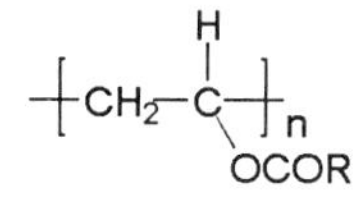

(2) Vinyl polymers

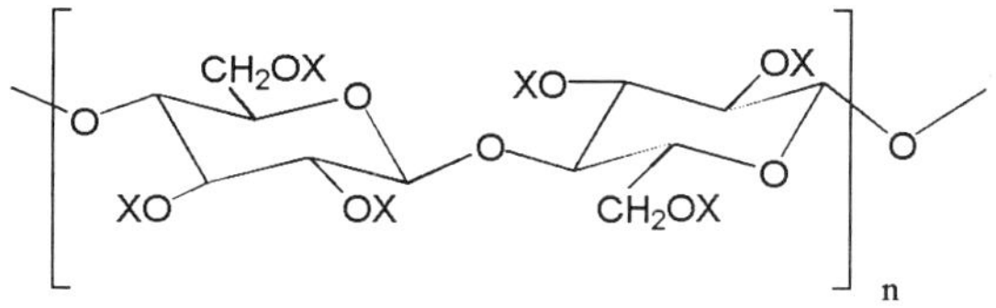

(3) Cellulosic Polymers

(3a) X = NO_2 cellulose nitrate (nitrocellulose)

(3b) X = R cellulose etherrs

(3c) X = COR cellulose esters

Fig. 4.5 The structures of polymers commonly used as film-formers in non-convertible surface coatings. Note: (i) in each case, R = alkyl group, commonly CH_3 to C_4H_9; (ii) cellulosic polymers (**3**) are derived from cellulose by replacement of the hydrogens of three hydroxy groups in the glucose rings by a group X. The structures are illustrated as if all three groups are substituted, although in practice the polymers are more likely to be mixtures of mono-, di- and tri-substituted products.

acrylic polymers (**1**), or, to a lesser extent these days, derivatives of cellulose (**3**).

A second, very important, group of non-convertible coatings are water-based latex (or emulsion) paints, such as those used for household decorative purposes and these are clearly much more environmentally-acceptable products. These paints consist of small solid particles of a high molar mass linear polymer, which is generally either of the acrylic (**1**) or vinyl (**2**) types, dispersed, together with the pigments, in an aqueous phase. The structures illustrated in Fig. 4.5 for these polymers are somewhat simplified since co-polymerisation with other monomers is commonly used to facilitate the film formation process and to modify the mechanical properties of the film. The aqueous phase contains a number of additives which are important to give the paint certain properties. For example, these include a water-soluble polymer, commonly a cellulose derivative, which assists in stabilising the dispersion of the polymer particles and the pigment, and is also used to control the rheological properties of the paint (see section 4.1.3), for example in providing the 'non-drip' qualities which are frequently desirable in this type of paint. Since the film-forming polymer does not actually dissolve in the water, these paints are two-phase systems, the polymer particles being referred to as the dispersed phase and the water as the continuous phase. After application of the paint to the surface, the water evaporates and the particles of polymer coalesce to form a continuous solid film. This coalescence, which is critical to the success of this approach to coating, involves overcoming the forces which were present in the wet paint specifically to stabilise the dispersion of particles. The driving forces for coalescence are thought to arise from capillary forces present as the particles come into contact with one another, and the reduction in surface energy as the very fine particles form a continuous film. Small amounts of water-miscible organic solvents (coalescing solvents) are also commonly incorporated into the formulation in order to facilitate the film formation process. Finally, the polymer chains become entangled with one another forming a coherent film. It is essential, therefore that the T_g of the polymer which is forming the film must be below ambient temperature so that the polymer molecules are capable of an appropriate degree of mobility. This is achieved from polymers **1** and **2**, by structural modification using co-polymerisation either to incorporate appropriately-flexible side chains or to increase the flexibility of the polymer main-chain or backbone. The polymer, however, does not change chemically as a result of this process. These water-based products are commonly referred to colloquially as 'emulsion paints', a term which derives from the fact that the film-former is prepared by a process in which an emulsion (i.e. a dispersion of droplets of one liquid in another) of liquid monomer in water is polymerised (referred to as emulsion polymerisation). However, since the polymer which is produced for incorporation into the paint is in the form of a dispersion in a liquid of tiny particles which are solid, the term 'emulsion' strictly no longer applies and the term 'latex' is more accurate.

Surface coatings which dry with chemical reaction

In these so-called *convertible* coatings, film formation involves as the first stage evaporation of the liquid, as is the case with non-convertible coatings, but afterwards the process is completed as the film-former undergoes a chemical reaction or a series of such reactions, generally involving polymerisation processes leading to a highly resistant cross-linked polymeric film. These coatings are generally used when the particular application demands good durability, since a crossed-linked polymer is generally more resistant to its environment compared with a linear polymer. A further advantage of this approach to coating is that the film-formers are relatively low molar mass materials which may be used in solution at high concentrations without giving rise to excessive viscosities, thus minimising the quantity of organic solvents which must be used, and consequently reducing the environmental impact.

In a convertible coating, after evaporation of the solvent, chemical reactions take place between functional groups on the film-former which result in the build-up of the molar mass of the system. An extremely important parameter in explaining film formation behaviour in the case of such systems is the *functionality* of the monomeric species. When there are two functional groups present (functionality 2) the molecules join together end-to-end forming a linear polymer as illustrated in Fig. 4.6(a). To form a cross-linked polymeric film, it is essential that the 'average' functionality of the system must be greater than 2. This is illustrated in Fig. 4.6(b) for the case of the reaction of a di-(A)-functional monomer with a tri-(B)-functional monomer. At a certain stage in the process, the reaction forms a polymeric network of effectively infinite molar mass, which is known as a *gel*. At the onset of gellation, major changes in the physical properties of the polymerising system take place. As a result of the dramatic increase in viscosity at the so-called gel point, processes such as flow and levelling are brought to an abrupt end. However, typically the glass transition temperature (T_g) of the system will remain below ambient temperature and the material will behave as a weak, soft rubber. Further cross-linking reactions must take place within the polymer after gellation before the criteria for complete film formation are met.

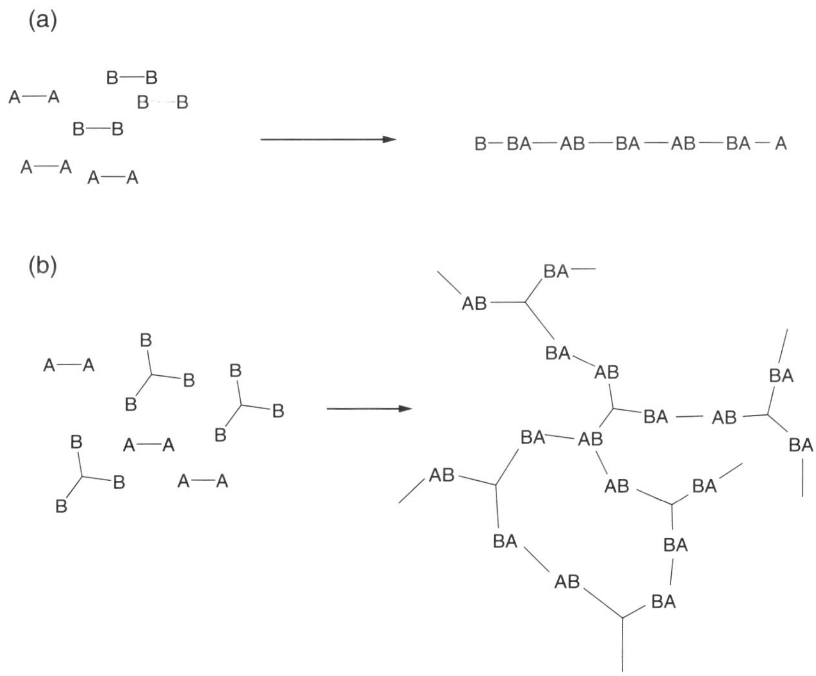

Fig. 4.6 Schematic representation of cross-linking reactions in convertible surface coatings. (a) Formation of linear polymer; (b) formation of cross-linked polymer.

The range of film-formers used in convertible coatings is remarkably diverse and volumes have been written on the subject. The chemistry of the cross-linking reactions occurring as the paint solidifies to a coherent film is generally extremely complex and in many cases incompletely understood and so it is dealt with only in outline and for selected cases in this text. In some cases, the polymerisation processes are due to the reaction of the film-forming ingredients with external factors such as the oxygen in the atmosphere, while in others there is no external involvement and polymerisation takes place only by reaction of the ingredients within the film. Traditionally, the film-formers used in convertible coatings are referred to as *resins*, although there is an argument that this term strictly refers to the final cross-linked structure of the film after chemical change due to the film formation process, and that the term *resin precursors* provides a more accurate description of the ingredient in the low molar mass form in which it is incorporated into the wet paint or ink.

Alkyd resins are the most extensively-used film-formers in decorative paints, particularly aliphatic hydrocarbon solvent-based gloss paints, and they are also used commonly in certain industrial paints. In fact, because of their extensive use, they are frequently referred to as the 'work-horse' of the paint industry. They provide tough, flexible finishes and good adhesion to surfaces. Most alkyd resins dry at normal temperatures by an air oxidation process. They are complex polymers consisting of a linear polyester backbone with long alkyl group side chains, commonly eighteen carbon atoms in length. A typical molecular structure of an alkyd resin is represented in Fig. 4.7, although it must be emphasised that such a structure is a gross simplification particularly in view of the fact that most commercial resins will be extremely complex mixtures of a series of structurally-related molecules. These resins are semi-synthetic in nature. The particular resin whose structure is given in Fig. 4.7 will have been derived from two main ingredients: triglycerides, which are components of some natural oils (such as linseed oil), and phthalic anhydride, the chemical ingredient which supplies the polyester backbone. The side-chains in the resin molecules contain a significant number of (C=C) double bonds, sometimes referred to as unsaturation, which are the reactive centres of the polymer molecules. A resistant cross-linked polymeric film is produced when the paint dries as a result of oxidation reactions between the unsaturated centres in the

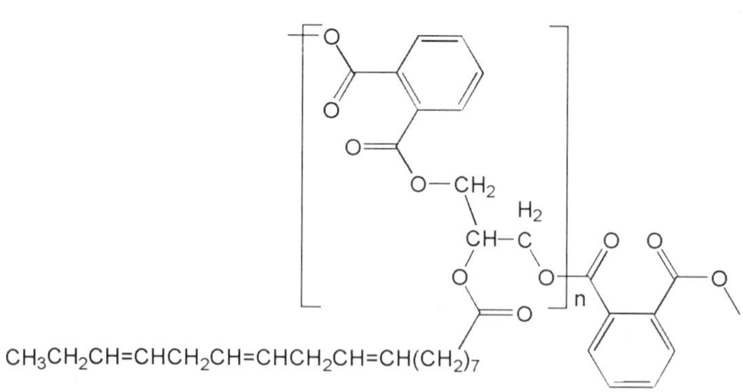

Fig. 4.7 Typical molecular structure of an alkyd resin (or resin-precursor) film-former.

polymer side-chains and oxygen in the atmosphere. A complex series of reactions involving free radical polymerisation is involved and, in the process, oxygen is taken up chemically into the film, forming peroxide (C—O—O—C) and ether (C—O—C) cross-links. The air oxidation process is promoted by the presence of small quantities of metal derivatives, generally salts of aliphatic carboxylic acids which are soluble in the hydrocarbon solvents used in the paint formulation. These components act as catalysts for the oxidation process. These catalysts were formerly one of the sources of lead in paints, but nowadays combinations of 'safer' metals such as cobalt, calcium and zirconium are used in preference.

There are a wide range of other film-formers which dry by chemical reactions in which the atmosphere plays no part. Such materials include amino resins, such as urea-formaldehyde and melamine-formaldehyde resins, epoxy resins, polyurethanes and phenolic resins. These coatings contain ingredients, often low molar mass polymers, which when subjected to a curing process undergo polymerisation to a cross-linked structure with high durability. Frequently in the drying of paints containing film-formers of these types, the final cross-linked polymeric film is formed from a polymerisation reaction involving the film-forming ingredients brought about by the application of heat. This type of industrial paint is used for example on car bodies or domestic appliances which are coated and then passed through ovens to produce rapidly a hard, durable finish. As an example, melamine-formaldeyde resins are commonly used in conjunction with certain alkyd resins which contain free hydroxy (OH) groups. When such a formulation is heated to temperatures in excess of, say, 140°C, a complex set of condensation reactions occur, for example liberating molecules of water or low molar mass alcohols, leading to a highly resistant cross-linked polymeric film.

Alternatively, normal temperature drying to a highly durable finish may be required, for example when the scale of application means that high temperature drying is not practical, such as on oil or gas pipelines or on the interior surface of the tanks in oil-tankers. One way in which this may be achieved is by the use of a 'two-pack' system in which two reactive ingredients are mixed just prior to application. Reaction takes place over a period of time to give a resistant cross-linked film. As an example, low molar mass epoxy resins, when mixed with certain polyamines and applied to a surface become touch-dry in a few hours and dry to a resistant coating in a few days. The curing involves an addition reaction of the amino groups in the polyamine component with the terminal epoxide groups of the resin as illustrated in Fig. 4.8. A highly resistant cross-linked polymer results.

4.2 Printing inks

The volume of material produced by printing, colour printing in particular, has increased dramatically in recent years. The wide range of substrates on to which printing is now carried out includes paper (in the production of

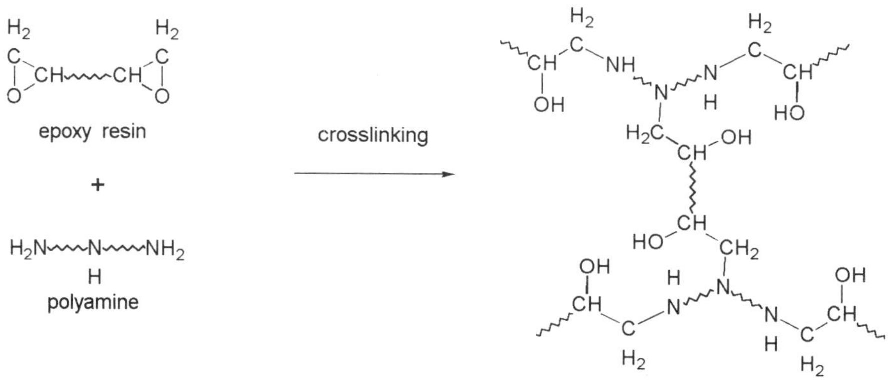

Fig. 4.8 Reaction scheme for the cross-linking reaction between an epoxy resin and a polyamine.

books, newspapers, magazines, banknotes, stamps, posters, etc.) and also board, plastic film and containers, metal foils and textiles. The printing ink is the fluid suspension which is printed as a surface coating on to the substrate and which then dries as a thin solid film. As is the case with paints, a printing ink must show good adhesion to the substrate to which it is applied and must retain its mechanical and colour properties for the expected lifetime of the printed product. However, unlike paints which have both a decorative and protective function, printing inks generally serve only the purpose of reproducing images or providing a decorative effect. Commonly this means that the durability requirements of inks are less than is the case with paints. Before considering the nature of printing inks, it is important to understand the nature of the various printing methods used, since the means by which the ink is transferred to the substrate is one of the most important factors to be taken into account when formulating the ink.

4.2.1 *Printing processes*

The printing industry, together with its associated industries such as publishing and packaging, represents a major component of most national economies. Printing involves the production of multiple copies of an original which may take a variety of forms, but is usually text or pictorial material. Printing is carried out on to a wide range of substrates including paper, board, plastic film, metal foils and textiles. Textile printing has many unique features and so it is considered in detail as a separate topic in Chapter 7. The substrates on to which the printed image is to be transferred may be presented to the printing press as sheets, i.e. sheet-fed systems, or from a continuous roll, referred to as web-fed systems.

There are five well-established major printing processes currently in use, namely lithography, flexography, gravure, letterpress and screen printing. The first four of these traditional techniques use printing plates to carry the image to the substrate. These plates may be either flat or, more commonly for continuous printing, cylindrical. On these plates there are two distinct physical areas: the image or printing areas, and the non-image or non-printing areas. In relief printing (as in the letterpress and flexographic processes), the image area of the printing plate is raised above the non-image area, while in gravure printing the reverse is true and the image area is recessed. In the lithographic process, currently the most important of the traditional methods, the image and non-image areas are in the same plane, and the process relies on these areas being chemically different. The printing plates may be manufactured using engraving processes or alternatively, and more commonly nowadays, using appropriate photographic methods. Screen printing is rather different from the other traditional printing processes in that it involves a stencilling process. Printing processes may be categorised as either direct or offset methods. Most gravure, flexographic, letterpress and all screen printing is carried out using the direct method in which the transfer of the ink takes place directly from the image carrier to the substrate. In offset printing, the image is transferred from the image carrier to an intermediate cylinder, from which it is subsequently transferred to the the substrate. Most lithography is carried out in this way.

The printing industry has for many years been at the forefront of technological change, and as a result a number of newer printing techniques have emerged for specific purposes many of these involving computer-assisted methods for reproducing the image. In addition to the traditional techniques, therefore, this chapter describes the techniques of ink-jet printing and toner printing which are used extensively for office printing purposes. Finally, there is a sub-section which outlines the way in which multi-coloured prints are obtained.

The offset lithographic process

Lithography has emerged as the most important printing technique, especially for general commercial literature, books, catalogues, greeting cards and packaging. In lithography, both the image and non-image areas are on the same plane on the printing plate. The process utilises both water (a highly polar solvent) and a viscous printing ink (based on a non-polar resin system) and makes use of the very different chemical characteristics of these two materials. The plate is treated chemically such that the non-image areas are hydrophilic (water-accepting) in character and thus attract water and repel the ink, which is hydrophobic or non-polar in character. The image areas, in contrast, are hydrophobic (water-repelling) and attract the ink (Fig. 4.9).

The principles of offset lithography are illustrated in Fig. 4.10. The process may be separated into a number of stages. A damping system supplies a carefully controlled amount of water, or fountain solution, to the plate or cylinder. The fountain solution contains varying quantities of water-soluble additives which are of importance to the printing process

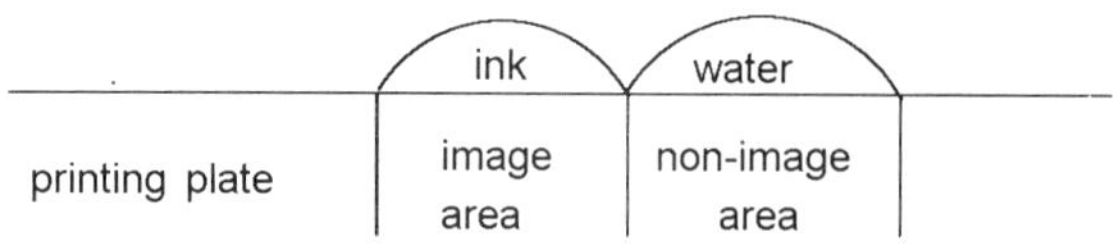

Fig. 4.9 Surface interaction on a lithographic printing plate.

such as film-forming agents, desensitisers, buffers and biocides. This solution forms a thin film on the water-accepting areas, but contracts into tiny droplets on the water-repellent areas. When the inked roller passes over the damped printing plate, the water film prevents ink contacting the the non-image areas, but the droplets of water on the image areas are replaced forming a continuous ink film. The process is called offset lithography because the plate does not print directly on the paper, but is first 'offset' on to a rubber blanket, wrapped around the offset cylinder. From there the image is transferred to the paper with the aid of pressure from the impression cylinder. Lithographic inks are generally highly viscous in order to maintain the integrity of the image under the pressure of printing and as an aid to water-repellency.

The popularity of lithographic printing is due to the fact that it is capable of high-speed operation, gives prints of good quality and is inexpensive, largely because the printing plates are easily prepared by a simple photographic process. The major difficulties of the process are associated with the maintenance of the water/ink balance.

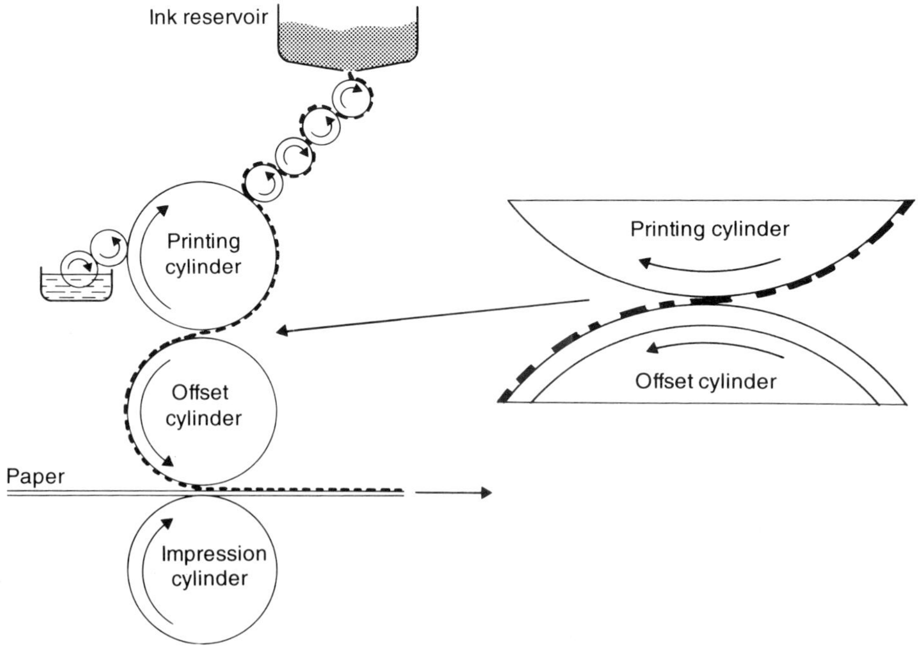

Fig. 4.10 Offset lithographic printing.

The flexographic process

Flexography is a process in which the printing image is raised above the non-image areas (Fig. 4.11). In this respect, the process is similar to letterpress printing (see below). The raised printing surface is generally made from a flexible material, such as rubber, mounted on a metal cylinder. This flexible surface is capable of transferring a good image to substrates with a wide range of surface characteristics. However, the process is especially well-suited to non-absorbent substrates such as plastic films, metal foils, and a range of non-absorbent paper products and so for printing on to such substrates it is frequently the method of choice. The printing unit consists of an inking unit, plate cylinder and impression cylinder. A system of rollers transfers the ink from the reservoir as a controlled thin film to the raised image areas on the printing cylinder. The substrate passes between the plate cylinder and the impression cylinder and receives the image. A low viscosity liquid ink is used, drying rapidly by solvent evaporation, in order to ensure adequate flow of ink through the distribution system from the ink reservoir to the substrate. This is achieved by a high solvent content, using carefully selected solvents such as alcohols, esters and ethers which do not attack the rubber surface of the printing plate. While much flexographic printing is still based on organic solvent-based inks, considerable progress has been made in recent years in the development of water-based inks and it is likely that these will increase in importance in years to come.

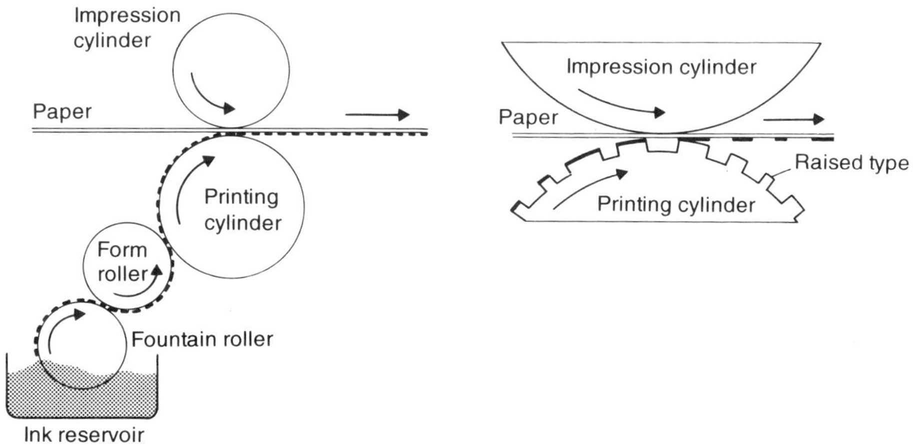

Fig. 4.11 Flexographic printing.

The gravure process

In the gravure process, the printing image is engraved into a cylinder (Fig. 4.12). The printing cylinder rotates in an ink reservoir and recesses which contain the image fill up with the ink. A doctor blade wipes clean any surplus ink from the non-image areas on the surface of the printing cylinder,

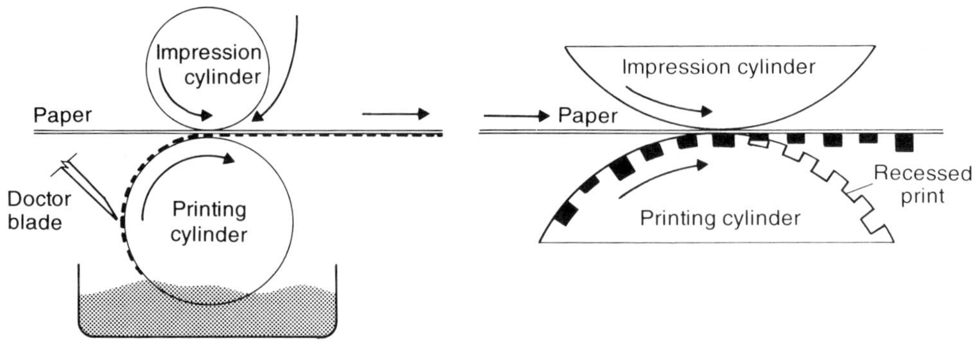

Fig. 4.12 Gravure printing.

leaving ink only in the recesses. Printing is achieved by passing the substrate between the printing cylinder and an impression cylinder under pressure. A liquid ink with a high solvent content is used so that the ink flows easily into the printing recesses, is subsequently attracted to the substrate by capillary action and dries rapidly. Gravure printing is frequently the method of choice for long-run printing where high quality prints are required, for example postage stamps, mail-order catalogues and travel brochures. The process is also used on a range of paper and board packaging materials and as an alternative to flexography for plastic films and metal foils.

The letterpress process

The letterpress process uses printing plates in which the image areas are raised above the non-image areas (Fig. 4.13), and in this respect resembles flexography (see above). In the letterpress process, ink is applied by a series

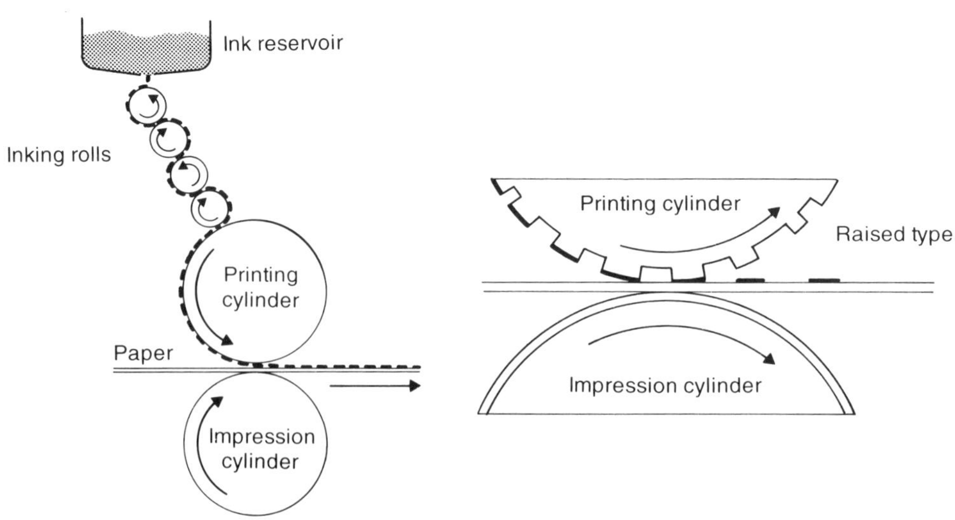

Fig. 4.13 Letterpress printing.

of rollers to the printing plate and a film adheres to the raised image areas. The substrate is brought directly into contact with the inked surface and sufficient pressure is applied to transfer the ink to the surface. The ink is of high viscosity, similar to lithographic inks, to ensure adhesion to the plate surface and to avoid spreading of the image under the very high printing pressures used. The letterpress process was formerly the dominant printing process, particularly for newspapers and other published printed materials, but its use has declined sharply in favour of the other methods, notably lithography (see Table 4.1). It is used still to a certain extent for short-run printing such as letter headings and business cards.

Table 4.1 A summary of the main features of the four traditional printing processes.

Printing process	Nature of image areas	Ink transfer method	Ink viscosity
Lithography	plane	offset	high (paste)
Flexography	raised	direct	low (liquid)
Gravure	recessed	direct	low (liquid)
Letterpress	raised	direct	high (paste)

Screen printing

Screen printing is a stencilling process in which ink is transferred to the substrate through a screen to produce the image. The process was formerly known as silk screen printing as the mesh which forms the image supported by the screen was frequently made of silk, although nowadays synthetic fibres or metal are more commonly used. In flat screen printing, the screen which carries the image is supplied with the ink and a squeegee blade forces the ink through the pores in the mesh under light pressure. At the same time the substrate is held in contact with the screen and receives the ink. For continuous production, rotary-screen printing is commonly used. In this technique, ink is fed to the inside of a cylindrical screen and is forced through the mesh using a stationary squeegee. Screen printing inks may be formulated to adhere to almost any surface and so the process is extremely versatile. Because of the simplicity of the process and the ability to print thick films, it is the most commonly used process for printing on to textiles. The principles of screen printing on to textiles are dealt with in Chapter 7. Screen printing is used for other substrates in cases where thick films of ink are required, and where simplicity of the design will allow. Examples include posters, road signs and PVC window stickers.

Ink-jet printing

Ink-jet printing is a non-impact means of generating images which involves directing small droplets of ink in rapid succession, under computer control, on to the substrate. There are a number of types of ink-jet printing methods.

Two of the principal types are the continuous jet method and the impulse or 'drop on demand' method.

In the first method, as the name implies, a continuous stream of ink droplets is forced out of a nozzle under pressure (Fig. 4.14). Some of the droplets are deflected and targetted at the substrate in such a way as to form the image, while the remainder are collected and either recycled or discarded. This selective deflection is achieved by first electrostatically charging some of the ink droplets as the jet passes through a charged electrode. The jet then passes between high voltage oppositely-charged deflection plates and the droplets which have acquired a charge are deflected by the field towards the substrate. The uncharged droplets continue undeflected and are collected in a gutter.

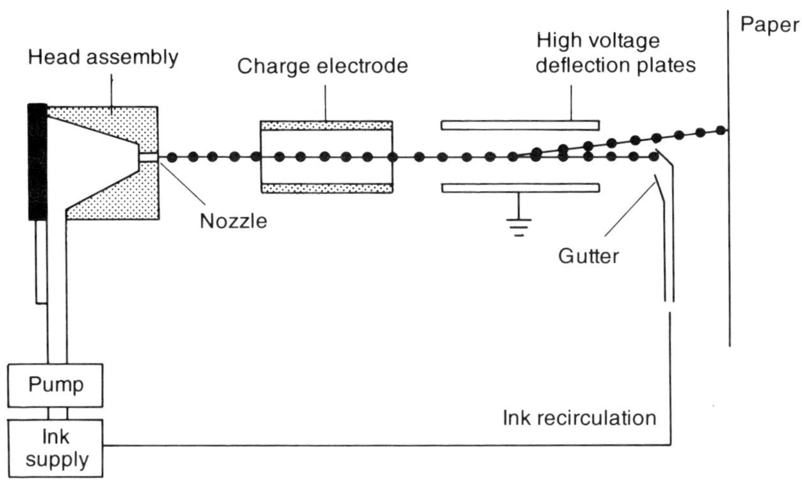

Fig. 4.14 Continuous ink-jet printing.

In the impulse or 'drop on demand' method (Fig. 4.15), the pressure on the ink is not continuous but is applied when the droplet is needed to form part of the image. An array of nozzles is used to generate the image so that no deflection system for the droplets is required and, since all the ink is used, there is no wastage. In drop on demand ink-jet systems, the print-head is required to be as close as possible to the substrate surface so as to produce an accurate image, while in continuous ink-jet systems the droplets can travel much further distances.

Ink-jet printing requires the use of inks which meet very stringent physical and chemical criteria. Inks using materials of extremely high purity principally to ensure very exact conductivity specifications are required. Ink-jet inks are low viscosity fluids with a remarkably simple composition consisting of a solvent and colorant together with other specific additives. In contrast to most other printing inks, the colorants used are mainly dyes because pigments, even when an extremely fine particle size is used, have a tendency to block the nozzles. For water-based inks, special dyes have been

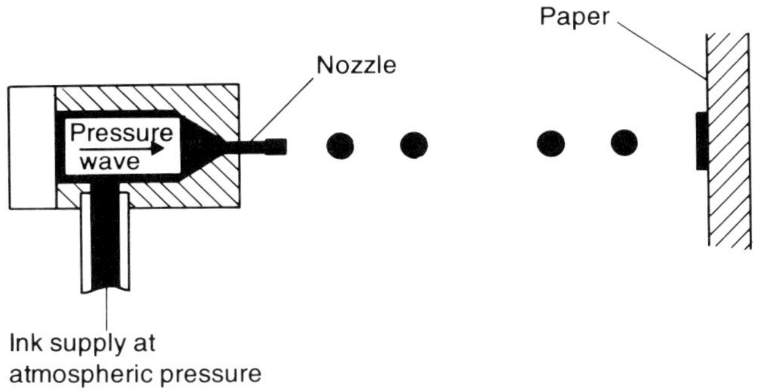

Fig. 4.15 'Drop-on-demand' ink-jet printing.

developed which are soluble at the slightly alkaline pH of the inks, but which are rendered insoluble by the weakly acidic pH conditions on the paper substrate. These provide prints which are reasonably resistant to wet treatments. Ink-jet systems are used extensively for home and office printing where high-quality text and graphics reproduction is required. The technology seems also set to make significant inroads into higher volume industrial printing, for example for packaging, textiles (see Chapter 7), wall-coverings and advertising displays.

Toner printing (photocopying and laser printing)

In these printing systems, the ink is a toner which is generally a powder consisting of pigment, charge control agent, and a low melting binder. Toner printing systems use optical or electrical methods to form an electrostatic latent image to which the toner is attracted and subsequently transferred to the substrate. There are a variety of toner printing methods including electrophotography, ion deposition, electrostatic, magnetographic and electrographic systems. Of these, electrophotography or xerography, as it is sometimes called, is the most widely used, as the basis of most photocopiers and laser printers. These two processes have distinct similarities so that it is convenient to discuss them in the same section, but but there are also some key differences.

The photocopying process, as illustrated in Fig. 4.16, may be separated into six stages. One of the key aspects of photocopying is a photoconducting material which, as the name implies, is a conductor of electricity in the presence of light but an insulator in the dark. In the first stage of the process, the dark photoconducting surface of a drum is given a uniform electrical charge. In the imaging step, the document to be copied is illuminated with white light. Where there is no print, light is reflected on to the photoconducting surface and this causes the charge to be dissipated. Where there is text, light is absorbed and does not reach the photoconductor. In this way, a latent electrostatic image is formed on the drum. The drum is then exposed to toner particles, which have been given an

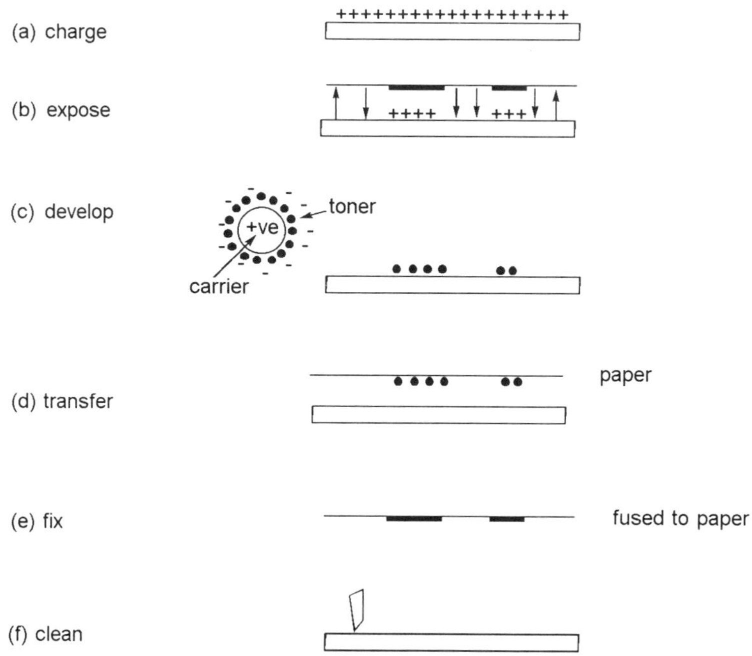

Fig. 4.16 The photocopying process.

opposite charge to that on the drum. The toner particles are thus attracted to the image areas of the photoconducting surface. The next step involves transfer of the image from the drum to the substrate, usually plain paper, by applying an electrical potential to the back of the paper of opposite charge to the toner particles. The next step fixes the image to the paper. This is achieved by a heat treatment which melts the toner resin and fuses it to the paper. Finally, the photoconducting surface is cleaned to make it ready for the next copying cycle.

In a laser printer, the first step also involves giving the photoconducting drum a uniform electric charge. However, in the second step a laser is used to write on the charged drum the information stored in the memory of the printer. In contrast to the photocopying process, therefore, the charge is dissipated in the image areas. In this case, the toner used carries a charge which is the same as that on the drum, so that it is repelled from the non-image areas on to the uncharged image areas. From this stage onwards, the principles of the laser printing process are similar to the photocopying process.

Colour printing

Black and white images are reproduced in a straightforward way by means of the various methods described in this chapter using inks generally containing carbon black pigments. Colour printing requires the accurate reproduction of images containing all the colours in the spectrum. Most colour reproduction is carried out using a four-colour printing process. In this technique, the coloured original is separated, generally using a photographic method, into four images each of which is printed from a separate printing plate or image-carrier with a different colour of ink to recreate the visual impression of the original. The four colours used are the subtractive primary colours, yellow, magenta and cyan, together with black (see Chapter 8 for an explanation of the concepts of primary colours). The different colours in the final print are produced by appropriate combinations of the yellow, magenta and cyan inks printed on to a white background. Although in principle a black ought to be produced when the three subtractive primaries are overprinted, the use of a black ink is required to give adequate reproduction of neutral colours, for example in shadows. This is because the coloured pigments used in the inks are not capable of providing perfect subtractive primary colours. The organic pigments commonly used in the inks are the Diarylide Yellows, metal salt azo magenta pigments and β-copper phthalocyanine which provides the cyan. The structures and properties of these pigments are described in Chapter 6. They provide the appropriate bright, intense colours and adequate fastness properties. In addition, the inks produced from these pigments are transparent which is essential for accurate colour reproduction. Carbon black is the major pigment used for the black inks.

4.2.2 The nature of printing inks

Printing inks are in many ways similar in composition to paints, as discussed in section 4.1. For example, the principal components of the inks are pigments, film-formers and a liquid, together with minor quantities of other additives which are incorporated for specific purposes. Each of the main components serves functions similar to its role in paint formulation and the reader is directed to the previous section for a detailed discussion of these. The film-former binds the pigment particles together, provides the mechanical properties of the dry film and gives adhesion to the substrate to which the print is applied. The liquid serves as a carrier of the other ingredients making the ink sufficiently fluid for the printing process. The pigment, of course, gives the ink the desired colour. In general, printing inks will have a higher coloured pigment content than paints. The reasons for this are the high intensity of colour required from a printing ink to give the desired visual impact, and the fact that printing inks are applied as much thinner films than paints. An understanding of rheology is as important with printing inks as it is with paints (see section 4.1.3). Printing inks are formulated to have a particular viscosity, which is determined by the nature of the printing process used, and the mechanism by which it is delivered by the printing process to the substrate. It is convenient to classify printing inks according to their viscosity as either paste inks, which as the name implies are of high viscosity (low solvent content), or liquid inks, which are of low viscosity as a result of a high solvent content. Letterpress and lithographic inks are paste inks, whereas gravure and flexographic inks are liquid inks.

The mechanisms by which printing inks dry, or solidify, for the most part involve similar physical and chemical principles to those discussed previously for paints (section 4.1.5), but with a few key differences. There are two distinct stages in the drying of a printed film of ink. In the first stage of drying an ink must 'set' to a consistency which will not result in smearing or transfer when subjected to light mechanical handling. Because of the delivery speed of continuous printing processes, setting must occur extremely rapidly to avoid 'set-off', for example, the transfer of ink from a printed sheet in a stack on to the back of the next sheet falling upon it. In the second stage of drying, which can take place over a somewhat longer period, the film acquires its final hard resistant form. As is the case with paints, printing inks may dry either with or without chemical reaction taking place, i.e. they may be considered as either *non-convertible* or *convertible* coatings. Although it is convenient to discuss the drying mechanisms in isolation, in practice it is normal for two or more of these to operate in combination.

Printing inks drying without chemical reaction

The two main printing drying mechanisms where no chemical reaction is involved are absorption and evaporation. Absorption involves penetration of the ink by capillary action into the pores of the substrate on to which the ink is printed. This process is frequently assisted by the pressure applied in the printing process. Inks which dry by the absorption mechanism alone provide rapid setting but because the ink remains as a fluid it generally leads to poor mechanical resistance. In the evaporation mechanism, the removal of a volatile solvent leaves a film of pigment and film-former. Flexographic and gravure inks generally dry by evaporation. The film-former is a hard linear polymer which is soluble in the ink liquid. Fig. 4.6 shows the structures of the range of film-formers used most often in non-convertible coatings. The most commonly used film-former in the case of these liquid inks is nitrocellulose (cellulose nitrate) (**3a**), although acrylic (**1**) and vinyl (**2**) polymers are also used. The solvents commonly used are alcohols, esters and ether-alcohols. However, because of the obvious economic and environmental advantages, liquid inks using water as the solvent with appropriately designed film-formers are becoming increasingly important.

Printing inks drying with chemical reaction

In printing inks where the drying process involves chemical reaction, i.e. convertible coatings, film-formers similar to those used in paint formulation are employed (see section 4.1.2). These reactions lead to resistant cross-linked polymeric films as a result either of air-oxidation of the film-former or reaction between film-forming ingredients. As with paints, the drying process may be accelerated by the application of heat. In cold-setting lithographic inks, alkyd resins such as that illustrated in Fig. 4.7 are commonly used. These resins dry by an oxygen-induced free radical polymerisation process which introduces ether and peroxide cross-links as a result of reaction with unsaturated centres in the side-chains (see section 4.1.5). In some quick-setting lithographic inks, a typical formulation ink might contain pigment, a non-drying hard phenolic resin, an alkyd resin drying by air oxidation and a relatively small amount of a high boiling hydrocarbon solvent. The initial phase of the drying of such an ink involves the absorption mechanism as the less viscous components of the ink (predominantly the hydrocarbon solvent) penetrate into the pores of the paper. This process which occurs rapidly produces a 'tack-free' or 'set' film as a result of an increase in the viscosity of the ink remaining on the surface. Subsequent oxidation of the alkyd resin converts the film to a chemically and physically resistant film over a period of some hours. Heat-setting lithographic inks consist of a resin combination in a high-boiling hydrocarbon solvent, drying principally by evaporation as the printed product, usually a continuous sheet, is passed through high-temperature ovens.

Ink-jet printing inks are very low viscosity materials as is required by the non-impact method of ink delivery. These inks, unlike most other printing inks, use soluble dyes as the colouring materials. These are used in preference to pigments as the particles of pigment tend to block the nozzles of the jets. Most ink-jet inks in current use are water-based. Toner inks, used in photocopying and laser printers, are generally powders often attached to carrier particles, such as iron powder. The toners consist of pigment (most commonly carbon black, although the production of multicoloured prints using pigments of the three subtractive primary colours, yellow, magenta and cyan is increasing in popularity), charge control agents which are materials assisting in the control of electrostatic charge applied during the printing process, and resin, a relatively low melting (60–70°C) polymer.

Recent years have seen the development of rapid-drying inks cured by the application of ultraviolet or electron-beam radiation. Of these, UV-curing inks have gained most commercial acceptance. These inks contain as the film-forming ingredients compounds (monomers or pre-polymers) capable of undergoing a polymerisation reaction, together with small amounts of materials termed photo-initiators which promote the reaction in the presence of UV radiation. The film-formers are themselves liquids, avoiding the need for an additional solvent in the formulation of the ink, which offers obvious potential environmental advantages. When the printed sheet is exposed to UV radiation, a free-radical polymerisation process takes place and a solid cross-linked polymeric film is produced almost instantaneously.

Bibliography

Outlines of Paint Technology, Vols 1 and 2, W.M. Morgan, Griffin, London (1984).

Surface Coatings, Vols 1 and 2, 2nd edn, Oil and Colour Chemists Association, Australia, Chapman & Hall, London (1983).

Introduction to Paint Chemistry and Principles of Paint Technology, G.P.A. Turner, Chapman & Hall, London (1980).

Organic Coatings: Science and Technology, Vols 1–6, G.D. Parfitt (ed.), Dekker, New York (1983).

Treatise on Coatings, Vols 1–4, R.R. Myers (ed.), Dekker, New York (1975).

The Chemistry and Physics of Coatings, A.R. Marrion, Royal Society of Chemistry, London (1994).

The Printing Ink Manual, R.H. Leach (ed.), 5th edn, Blueprint, London (1993).

Printing Inks for Lithography: Chemistry, Formulation, Application and Trouble Shooting, D.J. Owen, SITA Technology, London (1990).

Chapter 5
Application of Dyes

5.1 Selection of dye

The primary consideration in choosing a dye is that it can reproduce the desired colour in the material to which it is applied. Moreover, an intense colour is essential, so that the material can be dyed with as little dye as possible. The colour measurement and specification of dyes is the subject of Chapter 8.

However, colour itself is far from being the only consideration, and it is instructive to consider the application of a dye in mechanistic terms. It is clear that dye has first to reach the surface of the solid material to be dyed and then to be adsorbed by that surface. Thus, dye molecules have to be soluble in the dye liquor, so solubility is an important factor. Some types of dye, however, such as vat and sulphur dyes, are insoluble; these have first to be rendered soluble by reduction in alkali before application to the material. Disperse dyes too have only limited solubility, but it is sufficient – and essential – for their mode of action. Many disperse dyes are also capable of sublimation, so that individual molecules can be adsorbed by the material from the vapour phase, as in transfer printing.

Once dye molecules have been adsorbed by the surface of the material, they are normally required to penetrate into its bulk. This requirement is essential for dyeing textiles. Furthermore, once into the bulk of the material, they must remain there as dye molecules during the lifespan of the material. This is the basis of good fastness. As discussed in Chapter 2, retention of the dye can be achieved if there is sufficient intermolecular attraction between dye and material, a property which dyers term substantivity. The strongest attraction which can be realised between dyes and material is covalent bonding, and some dyes, known as reactive dyes, have functional groups able to react chemically with the material to be coloured.

An alternative method by which dye may be retained is through the formation of insoluble particles of colorant, trapped within the material. This approach is adopted in the application of vat and sulphur dyes; once the water-soluble reduced form has been incorporated into the material, it is oxidised back into the insoluble dye. Dyes of another class, azoic dyes, are also insoluble, but these are formed by chemical reaction within the fibres.

Levelness of colour too is important, a term which describes the even distribution of colour through the material. Some types of dye are able to migrate within the material to give levelness, but such dyes generally have poorer fastness properties. Less fugitive dyes have to be applied under more strictly controlled conditions, but their application may be assisted by suitable levelling agents. Levelling agents are discussed later in this chapter (section 5.6).

The choice of dye, therefore, depends on a wide range of fundamental criteria, and to achieve these criteria different types of dye are required for different materials. In Chapter 3, a number of materials were discussed, including textiles, leather and paper. Three different classes of textile material were also distinguished: cellulosic, protein, and synthetic and cellulose acetate fibres. Since by far the commonest use of dyes is in the coloration of textiles, this aspect will be considered first.

5.2 Cellulosic textiles

Before dyes are applied to any type of material, the material must be freed of impurity as far as possible. In the form of loose fibres and yarn, cellulosic textiles contain only those impurities found in the natural state. However, knitted fabric may also contain a small amount of lubricant, added to facilitate knitting. Woven fabric will also contain sizes, gelatinous substances applied to the yarn to impart strength for weaving. Starch sizes must be removed before scouring and dyeing; they can be broken down to soluble sugars by application of a suitable enzyme, such as diastase. Materials free of size are scoured in a bath with small amounts of detergent and alkali for 20–30 minutes at 80–90°C. The fibres are then thoroughly rinsed before dyeing. Undyed cotton is often also bleached with peroxides (or maybe hypochlorites), but for viscose rayon bleaching is usually unnecessary.

Several classes of dye can be successfully applied to cellulosic textiles, including direct, vat, sulphur, azoic and reactive dyes. The application of each is now considered in turn.

5.2.1 Direct dyes

Of the five classes, direct dyes are the simplest and most economical to apply. Being soluble in aqueous media, they are readily applied to almost all kinds of cellulosic textile, especially when salt (normally sodium chloride) is present in the dyebath. Their washfastness properties, however, are often only poor to moderate, although a suitable aftertreatment, once the dye has been applied, may give an improvement. Consequently, this class of dye is generally applied to cheaper cotton and viscose materials.

Direct dyes are ionised in aqueous media, the anions being coloured. Most are sulphonated azo compounds. A characteristic feature of direct dye molecules is their long, linear shape, which appears essential for substantivity to cellulosic fibres. An example of a simple monoazo direct dye is C.I. Direct Red 14 (**1**).

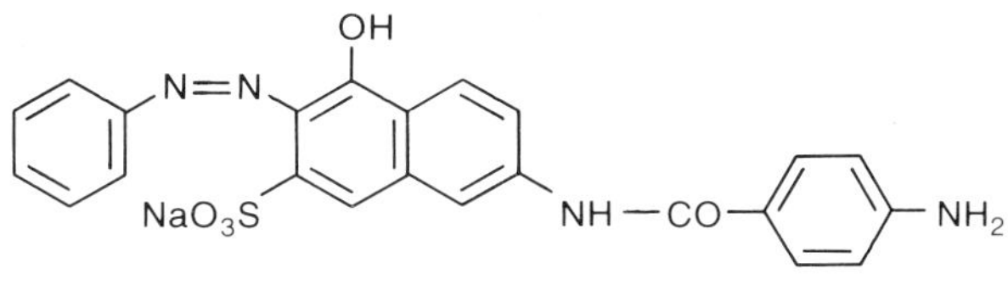

1

Some direct dyes contain as many as four azo groups per molecule, e.g. C.I. Direct Brown 44 (**2**).

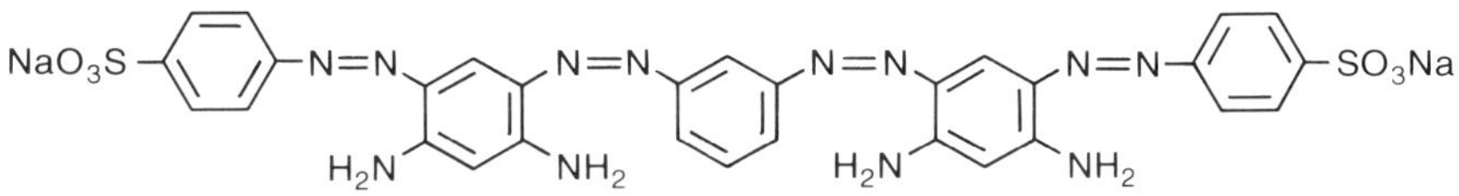

2

Application of direct dyes

A typical dyebath contains dye (or mixture of dyes), sodium chloride, and a wetting agent. A sequestrant may also be present (as described below).

Function of dyebath assistants

Sodium chloride: Sodium chloride is added to the dye liquor to promote exhaustion of dye, but the effectiveness of salt varies widely with different dyes. When immersed in water, cellulosic fibres assume a negative charge and so repel direct dye anions. By reducing the electrical potential on the cellulose fibres, electrolytes, such as sodium chloride, allow the dye ions to approach the fibres sufficiently closely for nonionic molecular interactions to become operative.

The degree of repulsion between cellulosic fibres and dye anions depends both on the overall negative charge carried by the cellulosic fibres and on the negatively charged substituents of each dye ion. Viscose rayon, which generates a greater negative surface charge than cotton, exerts a greater repulsion on dye anions. Similarly, the more negatively charged groups there are on the dye anion, the greater is the repulsion exercised by the fibres, and the more sensitive is the anion to the presence of salt.

Wetting agent: This is sometimes included to promote wetting of the goods.

Sequestrant: A few direct dyes are precipitated by hardness in water due to the presence of calcium and magnesium ions. Precipitation of these ions can be prevented by inclusion of a sequestrant in the dyebath. The sequestrant forms soluble chelate complexes with the metal ions.

Categories of direct dyes

The Society of Dyers and Colourists has established three main categories of direct dyes:

Class A – self-levelling. These dyes migrate well and give level dyeings, even when large concentrations of salt are present.
Class B – salt-controllable. These dyes have poor levelling properties, and exhaustion is effected through controlled additions of salt.
Class C – temperature-controllable. The exhaustion of these dyes cannot be adequately controlled through careful addition of salt alone. Careful control of temperature is also required.

With Class A dyes, salt can be present before the textile goods are introduced. With Class B dyes, however, salt has to be added gradually to the dyebath while the liquor and goods are being heated to the boil, and even afterwards at the boil. With Class C dyes, still greater care is necessary; the temperature is raised to boiling point very slowly indeed, and salt is gradually added only at the boil. If more than one dye is needed to obtain a desired shade, only members of the same category of direct dye should be mixed.

Aftertreatment of direct dyes

The poor wet-fastness properties of direct dyes may be improved by a suitable aftertreatment. Aftertreatments effect increases in the size and complexity of dye molecules after their incorporation into the fibres. The aim is to render the dye less soluble in water, and hence more resistant to wet treatments. Aftertreatments are always given once the dye has been applied to the fibres, for the resulting molecules would be generally too large to enter the fibres or, at best, would give highly unlevel dyeings.

Diazotisation and development

Direct dyes containing a primary amine group can be diazotised by treatment with sodium nitrite and hydrochloric acid, to produce a diazonium salt of the dye. This diazonium salt can then be coupled (or developed) with a phenol, naphthol or aromatic amine. For example, C.I. Direct Red 127 can be diazotised and developed with 2-naphthol, as illustrated in Fig. 5.1.

Development increases the size of the dye molecule, and washfastness is often improved. Development also changes the shade; different coupling components generally give different shades.

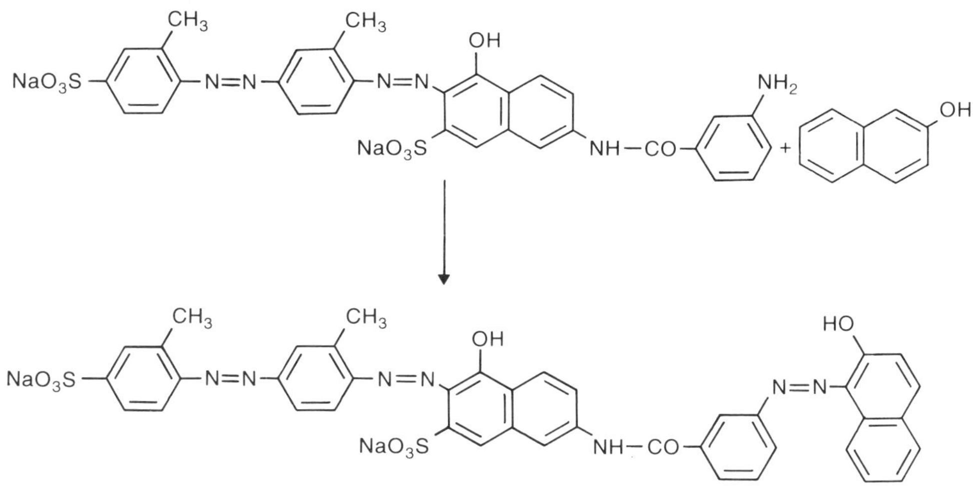

Fig. 5.1 Aftertreatment of direct dye, C.I. Direct 127, by diazotisation and development.

Coupling with a diazonium salt

The adsorbed dye may instead act as a coupling component, though this procedure is much rarer. The diazonium salt generally used is 4-nitroaniline. Thus, C.I. Direct Red 33 can be so treated (Fig. 5.2).

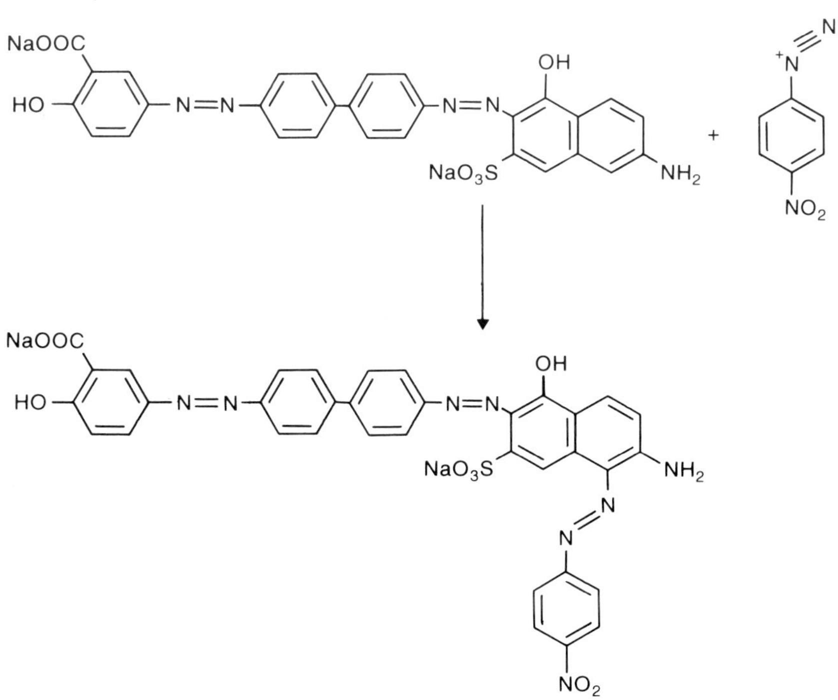

Fig. 5.2 Aftertreatment of the direct dye, C.I. Direct Red 33, with a diazonium salt.

A few direct dyes, such as C.I. Direct Red 155 and C.I. Direct Black 4, can be either diazotised and developed, or coupled.

Metallic salts: Aftertreatment can be effected using metallic salts, in particular copper sulphate, to form complexes of the metal and dye. The dye molecules must contain suitable groups (**3**):

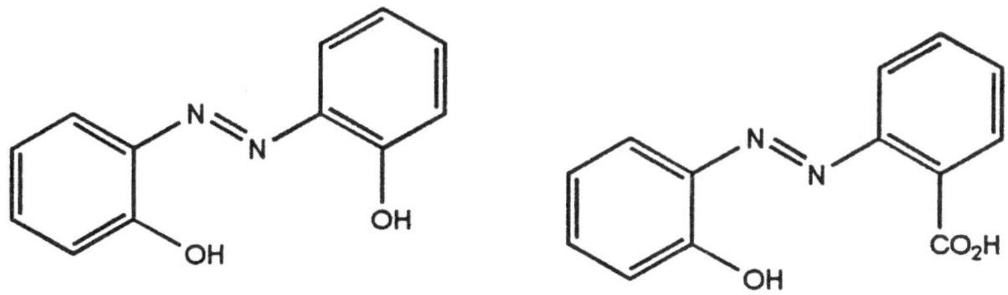

3

Washfastness is often improved, but duller hues are produced.

Formaldehyde (methanal): The mechanism involved in aftertreatments with formaldehyde is obscure. Dye molecules may be linked through methylene (—CH_2—) bridges. This type of aftertreatment is nowadays far less common, in view of the toxicological hazards presented by formaldehyde.

Cationic fixing agents: These improve wet-fastness by combining with the anionic dyes to form a complex, which is less soluble than the original dye. Although wet-fastness is improved, light-fastness is very often impaired, and changes of hue also often result.

5.2.2 *Vat dyes*

In contrast to direct dyes, vat dyes have better fastness properties. They are divided into two groups, the anthraquinonoid dyes and the indigoid dyes, both of which include a wide range of molecular structures. Although both groups possess good washfastness properties, especially where fastness to chlorine is essential, anthraquinonoid dyes are faster, most notably with respect to light. Examples of an anthraquinonoid and of an indigoid dye are shown in Fig. 5.3.

Despite their complex molecular structures, however, vat dyes do have one common feature: all contain carbonyl groups (>C=O). Vat dyes can, therefore, be reduced to leuco forms, whose molecules possess hydroxyl groups (≥C—OH). Although the leuco form is insoluble in water, it can be rendered soluble in alkali:

$$\geq C{-}OH + OH^- \rightleftharpoons \geq C{-}O^- + H_2O \tag{5.1}$$

The dissolved material can be oxidised back to the original vat dye.

Some dyes are marketed as solubilised vat dyes. These are generally sulphuric esters of the reduced form, e.g. C.I. Solubilised Vat Red 6, whose structure is shown in Fig. 5.3. These dyes dissolve in alkali and can be oxidised back to the parent vat dye by a mild oxidising agent.

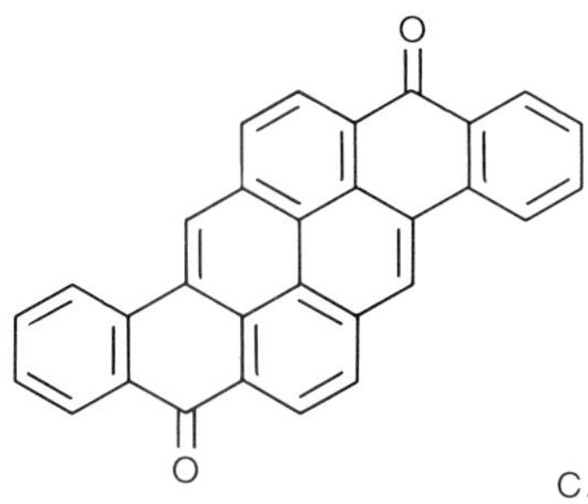

C.I. Vat Orange 9

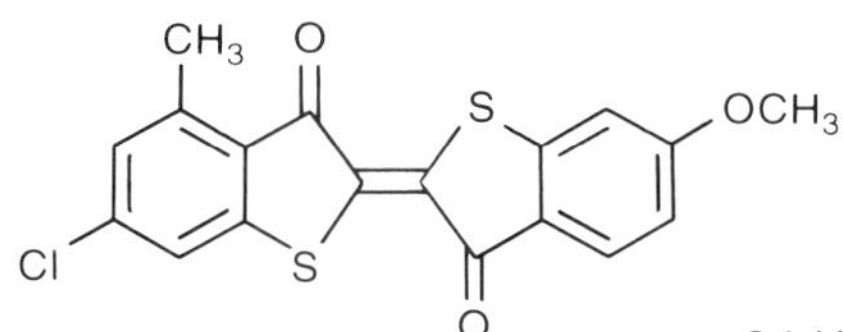

C.I. Vat Red 6

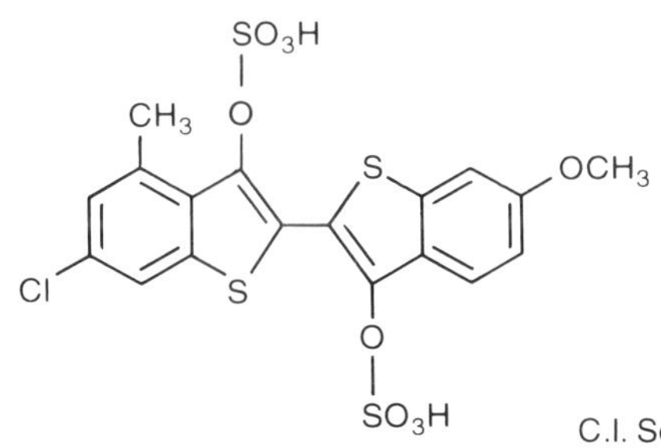

C.I. Solubilised Vat Red 6

Fig. 5.3 Examples of anthraquinonoid dye, C.I. Vat Orange 9, of an indigoid dye, C.I. Vat Red 6, and of a solubilised vat dye, C.I. Solubilised Vat Red 6.

Application of vat dyes

The application of vat dyes consists essentially of four stages:

(A) Vatting – preparation of the sodium derivative of the reduced (leuco) compound.
(B) Impregnation of the fibres with the leuco compound.
(C) Oxidation in the fibres to the insoluble vat dye.
(D) Soaping.

With solubilised vat dyes, step 1 is omitted.

Vatting: The required ingredients include dye (or mixture of dyes), sodium hydroxide and sodium dithionite. To prepare the leuco form, the vat dye is pasted with warm water, often containing a suitable wetting agent, and then reduced by sodium dithionite solution, made alkaline with sodium hydroxide. The amount of sodium dithionite required for reduction is influenced by the number of carbonyl groups in each dye molecule. In practice, this theoretical amount is well exceeded to allow for any oxidation, which may arise through air dissolved in the dye liquor, particularly in those machines which tend to aerate the liquor. The amount of sodium hydroxide required also depends on the number of carbonyl groups and is governed by the equilibrium in Eqn 5.1.

Vatting requires close control, for a number of undesirable effects can occur. If there is insufficient alkali, the leuco dye may be subject to undesirable molecular rearrangements. Careful control of temperature is also important; if the temperature is too low reduction may be incomplete, and if too high the reducing agent may react with the dye in another way. Moreover, some conditions promote the loss of halogen atoms from those vat dyes which contain them. Thus for reducing a given vat dye, there are well defined conditions of temperature, pH and concentration of reducing agent.

Impregnation of the fibres with the leuco compound: The leuco dye, once formed, is thus applied from a solution containing excess alkali and excess reducing agent. In many cases, sodium chloride is also present. The leuco form is anionic; leuco dyes are, therefore, similar to direct dyes in that anions are adsorbed by the cellulosic fibres. However, the concentration of electrolyte in the dyebath is particularly high, and consequently adsorption of dye is often rapid. On the other hand, migration is often slow. As with the application of direct dyes, electrolytes facilitate the approach of the leuco dye anions to the negatively charged cellulosic fibres, so that nonionic forces of interaction become operative.

Oxidation in the fibres to the insoluble vat dye: Once the leuco form has been adsorbed, it is oxidised to the parent vat dye. In many instances, atmospheric oxygen is sufficient. However, some leuco forms are more difficult to oxidise, and oxidising agents such as potassium dichromate or hydrogen peroxide are required. Oxidising agents may also be needed where the fibres are densely packed, as in package dyeing (see Chapter 9).

Soaping: After oxidation of the leuco form, the fibres are treated with boiling soap or detergent solution. Soaping has a two-fold function: it removes vat dye particles adhering to the fibre surfaces and it modifies the shade of the dye. On soaping, larger, better-defined crystals of dye are formed within the fibres, and indeed their lattice structure may be altered.

5.2.3 Sulphur dyes

Sulphur dyes are mainly restricted to black, brown, blue and green colours. The chief advantages of sulphur dyes are their cheapness and their general good fastness to washing. However, many are sensitive to chlorine and, consequently, have poor fastness to bleaches. The light-fastness of sulphur dyes varies from poor to good. In a number of cases, fastness properties may be improved by suitable aftertreatments.

The feature of sulphur dyes which characterises them as an application class is the presence of sulphur linkages in their molecules. However, their chemical structures are highly complex and ill-defined. They are insoluble in water but can be readily converted to a soluble form by reduction in alkaline solution. The reduction appears to involve the cleavage of disulphide linkages:

$$RSSR' \longrightarrow RSH + HSR' \tag{5.2}$$

After application, the soluble form can be oxidised back to an insoluble form.

Sulphur dyes often contain impurities, which remain insoluble after the reduction process. These insoluble impurities will be filtered in processes such as package dyeing and give rise to unlevelness. As a result, solubilised sulphur dyes have been developed. These types are prepared by reaction with sodium sulphite or sodium hydrogen sulphite, e.g.:

$$RSSR' + HSO_3^- \longrightarrow RSSO_3^- + R'SH \tag{5.3}$$

Solubilised sulphur dyes have no substantivity for cellulose. After initial conversion to the reduced form, their subsequent dyeing behaviour resembles that of normal sulphur dyes.

A third type, sulphurised vat dyes, also exists. These dyes resemble other sulphur dyes in being manufactured by sulphurisation processes, but they are applied in a similar manner to vat dyes. For reduction, it is necessary to use sodium dithionite rather than sodium sulphide.

Application of sulphur dyes

The application of sulphur dyes consists essentially of the following operations:

(A) Reduction to the soluble leuco form.
(B) Impregnation of the fibres with the leuco compound.
(C) Oxidation in the fibres to the insoluble form.

Some sulphur dyes are marketed already in the leuco form and are soluble in water. With these dyes, step 1 has already been carried out by the dye manufacturer.

Reduction to the soluble leuco form: The required ingredients include dye (or mixture of dyes), sodium sulphide and sodium carbonate. To prepare the soluble leuco form, the dye is pasted with sodium sulphide and sodium carbonate. The amount of sodium sulphide required varies very much from one dye to another. Boiling water is then added to the paste until dissolution is complete. Before contact with the fibres, solutions of leuco sulphur dyes often need straining, because of the insoluble impurities they contain.

Impregnation of the fibres with the leuco compound: The leuco form, once prepared, is applied to the fibres in the presence of sodium chloride and a penetrating agent. Often the liquor is brought to the boil in the early stages of dyeing, maintained at the boil for *c.* 30 minutes and then gradually allowed to cool. Although salt is sometimes initially present, it is often added in portions once the liqour is boiling, in order to reduce the risk of unlevel dyeings. The temperature is reduced in the later stages to promote exhaustion of the dye, but exhaustion may be low even in the presence of salt. Hence, low liquor ratios are important.

Oxidation: Once the leuco form has been adsorbed, it is oxidised in the fibres to the insoluble form. Exposure to air usually suffices, but in a few cases specific oxidising agents have to be used.

Aftertreatments

The fastness properties of many sulphur dyes may be improved by aftertreatments. Sulphur dyes appear to form complexes with some salts with consequent improvement in wet-fastness. Alkylating agents based on epichlorohydrin give rise to dyeings of improved fastness to severe washing treatments. Light-fastness is often considerably improved by treatment with copper sulphate.

5.2.4 *Azoic dyes*

Azoic colorants are unusual in being formed by reaction within the cellulosic fibres themselves. A solution of coupling component is first applied, whereupon the fibes are treated with a diazonium compound. The ensuing coupling reaction produces an azoic dye, which is insoluble in aqueous media. Nowadays, large ranges of coupling components and diazonium compounds are available, hence the number of azoic dyes which can be synthesised is extensive and represents a wide variety of bright shades. Most azoic dyes exhibit good all-round fastness properties.

The coupling component, applied first, can be a naphthol, phenol or aromatic amine. Many are arylmides of 2,3-hydroxynaphthoic acid derivatives (**4**) where Ar signifies an aromatic moiety. The diazonium salt is formed by treating a cold acid solution of a primary aromatic amine with sodium nitrite. As an example, the diazotisation of 4-chloro-2-nitroaniline (C.I. Azoic Diazonium Component 9) yields the diazonium salt (**5**).

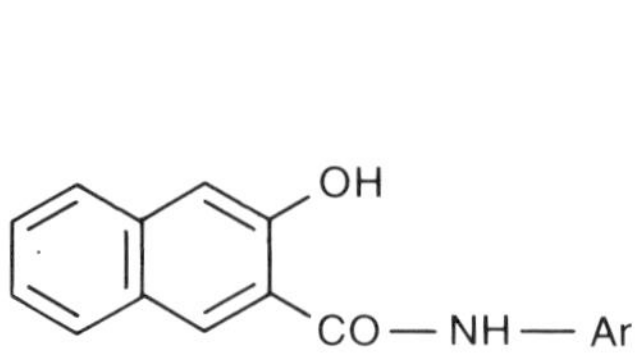

4

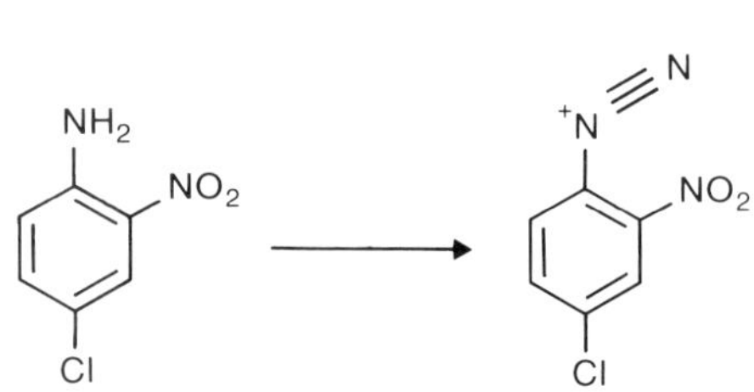

5

On reaction of the diazonium salt with the coupling component, C.I. Azoic Coupling Component 2 (**6**) a red azoic colorant (**7**) is produced:

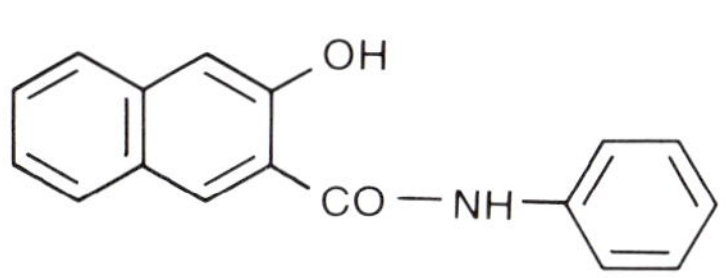

6

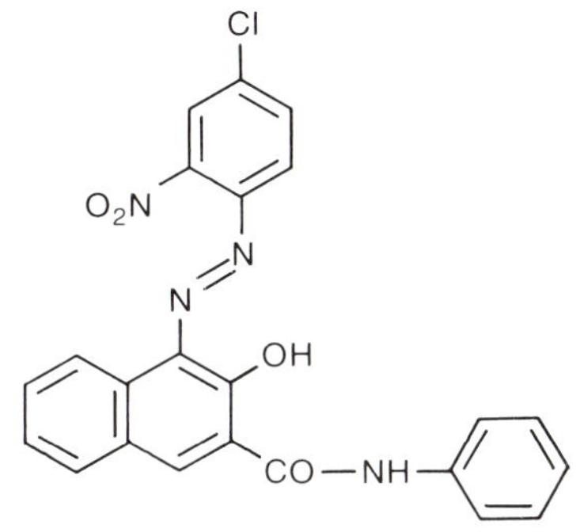

7

Many diazonium salts are unstable, so that stock solutions of them cannot be stored over long periods. Production of azoic dyes has, however, been facilitated by the introduction of stabilised diazonium salts, or fast salts as they often called. The dyer dissolves the stabilised salt in water to obtain a solution of diazonium compound.

Diazonium salts may be stabilised by a variety of methods. One widely practised method is to prepare a double salt, usually with zinc chloride, such as in C.I. Azoic Diazonium Compound 45 (**8**).

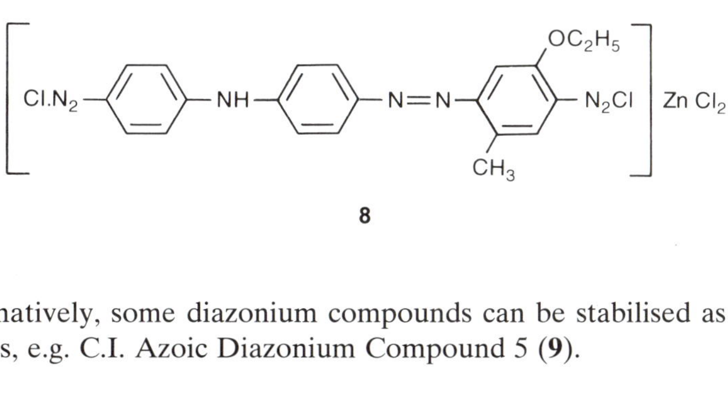

8

Alternatively, some diazonium compounds can be stabilised as arylsulphonates, e.g. C.I. Azoic Diazonium Compound 5 (**9**).

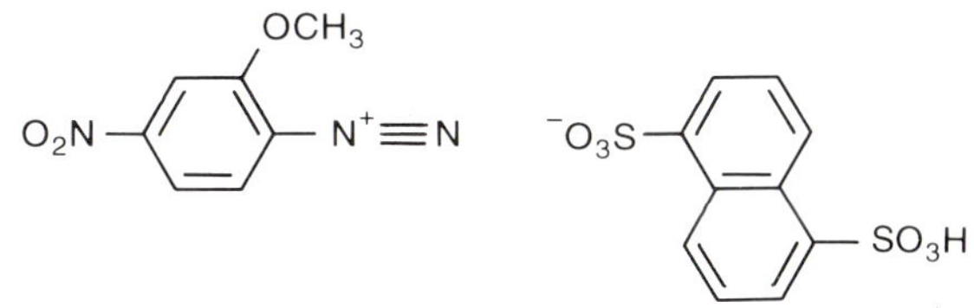

9

Application of azoic dyes

The application of azoic dyes consists essentially of three stages:

(A) Impregnation of the fibres with coupling component.
(B) Development of the colour with a suitable diazonium compound.
(C) Soaping.

Impregnation of the fibres with coupling component: Coupling components are insoluble in water but can be applied as sodium salts dissolved in alkaline solution. Hence, the coupling component is applied as an anion.

The choice of coupling component depends to a large extent on the liquor ratio used. The liquor ratio is the ratio of the mass of dye liquor to the mass of material to be dyed. In processes where the liquor ratio is high (as in many batch processes), coupling components with high substantivity for the substrate are desirable. Indeed, for application to cellulosic fibres, salt is often added to assist the uptake of coupling component, just as with direct dyes. In processes using very low liquor ratios (such as continuous processes), coupling components of much lower substantivity are applied, in order to avoid fast exhaustion and hence difficulties in controlling shade.

Close control of the uptake of coupling component is essential. First, the depth of shade produced by an azoic dye is determined by the amount of coupling component adsorbed by the fibres. Moreover, to obtain a level dyeing, it is essential that an even distribution of coupling component in the fibres be achieved. The adsorption of most coupling components decreases with rise in temperature, and so they are generally applied at 20–30°C. After impregnation, retained liquor is removed, to reduce the formation of loose colorant during coupling.

Development of colour with a suitable diazonium compound: The coupling reaction between the diazonium compound in solution and the adsorbed coupling component produces the insoluble azoic colorant. The rate at which coupling occurs is highly dependent on both pH and temperature. In alkaline conditions the diazonium compound tends to decompose, so the impregnated fibres have to be neutralised. Under strongly acidic conditions, however, the reaction is slow. Thus, there is an optimum pH for effecting a particular coupling reaction. Sodium chloride is usually present to inhibit migration of adsorbed coupling component into the bulk of the liquor.

The increase in the rate of coupling afforded by rise in temperature is limited by the stability of the diazonium compound, which may rapidly decompose in solution even at 20°C. Thus, the temperature of coupling is often well below 20°C.

Soaping: After development of the azoic dye, the fibres are treated at the boil with a solution of soap or detergent. As with vat dyeing, soaping both removes loose dye adhering to fibre surfaces and modifies the shade of the colorant.

5.2.5 Reactive dyes

Reactive dyes differ from other application classes in that they can react chemically with cellulosic fibres through formation of covalent bonds. This reaction is promoted in alkaline media. Unlike nearly all other application classes, reactive dyes do not have to be rendered insoluble in water to promote high fastness. Moreover, they do not have to be highly substantive to the cellulosic fibres. Hence, reactive dyes can be composed of smaller, simpler molecules than most other classes. Indeed, reactive dyes of comparatively low molecular weight are often advantageous, in order to achieve good penetration and levelness within the fibres prior to chemical reaction.

A particular problem with the application of many reactive dyes is the competing reaction with hydroxide ions in the dyebath to produce a hydrolysed, non-reactive form of the dye. The efficiency of dyeing is increased if reaction with the fibres can be promoted with simultaneous suppression of reaction with hydroxide ions. However, complete reaction with the fibres is never achieved.

Nearly all the fastness properties of reactive dyes are centred around the covalent linkages between dye and fibre. For reactive dyes to realise their fullest potential, the dye–fibre bonds must be resistant to normal domestic and industrial wet-treatments, and the removal of unreacted and hydrolysed dye from the fibres must be complete. In practice, very fast shades have been produced, which do not fade even after repeated wet-treatments. The light-fastness of reactive dyes is also good, and a large range of bright shades is available.

A feature common to all reactive dyes is that they are rendered soluble in aqueous media by incorporation of anionic solubilising groups, such as sulphonate groups. However, a wide variety of functional groups within dye molecules is available for chemical attachment, or fixation, to the fibres. For reasons of space, only a few can be discussed here.

Many of the reactive functional groups are based on heterocyclic rings containing nitrogen. These include the triazinyl, pyrimidinyl and quinoxalinyl dyes, examples of which are given in Fig. 5.4. They combine with cellulose under alkaline conditions. In the presence of alkali, some of the hydroxyl groups on the cellulose chains become ionised:

$$\text{Cell—OH} + \text{OH}^- \rightleftharpoons \text{Cell—O}^- + H_2O$$

The dye reacts with the cellulosate ions by a nucleophilic substitution mechanism, e.g.:

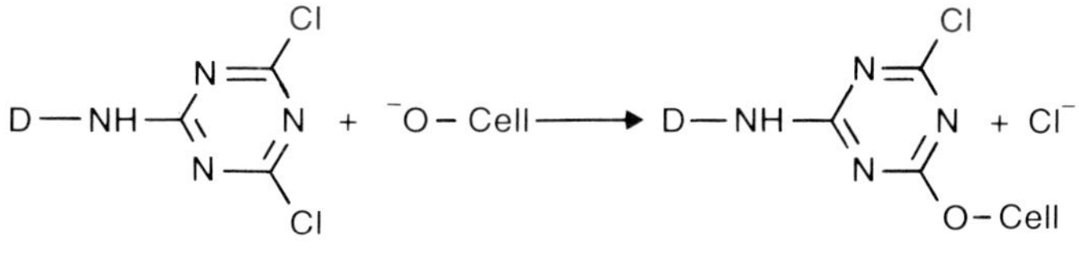

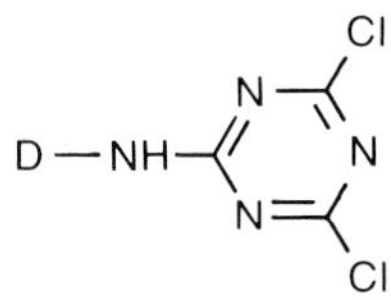

dichlorotriazinyl dyes

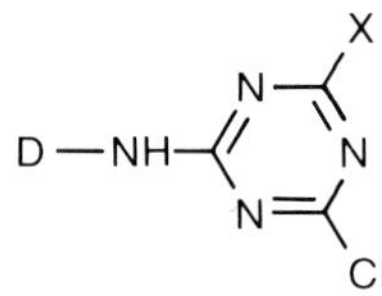

monochlorotriazinyl dyes

X represents a funtional group other than a halogen

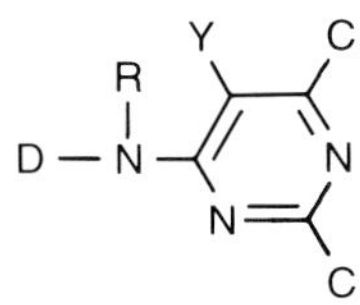

dichloropyrimidinyl dyes

Y ≡ $-NO_2$,–CN,–Cl or H

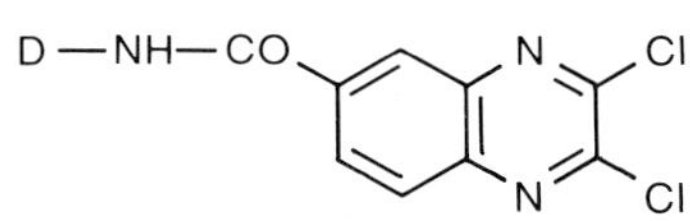

dichloroquinoxalinyl dyes

Fig. 5.4 Examples of reactive dyes based on heterocyclic rings containing nitrogen. D represents the remainder of the dye molecule, including the chromophore and the solubilising group.

In the case of dichlorotriazinyl dyes, containing two chlorine atoms per molecule, the second chlorine substituent may also be replaced by a cellulosate ion. For these dyes, temperatures of up to only 40°C are normally sufficient for reaction. The monochlorotriazinyl dyes, which possess only one chlorine atom per molecule, require higher temperatures, usually 70°C or above. In all cases, reaction with hydroxide ions in solution competes with the bonding to cellulose, so in practice complete fixation is never achieved.

Vinyl sulphone dyes are also important for dyeing cellulosic fibres and these dyes are included in Fig. 5.5. Strictly, these compounds are precursors, from which vinyl sulphone derivatives:

$$D{-}SO_2{-}CH{=}CH_2$$

can be formed in the presence of mild alkali. Reaction with cellulose then yields:

$$D{-}SO_2{-}CH_2{-}CH_2{-}O{-}Cell$$

$$D—SO_2—CH_2—CH_2—O—SO_3Na$$

$$D—SO_2—CH_2—CH_2—Cl$$

Fig. 5.5 Examples of vinyl sulphone reactive dyes. D represents the remainder of the molecule.

The reactivities of vinyl sulphone dyes lie between those of the mono- and dichlorotriazinyl dyes. Vinyl sulphone dyes are often fixed to cellulosic fibres at 40–60°C. These dyes can also be hydrolysed in alkaline media.

The different levels of reactivity associated with the chlorotriazinyl and vinylsulphone groupings can e.g. even be incorporated into the same dye (**10**).

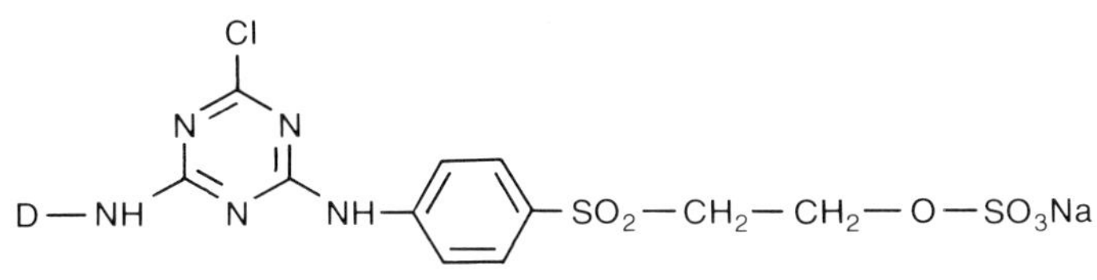

10

The aim of this approach is to give greater flexibility of application by reducing dependence on pH and temperature.

Application of reactive dyes

Dyeing with reactive dyes essentially involves the following stages:

(a) High, uniform penetration of dye into the textile fibres.
(b) Covalent bonding (fixation) of the adsorbed dye to the fibres.
(c) Removal of any unbound dye from the fibres by soaping.

To obtain a level dyeing after fixation, dye must be uniformly distributed throughout the fibres before fixation. Levelness is best achieved with dyes of low substantivity, but adsorption of such dyes from the liquor is liable to be poor.

Reactive dyes are initially applied from neutral solution and at this stage behave very much like direct dyes. Thus, adsorption from the liquor is considerably enhanced by adding salt to the dyebath. Generally, much higher concentrations of salt are needed than those required for the application of direct dyes.

Once uniform penetration of the fibres has been achieved, the dye can be fixed. Fixation occurs on addition of alkali, usually sodium carbonate or bicarbonate. After fixation, the dyed fibres are soaped and rinsed to remove hydrolysed (and perhaps unreacted) dye.

5.3 Protein textiles

Protein textiles generally contain a high content of impurity. As much as 50% of raw wool contains grease, oil, dirt and other matter, and these impurities have to be removed by thorough scouring. Scouring is carried out with a mixture of alkali and detergent, but the concentration of alkali must be carefully controlled in order to avoid degradation of the fibres. The fibres are then rinsed. If pale shades of dye are to be applied, the wool must also be bleached, usually with hydrogen peroxide, to remove yellowness.

The soft quality of silk is obtained by removal of the sericin gum, by scouring in alkaline detergent. Raw silk contains 15–25% sericin, and different qualities of silk are obtained according to the amount of sericin removed. The more sericin removed, the finer the quality of the silk. Like wool, silk is normally bleached using hydrogen peroxide.

The classes of dye applied to protein textiles are acid, metal-complex (premetallised) and reactive dyes. Mordant dyes can also be applied to wool.

5.3.1 Acid dyes

Acid dyes are conveniently divided into three groups, according to their pH of application, levelling properties and degree of wet-fastness in the fibres, as shown in Table 5.1.

Table 5.1 The three groups of acid dyes.

Group	pH in dyeing	Wet-fastness	Levelling properties
Levelling (or equalising)	2–4	Poor	Good
Milling	4–6	Good	Moderate–poor
Supermilling	6–7	Very good	Very poor

This classification was originally devised for dyeing wool. There is no sharp distinction between the groups, and indeed some acid dyes are difficult to categorise.

Right from the commencement of dyeing, acid dyes are seldom adsorbed evenly. The degree of unlevelness developed during dyeing is largely governed by the 'strike', the extent to which dye is initially adsorbed. With levelling dyes, which exhibit good migration, levelling can eventually be improved by movement of dye molecules from heavily dyed to more lightly dyed parts of the substrate. Milling dyes, however, exhibit poorer migration, and supermilling dyes migrate only to a very limited extent.

Acid dyes are anionic in aqueous media, the anion being coloured. Most acid dyes are sodium salts of aromatic sulphonic acids, but some contain carboxyl and phenolic groups. There are chemically many different types, the most common being azo, anthraquinone and triphenylmethane compounds. Some examples are given in Table 5.2.

It will be noted that the three groups of acid dye also differ extensively in molecular size and complexity. Levelling dyes tend to have the lowest molar mass and the highest solubilities in aqueous media. Milling dyes generally consist of more complex molecules of lower solubility in aqueous media.

The molecules of supermilling dyes are more complex still and many contain long chains of aliphatic hydrocarbon, which confer high affinity for the fibres.

Application of acid dyes to wool – levelling dyes

A typical dyebath contains dye (or mixture of dyes), sulphuric acid (or sometimes formic acid), and Glauber's salt (hydrated sodium sulphate, $Na_2SO_4.10H_2O$).

Function of dyebath assistants

Sulphuric acid: When wool is immersed in strongly acid solutions, such as sulphuric acid solution, its carboxyl groups lose their negative charges and an overall positive potential is established on the fibres:

$$\wr\!-NH_3^+ + {}^-OOC\!-\!\wr + H^+ \rightleftharpoons \wr\!-NH_3^+ \ HOOC\!-\!\wr$$

Anions, such as sulphate and dye ions are then attracted towards the positively charged amino groups.

Glauber's salt: If Glauber's salt is included in the dyebath, the number of sulphate ions present is raised. The competition between sulphate and dye anions for the positive sites on the wool increases. The sulphate ions, being smaller, reach the positive sites more easily, but dye anions are eventually preferentially adsorbed because of their much higher affinity. This competition leads to good migration, but also to incomplete adsorption from the dyebath.

Summary of dyeing procedure

The wool is often added at 40°C or so to a dyebath containing 3–5% (on mass of fibres, omf) sulphuric acid and 10–20% (omf) Glauber's salt. After 10 minutes, the liquor is brought to the boil for 30–45 minutes and is maintained at the boil for an equivalent time. Treatment at the boil is essential so that levelness can be achieved.

Table 5.2 Some examples of acid dyes

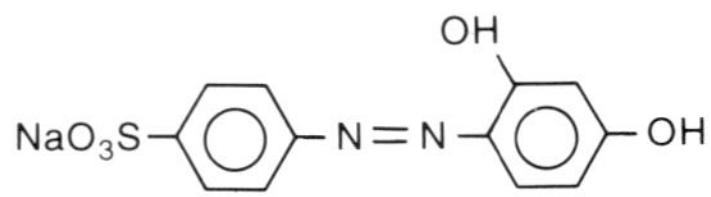

C.I. Acid Orange 6

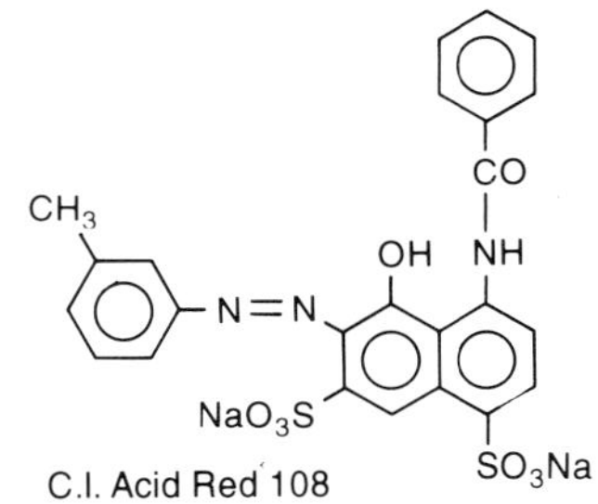

C.I. Acid Red 108

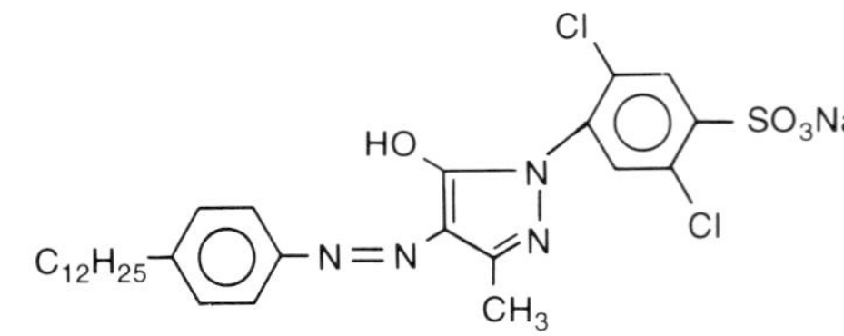

C.I. Acid Yellow 72

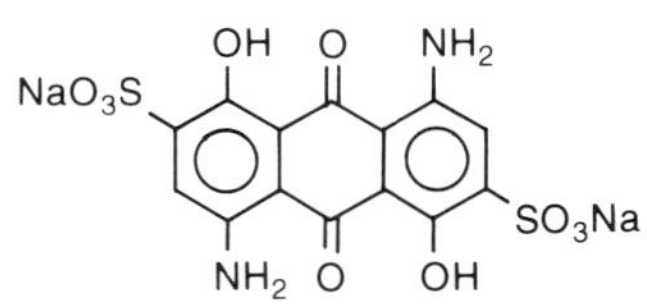

C.I. Acid Blue 45

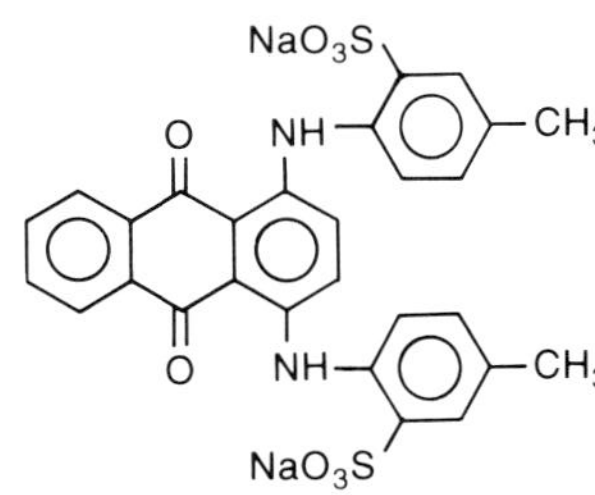

C.I. Acid Green 25

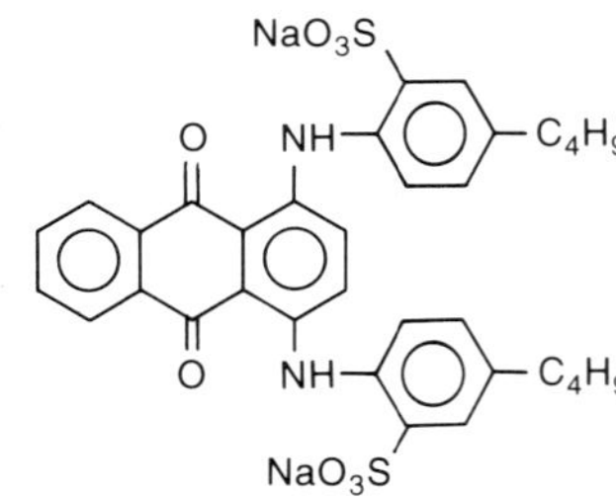

C.I. Acid Green 27

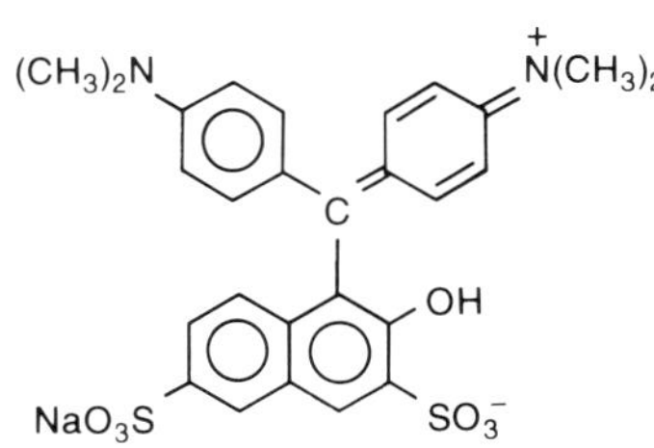

C.I. Acid Green 50

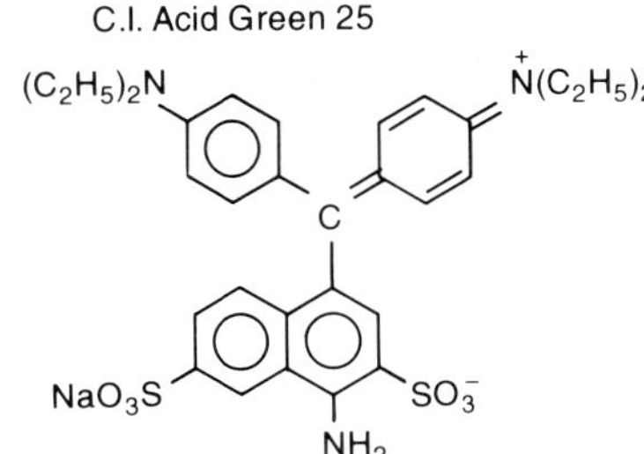

C.I. Acid Blue 108

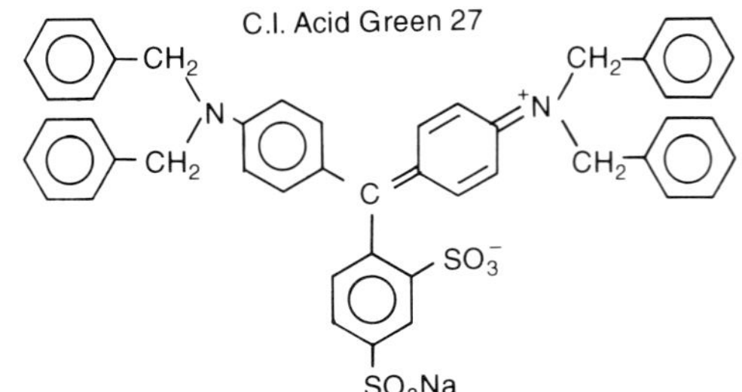

C.I. Acid Blue 103

Milling dyes

A typical dyebath contains dye (or mixture of dyes), acetic acid and Glauber's salt.

Function of dyebath assistants

Acetic acid: Milling dyes, which generally have higher molar masses and lower solubilities in aqueous media than levelling dyes, are also more substantive to wool. Milling dyes, therefore, depend to a much smaller extent on ionic attraction for complete adsorption, and fewer positively charged sites on the wool fibres are required. Thus, a dyebath of higher pH is desirable, and this is conveniently obtained by using acetic acid rather than sulphuric acid. Indeed, if the pH is low, the attraction between dye and fibres is too high; the adsorption of dye is rapid and the dyeing is unlevel.

Glauber's salt generally still promotes levelling. However, its effectiveness is reduced, since the anions of milling dyes have higher affinity and there are fewer positively charged sites on the wool fibres.

Summary of dyeing procedure

The wool is added at *c.* 40°C to a dyebath containing 3–5% (omf) acetic acid and *c.* 10% (omf) Glauber's salt. The temperature is raised to 100°C over 30 minutes or so and maintained there for a further 30 minutes. If adsorption of dye is not sufficient by this time, some further acetic acid may be added.

Supermilling dyes

A typical dyebath contains dye (or mixture of dyes) and ammonium acetate or sulphate.

Function of ammonium salt

Supermilling dyes possess the highest affinity of the acid dyes. They require virtually no ionic attraction for good adsorption and so are applied from an almost neutral dyebath. Significant acidity in the dyebath leads to very rapid, unlevel adsorption of dye.

The ammonium salt serves to maintain the pH of the dyebath, by neutralising any alkali in the wool residual from scouring. In addition, it can slowly liberate acid as dyeing proceeds, e.g.:

$$NH_4OOCCH_3 \rightleftharpoons NH_3 \uparrow + CH_3COOH$$

The pH of the bath will fall as the ammonia escapes, and adsorption of the dye will be assisted. However, if the dyebath is enclosed, ammonia cannot readily escape and the liberation of acid is then inhibited.

Summary of dyeing procedure

The wool is added at 40°C or so to a dyebath containing 2–4% (omf) ammonium acetate or sulphate. The temperature is gradually raised to the boil (*c.* 45 minutes) and held at the boil for the same length of time. Addition of acetic acid may be necessary after some time at the boil to assist adsorption of dye.

Application of acid dyes to silk

Acid dyes are generally applied to silk under neutral or mildly acidic conditions. Thus, supermilling or milling dyes are required. A levelling agent is also normally present, and sometimes Glauber's salt. The silk is entered into the dyebath at 30°C, whereupon the bath is heated to 70–80°C and then maintained at this temperature for 45–60 minutes.

5.3.2 Chrome mordant dyes

Before the advent of synthetic dyes, many dyes were applied to textiles which had first been impregnated, or mordanted, with a metal hydroxide. The coloured product formed within the fibres was largely insoluble in water. Thus, mordanting conferred a degree of wet-fastness upon the dye; it also tended to improve the light fastness. Although many metals were used as mordants at one time, nowadays only chromium is used to any extent. However, in practice, chromium mordants have to be used with great care, to prevent undesirable discharges of chromium (see Chapter 12). Dyes capable of combining with chromium are still commonly applied to wool, their mode of action relying on the high content of the aminoacid, cystine, in wool keratin chains. Chrome mordant dyes have generally good light- and wet-fastness properties, but they nearly all possess dull shades.

Chromium dyes are anionic, for they contain sulphonate or even carboxylate groups. Most of them contain azo groups, but others are anthraquinone and triarylmethane dyes (Fig. 5.6).

All chrome mordant dyes are capable of forming coordination complexes with chromium, but the nature of the complexes is far from completely established. It is known that chromium exists in the trivalent state in these complexes and it has been thought probable that 1:2-complexes are mostly formed. (Such complexes would contain one equivalent of chromium to two equivalents of the parent dye.) However, it is likely too that multiple bonds between the wool keratin, chromium and dye can also be formed.

Application of chrome dyes

Both dye and mordant have to be applied under conditions which promote their combination within wool fibres. The form of chrome used is nearly always potassium (or sodium) dichromate. Potassium chromate is some-

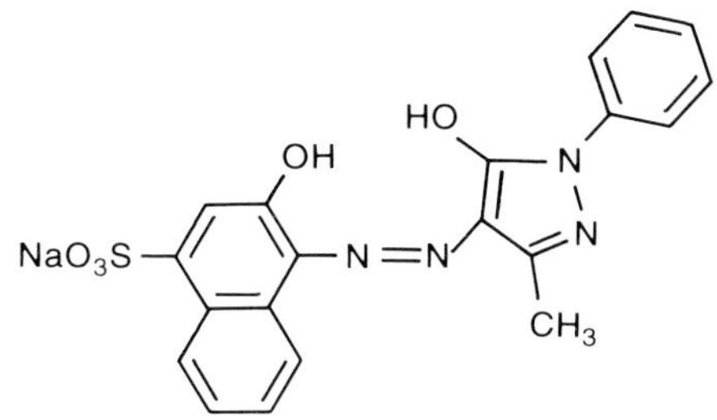

C.I. Mordant Red 7

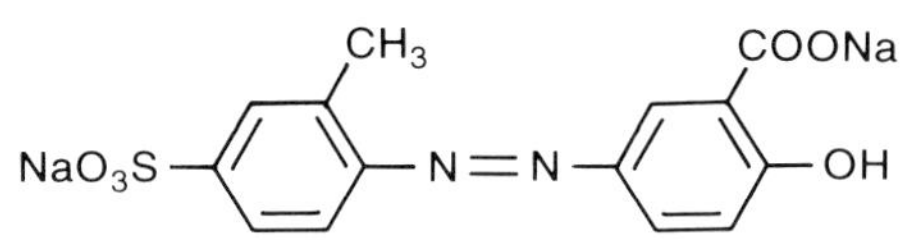

C.I. Mordant Yellow 60

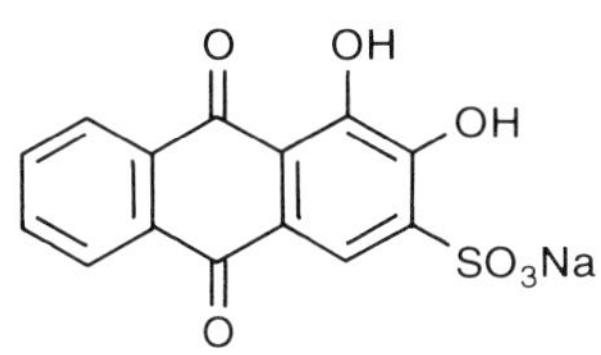

C.I. Mordant Red 3

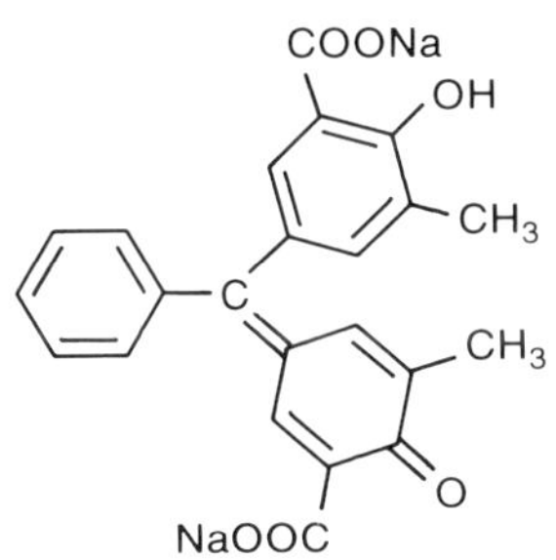

C.I. Mordant Blue 1

Fig. 5.6 Examples of chrome mordant dyes.

times used instead of dichromate, but in solution one species changes to the other on adjusting the pH in the range 6–8:

$$Cr_2O_7^{2-} + 2OH^- \rightleftharpoons 2CrO_4^{2-} + H_2O$$

$$2CrO_4^{2-} + 2H^+ \rightleftharpoons Cr_2O_7^{2-} + H_2O$$

The mechanism of mordanting is highly complex and is still not yet fully understood. The first stage, the interaction of dichromate ions with amino groups in the wool, is similar to the adsorption of acid dye anions:

$$2 \wr\!\!-NH_3^+ \ {}^-OOC-\!\!\wr + Cr_2O_7^{2-} + 2H^+ \rightarrow (\wr\!\!-NH_3^+)_2\ Cr_2O_7^{2-} + 2HOOC-\!\!\wr$$

The dichromate ions, which contain hexavalent chromium, are then reduced by a complex route to trivalent chromium ions at the expense of cystine present in the wool. Some of the cystine is oxidised with consequent rupture of its disulphide bonds. The trivalent chromium ions combine with molecules of the parent dye and interact with functional groups in the keratin chains.

The great complexity of the chemical reactions which appear to take place within the fibres suggest that a number of different products may well be formed. The variations in shade resulting from dyeing the same dye by different methods (see below) support this suggestion.

Dyeing methods

There are three ways of applying chrome dyes:

(a) Chrome mordant method; dichromate is added before dyeing.
(b) Afterchrome method; dichromate is added after dyeing.
(c) Metachrome method; dichromate and dye are applied simultaneously.

Chrome mordant method: This is the oldest method. However, it is nowadays least used, for it requires two baths, one for mordanting and the other for dyeing. It is, therefore, often uneconomical in comparison with other methods, although some chrome dyes can be applied only by this method.

The goods are first added at *c.* 60°C to a mordanting bath, containing dichromate solution and often either sulphuric or an organic acid. The temperature is raised to 100°C over 45 minutes and held there for a further 60–90 minutes. Before dyeing, the mordanted wool is thoroughly rinsed. It is then added to a dyebath containing acetic acid at 50°C. The bath is brought to the boil over 45 minutes and maintained at the boil for 60–90 minutes longer.

Afterchrome method: This is the most important way of applying chrome dyes. The dye is applied like an acid milling dye. Once the dye has been adsorbed, the dyebath is cooled to 70°C, whereupon dichromate is added. The lower temperature helps to distribute the dichromate ions evenly within the wool fibres, as the chromium is not significantly reduced to the trivalent state at this temperature. The temperature is then raised to 100°C and held there for 30 minutes, while the dye is mordanted.

The method has a number of advantages. It requires only one bath. The dye can be initially applied in a level manner, yet after mordanting, it does not migrate. The dye, however, must be completely adsorbed before the addition of dichromate. If dye and dichromate are in the bath together, a chromium complex of the dye may be precipitated.

Metachrome method: Some chrome dyes can be applied simultaneously with mordant. The dyebath contains a mordant consisting of chromate and ammonium sulphate and is made very slightly alkaline with ammonia. The wool is added at 50°C; the temperature is then raised to 100°C over 45 minutes or so and held there for a further hour.

The use of the metachrome method is restricted by several factors. The dye must, for example, resist oxidation by the chromate in the bath, and it must have good substantivity up to pH 8. Thus, the range of shades is limited, especially in comparison with the range which can be produced by the afterchrome method.

5.3.3 Premetallised dyes

In the application of mordant dyes, chromium and dye are combined in the wool fibres. Alternatively, chromium-dye complexes can be prepared before dyeing by the dye manufacturer; these are the premetallised (metal-complex) dyes. There are two main types, the 1:1- and 1:2-chromium complexes, but the latter are commercially far more important.

The 1:1-complexes are derived from monoazo dyes, whose molecules also contain chelating groups. Examples are C.I. Acid Blue 158 and C.I. Acid Red 180, whose structures are shown in Fig. 5.7. Solubility in aqueous media is imparted to these dyes by the presence of one or two sulphonate groups. If there are two sulphonate groups, the dye has an overall negative charge; if there is only one such group, the dye behaves as a zwitterion, in which different parts of each molecule possess equal and opposite charges.

The 1:2-complexes are also mostly derived from monoazo dyes containing chelating groups. A few dyes possess azamethine, —CH=N—, groups instead. An example of the 1:2-complex, C.I. Acid Red 296, is shown in Fig. 5.7. Each molecule of 1:2-complex has a single negative charge. Sulphonate groups are generally absent. Instead, low, but adequate solubility in aqueous media is conferred by incorporation of highly polar nonionic groups, such as methylsulphonyl (—SO_2CH_3) and sulphamoyl (—SO_2NH_2) groups. The absence of sulphonate groups avoids tippy dyeing (see section 3.2.2), to which the corresponding sulphonated dyes often give rise. The problem is less serious with deep shades and indeed some sulphonated 1:2-complex dyes are now marketed specifically for producing deep colours.

Application of premetallised dyes

1:1-Complexes are applied under strongly acidic conditions (pH = 2). Several types of bonding between dye and wool fibres are possible: at pH = 2, there is strong attraction between the sulphonate groups in the dye and the positively charged amino groups in the wool, and van der Waals interactions between dye and fibres must also be taken into account. (At higher pHs, the positively charged chromium in the dye will be attracted towards ionised carboxylic acid groups in the wool. Moreover, uncharged amino groups in the wool may coordinate with chromium in the dye.) Thus, the

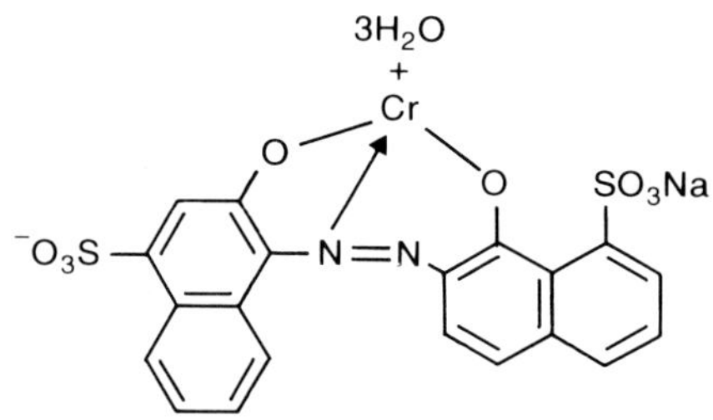

C.I. Acid Blue 158

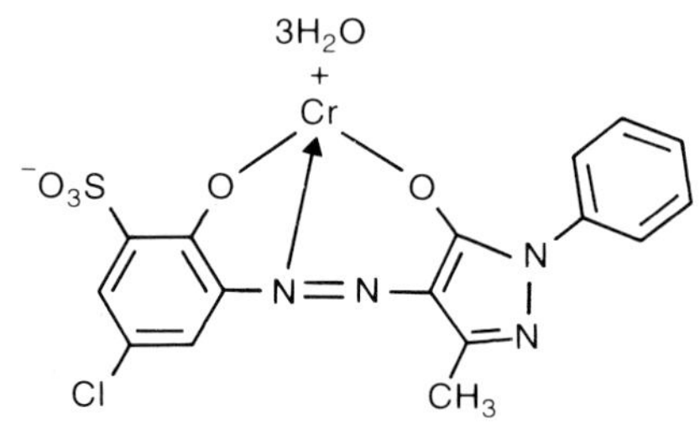

C.I. Acid Red 180

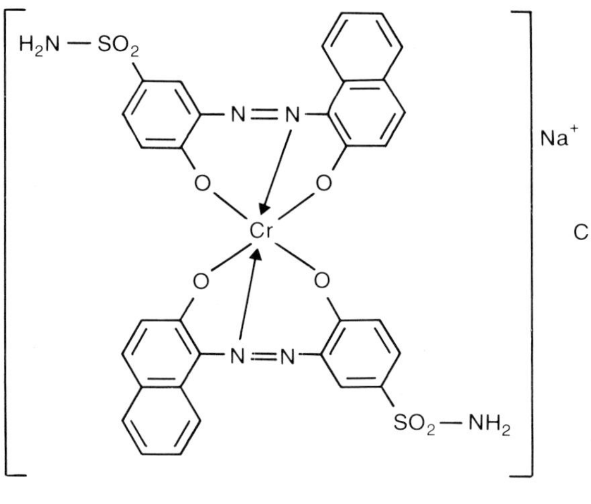

C.I. Acid Red 296

Fig. 5.7 The structures of some premetallised dyes.

application of 1:1-complexes resembles that of acid levelling dyes, although in practice there are some differences. Except in the presence of suitable levelling agents, 1:1-complexes are much more prone to give tippy dyeings. Furthermore, complete development of shade requires prolonged boiling, and Glauber's salt imparts virtually no levelling action even at a pH as low as 2.

1:2-Complexes are applied under nearly neutral conditions in the presence of ammonium sulphate or acetate. They behave very much as supermilling dyes. However, sensitivity to pH is much less marked with 1:2-complexes, perhaps because the negative charge is far less localised.

Summary of dyeing procedures

In the application of 1:1-complexes, the wool is entered at 60°C into a dyebath containing 2–3% (omf) dye and 8% (omf) sulphuric acid. The bath is brought to the boil over 30 minutes and maintained at the boil for a further 90 minutes. With 1:2-complexes, the dyeing process is begun at 40°C in a bath containing 2–3% (omf) dye and 5% (omf) ammonium sulphate or acetate. The bath is brought to the boil over 45 minutes and maintained at the boil for 30–60 minutes.

5.3.4 Reactive dyes

The main features of these dyes have already been described in section 5.2.5. In wool keratin, there are a variety of functional groups with which reactive dyes may react. These groups include the amino (—NH_2), hydroxyl (—OH) and thiol (—SH) groups. However, alkali is not used to induce reaction, lest the wool be degraded. Instead, the pH is maintained at 4–6, and reaction is induced by an increase in temperature. Silk, however, is more resistant to alkali, and alkaline conditions can be used to fix reactive dyes to silk. Much of the fixation is through the hydroxyl groups on some of the constituent aminoacids.

A wide range of reactive dyes is commercially available for application to protein fibres, including the chlorotriazinyl dyes, already discussed (section 5.2.5). Also prominent are the α-bromoacryloamido, the fluoro-chloropyrimidine (FCP) and vinyl sulphone dyes, whose general formulae are given in Fig. 5.8.

Application of reactive dyes to wool

The application involves the same three stages required for dyeing cellulosic textiles (section 5.2.5), although fixation of the dye is achieved by an

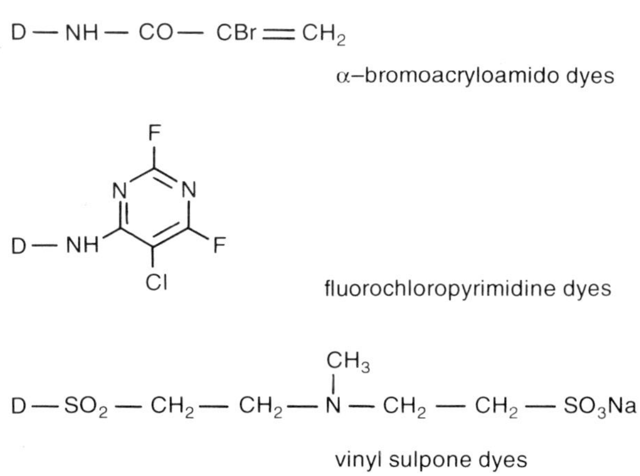

Fig. 5.8 Examples of reactive dyes applied to wool.

increase in temperature, instead of addition of alkali. The conditions of dyeing are initially arranged so that the dye is evenly adsorbed under weakly acidic conditions (pH 4–6), usually at a temperature well below boiling point. Glauber's salt is also present in the dyebath. At this stage, the dye behaves like an acid milling dye. A levelling agent is often included amongst the dyebath ingredients.

The temperature of the dyebath is then gradually raised to the boil and subsequently kept there. The adsorbed dye now reacts with the wool to form covalent linkages. Thus, provided fixation is sufficiently slow, the dye can migrate in the early stages of dyeing below the boil. As the temperature increases, however, so the rate of reaction increases, and the dye consequently becomes fixed. Finally, unattached dye is removed by soaping.

Application of reactive dyes to silk

The silk is entered into a neutral dyebath containing dye and Glauber's salt at 20–30°C. The temperature is raised over 30 minutes or so to *c.* 50°C. Sodium carbonate is then added and the temperature is held at 50°C for 45 minutes. The dyed silk is then soaped, to remove unattached dye.

5.4 Synthetic and cellulose acetate textiles

Synthetic and cellulose acetate textiles are generally freer of impurity than natural textiles. Nevertheless, some preliminary scouring is often advisable, particularly to remove traces of dirt and lubricants. Moreover, those fibres which are slightly yellow, such as acrylics, require bleaching e.g. with peroxides.

The commonest types of synthetic and cellulose acetate textile have already been described in Chapter 3, and they clearly represent a wide variety of macromolecular structures. However, there are some common features. One feature, an undesirable one, is the propensity, after being dyed, to reveal stripes of lighter and darker colours. This barriness, as it is called, results from variations, arising during manufacture, in chemical and physical stucture along the yarn. The yarns and fabrics of many of these textiles can also be heat-set: a desired shape can be obtained by heating, usually under tension, to 20–30°C above the temperature likely to be encountered during use, and then cooling. However, although dimensional stability is conferred, the fibres may become less receptive to dyes. Indeed, in some cases, it is almost essential to heat-set after dyeing.

Nearly all these textiles can be dyed with disperse dyes, and this class of dye is particularly important for dyeing polyester and cellulose triacetate. Polyamides, however, can be successfully dyed with acid dyes, and to a lesser extent with metal-complex and even chrome mordant dyes, though the latter have largely been superseded. These classes of dye can be applied, because polyamide chains are terminated by primary amine (—NH_2) and carboxylic acid (—COOH) groups and, in this respect, are similar to the protein chains of wool and silk. Moreover, acrylics, because they include anionic groups in their macromolecular chains, are commonly dyed with cationic (basic) dyes. Other classes of dye, such as vat and azoic dyes, are also very occasionally used to dye synthetic textiles.

Polypropylene is exceptionally difficult to dye, but advantage can be taken of the additives which are sometimes incorporated. Disperse dyes can be used but they have to be specially selected for application to polypropylene. Some types of polypropylene contain cationic groups, to effect the application of acid dyes. An interesting variation arises from the incorporation of nickel compounds to stabilise polypropylene against photodegradation. Fibres containing these compounds can be dyed with disperse mordant dyes, which are applied like mordant dyes on already mordanted fibres.

5.4.1 Disperse dyes

Disperse dyes are unusual in that they are dispersed, rather than dissolved, in the dye liquor, where they exist as finely divided particles of *c.* 1 μm cross-section. They are thus not very soluble in aqueous media, although such solubility that they do possess is essential to their mode of action. They cover a wide range of colours, but there are few brilliant shades amongst them. They exhibit a wide range of fastness properties; their heat- and wash-fastness properties, in particular, are extensively governed by molar mass. There is generally an inverse relation between ease of application and heat fastness, and indeed some dye manufacturers use this relation to classify disperse dyes.

Molecules of disperse dyes contain no charged groups; they are nonionic. Three important types are azo, anthraquinone and nitrophenylamine dyes, examples of which are shown in Fig. 5.9. Thus, although they are nonionic, they do contain a number of polar substituents. Since, in addition, their molecules are quite small, they have a low, but nonetheless finite solubility in water.

It is now generally accepted that disperse dyes are transferred from the solid phase to the fibre as individual molecules. Although disperse dyes have only a low solubility in water, they are highly substantive in the dissolved state. Thus, the solid phase serves as a reservoir to maintain a saturated solution.

For successful dyeing, the dye must be presented to the textile fibres as a fine, stable dispersion. Dispersions are often prepared by grinding crystals of dye in water in the presence of a dispersing agent.

Application of disperse dyes

A typical dyebath contains dispersing agent and dye (or mixture of dyes). The bath may also contain carrier and, in dyeing polyester, acetic acid. The use of carrier, however, is now declining.

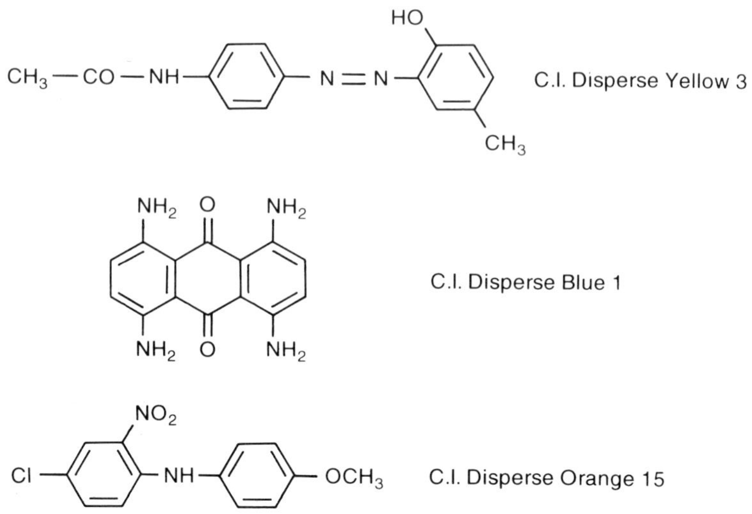

Fig. 5.9 Examples of disperse dyes.

Function of dyebath assistants

Dispersing agent: The functions of a dispersing agent are to assist dispersion of the dye and then to maintain the dye in a finely divided state. It also effectively increases the solubility of the dye. Dispersing agent thus reduces the affinity of the dye molecules for the textile fibres and so assists levelling.

Carrier: Carriers accelerate the rate of dyeing. They are especially useful in dyeing fibres of very high crystallinity, such as polyester, which may otherwise be difficult to dye.

There are several chemical types of carrier in use. Different textiles usually require different carriers. For polyesters, *o*-phenyl phenol, diphenyl and chlorinated benzenes are generally used, whereas tripropyl phosphate is an example of a carrier for dyeing cellulose triacetate. Amongst the range of carriers in use, most are insoluble in water and are supplied as emulsions, but some, like benzoic acid, are soluble at the temperature of dyeing.

The mechanism of the action of carriers is yet to be fully understood. Evidence is available to suggest that they assist dyeing by increasing the overall solubility of the disperse dye in the dye liquor and hence assisting transfer of dye molecules to the fibres. There is strong evidence too, however, that carriers modify the physical structure of the fibres in such a way that uptake of dye is accelerated.

Acetic acid: Acetic acid is especially important for dyeing polyester, to ensure that no trace of alkali is present during dyeing. Alkali degrades polyester.

Dyeing procedure

The textile is added to the other contents of the dyebath, often at 60°C, though nearer to room temperature in some cases. The dyebath is heated to a suitable temperature for dyeing and maintained at this temperature for 45–60 minutes. The temperature of dyeing depends on the textile; for example, polyester is commonly dyed under pressure at *c.* 130°C, whereas secondary cellulose acetate is dyed at only 60–70°C.

The fibres can then be scoured to improve rub-fastness, by removing any colour deposited on the fibre surfaces. When medium or deep shades have been dyed on polyester, scouring is usually inadequate, and a 'reduction-clearing' treatment is given instead. The fibres are treated for 20–30 minutes at 50–80°C with a solution containing sodium hydroxide and sodium dithionite. Under these conditions, there is no penetration of the solution into the bulk of the fibres.

Continuous dyeing of cloth, notably polyester, may be effected by thermofixation. In this method, the cloth is first padded with a suspension of the disperse dye and then dried, so that the dye adheres to the fibre surfaces. The cloth is then passed through a heated chamber, whereupon dye is transferred into the bulk of the fibre by sublimation. Dyes of lower molar mass tend to sublime more readily, but they also suffer from low wash-fastness and poor resistance to heat treatments. Dyes of higher molar mass have better fastness properties but are more difficult to apply. After thermofixation, scouring or even a 'reduction-clearing' treatment is necessary to remove any dye remaining on the fibre surfaces.

5.4.2 Acid dyes

Acid dyes are often applied to polyamides. These dyes are likely to be adsorbed through interaction with terminal amine ($—NH_2$) groups in the polyamide chains and also through nonionic interactions with groups along the chains. However, the number of these amine groups in a given polyamide fibre is limited: thus, Nylon 6,6 has 35–37 mmol g^{-1}, whereas wool has 800–900 mmol g^{-1}. Polyamides, therefore, become saturated with acid dyes at only moderate depths of shade. In some types of polyamide, known as deep-dyeing Nylons, the number of terminal amine groups has been significantly increased (up to 95 mmol g^{-1}), to enable greater depths of shade to be produced.

The affinities of acid dyes for polyamides are generally higher than for wool, so these dyes are seldom applied to polyamides under strongly acidic conditions. Most acid dyes are applied in the pH range 3–5, in the presence of formic or acetic acid. Under these conditions, they can migrate sufficiently well to mask physical variations in the fibres and prevent barriness. Other acid dyes, of more complex molecular structures, are applied under nearly neutral conditions, in the presence of ammonium acetate or sulphate. These dyes have higher affinities but poorer migration properties.

Summary of dyeing procedure

The polyamide is normally added to the dyebath at 40°C and the temperature is gradually raised to *c.* 95°C. Dyeing is then continued for a further 45–60 minutes. Slow exhaustion of the dye is desirable to facilitate level dyeings and suitable levelling agents are often used.

5.4.3 Premetallised dyes

The strongly acidic conditions required for the application of 1:1-complexes can result in degradation of polyamide fibres. Thus, the use of these dyes is limited. By contrast, the 1:2-complexes enjoy more extensive use; they are applied at nearly neutral pH, in the presence of ammonium acetate or sulphate, and have very good light- and wet-fastness properties. They can, however, highlight any barriness, but this difficulty can generally be remedied by careful application.

5.4.4 Cationic dyes

Cationic dyes – or basic dyes, as they are traditionally named – were originally applied to many types of cellulosic and protein textiles and were notable for their brilliant shades. However, because of deficiencies in fastness, particularly in light-fastness, their application to these textiles gradually declined. Nowadays, cationic dyes are applied extensively to commercial acrylics, for which they have high substantivity. Acrylic fibres contain anionic groups (see section 3.2.4), to which cationic dyes become strongly attached. However, the high substantivity suggests that nonionic forces of attraction must also be significant.

Cationic dyes are now marketed with good light- and wash-fastness on acrylics, yet still retaining brilliance of shade over a wide range of colours. Although these dyes are predominantly applied to acrylics, an affinity for them can also be imparted to polyamide and even polyester fibres by the introduction of monomers containing anionic groups, usually sulphonate groups.

Cationic dyes are ionised in aqueous media. The cation is coloured and generally contains an azo or anthraquinone group. Cationic dyes are often marketed as chlorides, though sometimes as oxalates and other salts. Some examples are shown in Fig. 5.10.

Application of cationic dyes

A typical dyebath contains dye (or mixture of dyes), acetic acid, sodium acetate, cationic retarder, Glauber's salt and nonionic surfactant.

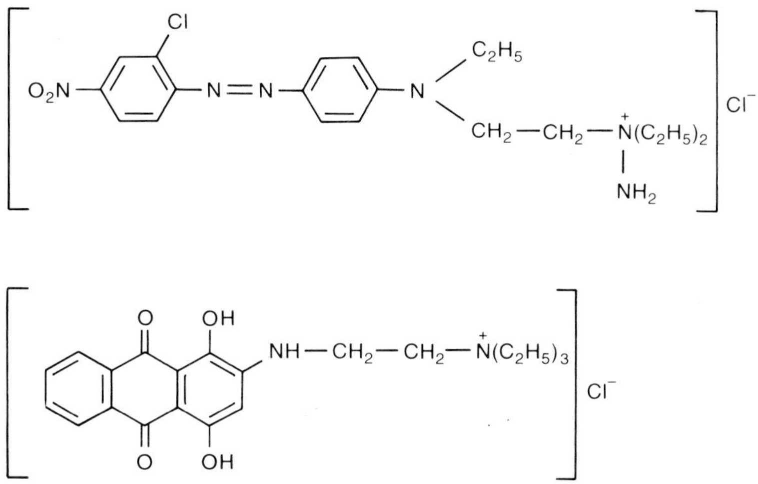

Fig. 5.10 Some examples of cationic dyes.

Function of dyebath assistants

The function of many of the dyebath assistants is to improve levelness of dyeing.

Acetic acid and sodium acetate: Acetic acid is included to give a pH to the dye liquor of *c.* 4–5, often with sodium acetate to provide a buffered system. The ionisation of some of the anionic groups on the acrylic fibres is suppressed, and their affinity for cationic dyes is consequently modified. Moreover, above pH 6, the dye may be prone to decomposition.

Cationic retarder: The retarder is colourless and, by competing with the dye for anionic groups on the acrylic fibres, reduces the rate of adsorption of dye. It permits a more uniform distribution of dye and so acts as a levelling agent (see below in section 5.6.2).

Glauber's salt: Glauber's salt can be included as a mild retarder and to assist levelling. However, it can also inhibit the action of other levelling agents.

Nonionic surfactant: This assists in wetting the substrate and in preventing precipitation of dye.

Dyeing procedure

The dissolved dye is added to the other contents of the dyebath at 60°C. The temperature is initially raised rapidly to *c.* 80°C and then much more slowly to 100°C. When boiling point has been reached, adsorption of the dye should be complete. After a further 10–15 minutes, the dyebath is slowly cooled to 50°C, to prevent undesirable changes in the texture of the fibres. The fibres are then rinsed.

The glass transition temperature of most commercial acrylic fibres is *c.* 80°C, so below 80°C the temperature of the dyebath can be raised quickly because little dye is adsorbed. However, at the glass transition temperature, the rate of diffusion of dye into the fibres suddenly increases. Small rises in temperature above 80°C accelerate diffusion even more and increase the rate of dyeing considerably. The high affinity of cationic dyes for the anionic groups in the fibres and the sudden increase in uptake of dye above 80°C both demand careful control to achieve level dyeings.

5.5 Dyeing fibre mixtures

In practice, many textiles consist of two (or even three) different types of fibre. The constituent fibres may have been spun together into the yarn to form a blend; or yarns of a different type may have been woven or knitted together to form a fabric, called a union. Nowadays, the term, fibre blend, is commonly used to refer to any textile, however formed, containing more than one type of fibre.

One approach to dyeing fibre mixtures is to dye each fibre component separately before blending. However, it is often more realistic to apply colour once the fibres are in fabric form, to give greater flexibility in meeting the changing demands of fashion. Depending on their chemical nature, the different components in a blend may require different classes of dye, and it is often essential that the colours of the different dyes match well. Moreover, a number of dyed effects can be achieved in blended fibres. A solid dyeing may be required, in which the depth and hue of colour in each component should be identical. Alternatively, one fibre component may remain undyed to give a reserve effect, or one component may be dyed more deeply than the other, a tone-in-tone effect. Finally, a cross-dyeing effect is possible too, in which there are contrasting hues in the two components.

Many different types of blended textile have been produced commercially, and it would clearly be impracticable to discuss the coloration of every one here. The foremost blends nowadays are those consisting of mixtures of natural and synthetic fibres; the natural fibres provide the desired comfort and the synthetic fibres provide the strength and durability. Three different types of blend are discussed here. Not only are they important in a commercial sense, but the principles underlying their coloration form a basis for the application of dyes to most other types of fibre blend.

5.5.1 Wool–polyamide blends

These blends have a wide variety of commercial applications, including carpeting, hosiery and sportswear. Both components can be dyed with acid, mordant or metal-complex dyes. However, care has to be taken that the dye is equally distributed between the two, or else the desired uniformity of colour is not achieved. For example, at the start of dyeing most levelling dyes will colour the polyamide component preferentially, but as the depth of colour is increased, more dye begins to be adsorbed by the wool component and eventually a solid colour is yielded. Once the polyamide component becomes saturated with dye, further dye is adsorbed only by the wool component.

5.5.2 Wool–polyester blends

The addition of polyester to wool gives improved resistance to wear and creasing. Dyeing of wool–polyester blends requires two types of dye: supermilling or 1:2-premetallised dyes for colouring the wool component and disperse dyes for colouring the polyester component. However, the wool component is readily stained by the disperse dye at lower temperatures, though at higher temperatures some of it is transferred to the polyester component. Nevertheless, a reduction-clearing treatment is often advisable. The presence of a suitable carrier is also usually advisable, for application of disperse dye at *c.* 130°C can damage the wool component.

5.5.3 Cotton–polyester blends

These blends commercially are perhaps the most important of all and are used in a wide range of garments. Disperse dyes are again used for application to the polyester component and are normally applied first. The extent of staining of the cotton component by these dyes can be lessened by subsequent reduction-clearing. The cotton component can then be dyed by almost any class of dye suitable for cellulosic fibres, though in practice vat and reactive dyes are generally used, in view of their superior fastness properties.

5.6 Levelling agents

Reference has been made to the desirability of applying dyes in a level fashion. In the application of a dye with a high substantivity for a textile, achieving levelness can be difficult, for these dyes have poor migration properties. Moreover, as with polyamides, chemical and physical variations along the fibres can produce an unlevel effect. Problems in achieving levelness can be ameliorated by the use of levelling agents during dyeing. These agents are colourless compounds and there are essentially two modes of action by which they can function. For example, a levelling agent may contain chemical groups similar to those by which the dye is attached to the fibres. This results in competition between the dye and levelling agent for adsorption by the fibres. Alternatively, the levelling agent may be capable of forming a weak complex with the dye. As dyeing proceeds, the complex gradually breaks down, allowing the dye to be adsorbed slowly by the fibres.

Whichever mode of action is operating, the levelling agent is able to reduce the rate at which the dye is adsorbed. Dye can, therefore, be applied in a more controlled manner. However, the use of levelling agents requires great care. If the wrong levelling agent is used for a particular dyeing process, dye may be insufficiently adsorbed by the textile fibres; indeed it may even be precipitated in the dyebath.

Levelling agents can be grouped according to their ionic character, and there are many ionic types. Quite often more than one type is present during dyeing. A brief discussion of some of them is now presented.

5.6.1 Anionic levelling agents

Anionic levelling agents are used extensively in dyeing wool and polyamides and, apart from being colourless, behave similarly to acid dyes. Indeed, anionic dyes are generally sulphonates. Molecules of these levelling agents compete with molecules of dye for adsorption by the textile fibres and so retard dyeing. However, it is essential that the dye has greater affinity than the levelling agent for the fibres, for eventually levelling agent must be completely replaced by dye. Thus, not all anionic levelling agents can be used successfully with all acid dyes. Some agents are suited only to some dyes; if used with other acid dyes, unlevel dyeings are likely to result. Anionic levelling agents can also sometimes be used in the application of disperse dyes to polyester.

5.6.2 Cationic levelling agents

Cationic levelling agents are used mostly as retarders in dyeing acrylic fibres, for these agents contain chemical groups similar to those in cationic dyes. Many cationic levelling agents are quaternary ammonium salts. Competition between dye and levelling agent retards the dyeing process. However, it is the dye which is eventually retained on the acrylic fibres, as it has the greater affinity for them. Cationic levelling agents are also sometimes used in the dyeing of wool.

5.6.3 Nonionic levelling agents

Nonionic levelling agents act by forming loose complexes with dyes and are used mainly in the application of acid and metal-complex dyes to wool. They also assist penetration of the wool by the dye and improve migration of dye in the fibres. However, the effectiveness of a nonionic levelling agent is very dependent on the dye in use. If the complex is not sufficiently soluble, dye is precipitated in the dyebath. If the complex is too stable, dye is retained in the dyebath, with negligible adsorption by the fibres. Thus, like other types of levelling agent, nonionic agents have to be used with care. Some nonionic agents are also used to assist the level application of disperse dyes to polyester and even of direct dyes to cotton. Many nonionic levelling agents have the general formula:

$$R—(O—CH_2—CH_2)_n—OH$$

where R represents a hydrocarbon moiety.

5.6.4 Nonionic/cationic levelling agents

These agents have weakly cationic character and find most use in the application of anionic dyes to wool. In particular, they reduce the degree of skittery dyeing (where there is an unequal distribution of dye between fibres in a yarn) and tippy dyeing. They also reduce the sensitivity of the dye to the pH of the dye liquor. They act by forming loose complexes with anionic dyes such that, as dyeing proceeds, a complex gradually dissociates and dye becomes adsorbed by the textile fibres. By forming loose complexes, these agents also limit the aggregation of dye molecules in the dyebath. However, many of these agents can be used only with specific dyes. The structures of these levelling agents are derived from cationic and nonionic levelling agents and can be represented by a general formula (**11**).

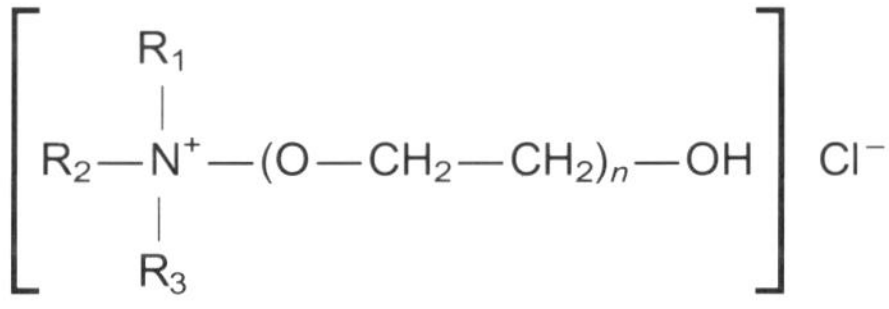

11

5.6.5 Amphoteric levelling agents

Amphoteric levelling agents contain both anionic and cationic groups. At low pH these agents carry a net positive charge, whilst at high pH they carry a net negative charge. They have found most use in the application of reactive dyes to wool at pH 4–6, and it appears that they can interact with both dye and fibres. Not only do they promote levelling, but they often improve the adsorption of dye as well. More recently, amphoteric agents have been used with acid milling dyes and those 1:2-premetallised dyes which contain anionic sulphonate groups.

5.7 Paper

As discussed in section 3.3, paper is essentially a cellulosic material, although many grades have a significant content of lignin as well. Dyes are used to colour a wide variety of papers; these include the cheaper grades, such as soft tissue products, and good quality uncoated printing and writing grades. Where superior colour fastness is required, as in cover papers, organic pigments may also be incorporated. Indeed, pigments may well be the sole colorants where very high light-fastness is necessary.

Dyes applied to paper include acid dyes and cationic dyes, but direct dyes are the most extensively used class. In addition, a new class, cationic direct dyes, has been introduced, specifically for application to paper.

Although paper is primarily cellulosic, there are some clear differences between dyeing paper and cellulosic textiles. Paper is normally dyed at temperatures below 40°C and within a pH range of 4–8. Moreover, fluctuations in pH during dyeing can affect dye performance in terms of strength, shade and even differences between the two sides of the paper (known as two-sidedness). Processes involving large additions of salt are not used in dyeing paper. Hard water, of itself, can increase the affinity of direct dyes, especially in the coloration of tissue grades.

Paper contains fillers, such as china clay, chalk or even titanium dioxide, to promote opacity, and the depth of dyeing must be equal in both filler and paper fibres. Direct dyes have only poor to moderate affinity for fillers, so that increased amounts of dye are required to achieve a desired depth of shade. Two-sidedness may also be a problem. Nevertheless, as with dyeing cellulosic textiles, high adsorption of dye and good colour fastness properties are important.

5.7.1 Acid dyes

Traditionally, acid levelling dyes were extensively used, but they have poor affinity for paper. Their retention on paper can be improved by after-treatments with fixatives. However, the application of acid dyes to paper is becoming increasingly rarer.

5.7.2 Direct dyes

Direct dyes are much better suited for application to paper, because of its high cellulosic content. They are used especially for tissues, toilet papers, wrapping papers, viscose film and printing and wrapping papers. Some direct dyes have poor wet-fastness properties in paper and, if possible, care has to be taken to ensure that dyes of improved wet-fastness are used. If no suitable dye is available for a required shade, an organic pigment may well be applied instead. Wet-fastness can, however, be improved by the addition of a fixing agent before or after application of the dye, depending on the dye being applied. The fixing agents used are polymers containing quaternary ammonium groups, which confer cationic character. Thus, salt-links can be formed with the anionic dyes. However, if too much fixing agent is applied, dye may be precipitated in the dyebath. Cationic fixing agents also reduce light-fastness.

5.7.3 Cationic dyes

Most cationic dyes have affinity only for those papers with an appreciable content of lignin, with whose anionic groups they can form salt-links. With decreasing lignin content in paper, the affinity of cationic dyes is accordingly reduced. The use of fixing agents has little effect on the retention of cationic dyes.

5.7.4 Cationic direct dyes

Cationic direct dyes resemble conventional direct dyes in that they consist of long, linear molecules. However, unlike their traditional counterparts, they possess cationic character, which arises from a positively charged functional group attached to the chromophoric part of each molecule. The functional group usually possesses a quaternary ammonium structure. An example of such a cationic structure is shown in Fig. 5.11.

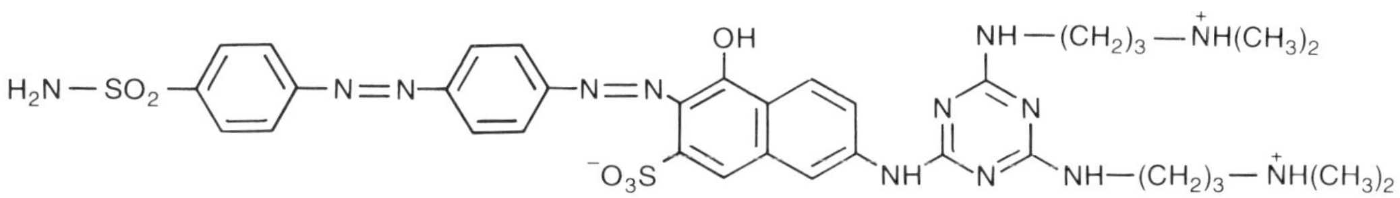

Fig. 5.11 An example of a cationic direct dye structure.

When compared with conventional direct dyes, cationic direct dyes are adsorbed more readily, have higher substantivity for all grades of paper and generally exhibit improved fastness properties.

5.8 Leather

As discussed in section 3.4, leather is composed predominantly of the protein, collagen. The physical properties of leather can present particular difficulties for dyeing. Leather is not a thin material, so that good penetration of dye (where required) may be difficult to achieve. The temperature of dyeing must be below 60°C. Moreover, a hide exhibits considerable variation: different layers in a hide (see section 3.4) have quite different physical properties. For example, the grain side of leather differs from the flesh side in fineness and density, and the resulting differences in porosity between the two sides make leather a material with two surfaces of different accessibilities and even different affinities to dyes. In addition, hides are likely to contain blemishes, which can render level application of dyes very difficult and even impossible. The assistance of suitable levelling agents, therefore, can be very important.

However, perhaps the chief influence on the application of dyes is the type of tannage used to preserve the leather. Collagen, like all proteins, is amphoteric. Immediately prior to tanning, the isoelectric point of collagen is 5.0–5.5, but tanning alters the isoelectric point. Thus, after chrome tanning, it rises to 6–7. Below this pH range, therefore, chrome-tanned leathers carry an overall positive charge and are hence more readily dyed by anionic dyes. (Above pH = 7, collagen is prone to damage). The pH of chrome tanned leather is generally within the range 3.0–3.5, but such a low pH would result in too rapid an adsorption by anionic dyes with the consequent risk of poor levelness. In practice, the pH is increased to *c.* 5.5, in order to provide a controlled application of dye. The increase in pH is effected by the addition of a so-called 'neutralising agent', usually the salt of an aromatic sulphonic acid or aliphatic dicarboxylic acid. Later in the dyeing process, the pH may be reduced to accelerate dye adsorption.

After vegetable tanning, the isoelectric point is lowered to *c.* 3.5, so very acid conditions are required for the application of anionic dyes. Cationic dyes can, therefore, be applied instead, at a pH of *c.* 6.

In addition to the type of tannage, the selection of dyes also depends on the required degree of penetration of dye into leather. Only a surface application may be necessary, as traditionally with shoe upper leathers. Alternatively, dyes may be required to penetrate partly or completely through the whole hide, as with clothing and upholstery leathers. Nowadays, good penetration of dye is preferred for nearly all types of leather.

Dyes applied to leather embrace a wide range of classes. Traditional amongst the anionic dyes are acid and direct dyes, but nowadays 1:2-premetallised dyes are extensively applied, for these give superior fastness properties. Soluble sulphur dyes also find some use in leather dyeing and are applied as anionic dyes. They penetrate leather well and possess good wet-fastness properties. Cationic dyes are very often applied after acid dyes as 'topping dyes', to obtain strong shades on heavily retanned leather. Even reactive dyes are finding some favour, although the use of alkali to achieve fixation can lower the quality of the leather, since collagen is susceptible to damage by alkali. Many of these classes of dyes have been developed specifically for application to leather, and some are now briefly discussed.

5.8.1 Acid dyes

Although at one time extensively applied to leather, acid levelling dyes are now considered to possess inadequate fastness properties. By contrast, a number of acid dyes of milling type are extensively used to dye chrome tanned leather. Acid dyes with more complex structures are also quite widely applied to leather. Many of these dyes contain chromium and bear some resemblance to 1:1-premetallised dyes. They readily cover leather surfaces and penetrate well.

5.8.2 Direct dyes

Direct dyes have little substantivity for leather but are still sometimes used for selected shades. Some are also applied as 'topping dyes' after the application of acid dyes. Although direct dyes provide good surface coverage of leather, their penetration is poor.

5.8.3 *1:2-Premetallised dyes*

Sulphonated 1:2-premetallised dyes are widely used for colouring leather on account of their superior fastness properties. In particular, they provide dyeings of the good fastness properties required for leather clothing. It is considered that the chromium in the dye can form complexes with collagen chains in the leather. Ionic and secondary forces of interaction supplement the formation of these complexes. The application of non-sulphonated 1:2-premetallised dyes requires care, for they can have poorer fastness properties, particularly with respect to dry cleaning.

5.9 Application of dye mixtures

In practice, mixtures of dyes, rather than individual dyes, are more frequently applied to achieve a desired colour. However, the application of mixtures of dyes can present additional problems. For example, where adsorption of dye is limited by the number of available dye sites in a substrate, there is competition for these sites between the component dyes in the mixture. This can be a problem in applying mixtures of acid dyes to polyamide textiles, because their fibres become saturated at only moderate depths of shade (see section 5.4.2). The dye with the greatest affinity for polyamide tends to be preferentially adsorbed. With increase in concentration of dye mixture in the dyebath, this particular dye progressively dominates the colour of the dyed fibres. This effect is known as blocking. Competition also becomes more marked when dyes with different degrees of sulphonation are present in the mixture, since increased sulphonation tends to reduce affinity and promote solubility in the dyebath.

Care also has to be taken in applying mixtures of cationic dyes to acrylic textile fibres. Not only is the number of dye sites in each acrylic fibre limited, but very many cationic dyes also show little migration on account of their high substantivity for acrylics. It is, therefore, essential that all the component dyes are adsorbed at nearly the same rate. Accordingly, the Society of Dyers and Colourists has introduced the concept of a compatibility value, K, ranging from 1 to 5. K is related primarily to affinity and, to a lesser extent, to diffusion. Dyes with lower values of K are adsorbed more strongly at the fibre surface, whilst dyes with higher values of K diffuse more readily into the interior of each fibre. Dyes for which the value of K varies by no more than 0.5 are considered to be sufficiently compatible in practice.

In the application of dyes to leather, differential penetration of the leather may be a problem. Thus, one component of a dye mixture may penetrate readily into leather, whilst another component remains largely on the surface. The bulk of the leather will then possess a different colour from its surface.

Dyes in a mixture may also interact with one another in the dyebath to form complexes. Problems associated with the formation of complexes are highlighted in the application of mixtures of direct dyes to cellulosic textiles, particularly if the mixture comprises direct dyes of different classes (see section 5.2.1). A complex is far less readily adsorbed than any of the component dyes because of its larger size and high charge. Moreover, if one component dye is in excess and hence not completely associated in a complex, it will be preferentially adsorbed. The desired colour will not be obtained.

There may even be problems once a dye mixture has been applied. One adsorbed dye may catalyse the gradual degradation of another within a substrate, a process known as catalytic fading. For example, some yellow vat dyes, notably C.I. Vat Yellows 2 and 33, can degrade blue vat dyes in cellulosic textiles. The result is fading of the green colour produced by mixing the blue and yellow dyes and a progressive change from green to yellow. By contrast, some blue reactive and direct dyes can degrade yellow dyes of the same class, a problem that is particularly serious in textiles which are frequently wetted, such as towelling and swimwear.

Bibliography

Colour Index, Vols 1–9, Society of Dyers and Colourists, Bradford, 3rd revision (1988).

Cellulosics Dyeing, J. Shore (ed.), Society of Dyers and Colourists, Bradford (1995).

Wool Dyeing, D.M. Lewis (ed.), Society of Dyers and Colourists, Bradford (1992).

The Dyeing of Synthetic-polymer and Acetate Fibres, D.M. Nunn, (ed.), Dyers Company Publications Trust (1979).

The Theory of Coloration of Textiles, A. Johnson (ed.), Society of Dyers and Colourists, Bradford (1989).

The Chemistry and Application of Dyes, D.R. Waring and G. Hallas (eds), Plenum Press, New York and London (1990).

Colorants and Auxiliaries, J. Shore (ed.), Society of Dyers and Colourists, Bradford (1990).

Review of Progress in Coloration and Related Topics, Vol. 14, Society of Dyers and Colourists, Bradford (1984).

Chapter 6

Pigments and their Application

The ways in which pigments differ from dyes as colouring materials have been discussed previously in Chapter 2 of this book. Essentially, a pigment is a finely-divided solid colouring material which is insoluble in its application medium, and it is incorporated into this medium as an intimate dispersion of solid particles. Pigments are generally applied by dispersion into a liquid medium. The pigmented medium then usually solidifies either by solvent evaporation, physical solidification or by polymerisation and the pigment particles become fixed mechanically in the solid polymeric matrix. The most important applications of pigments are in paints, printing inks, and plastics. In addition, however, they are used to a certain extent in an extremely diverse range of applications, including the coloration of paper, textiles, elastomers, glass, ceramics, cosmetics, crayons, and building materials such as cement and concrete. In contrast to textile dyes where the individual dye molecules are strongly attracted to the individual polymer molecules of the fibres to which they are applied, pigments are considered to have only a weak affinity for their application medium, and only at the surface where the pigment particle is in contact with the medium.

This chapter focuses on the chemistry of application of pigments. In this sense, the chapter complements the discussion of the chemistry of textile dyeing which the reader will find in Chapter 5. The first section in the present chapter deals with the functions performed by pigments in their application, which can extend beyond simply providing an appropriate colour. The second section considers the requirements of pigments for their particular applications, in terms both of their optical properties and their technical performance. The chapter concludes with an overview of the chemical structures and properties of the most important products in the commercial range of pigments. It is worthy of note that there are a wide variety of chemical types not only of traditional coloured pigments, but also important pigments which are white, black, metallic or pearlescent. Pigments are broadly classified into inorganic and organic types. The description of the structures of organic pigments contained in this chapter complements the discussion in Chapter 2 of this book on how, for selected cases, organic pigment molecules are designed to meet certain application requirements.

6.1 The functions of pigments in their application

Textile materials are dyed mainly to impart an appropriate colour, making the article more attractive to the customer and therefore more saleable (see Chapter 5). For many applications, the ability to provide the desired colour is also arguably the most important function of a pigment in its application. However, the optical role of a pigment can extend wider than simply providing colour, since it plays a decisive part in determining whether the medium is opaque or transparent and it may have a role in modifying gloss. In addition, in many applications the pigment can assume a range of other important functions.

6.1.1 The optical functions of pigments

Pigments have a profound influence on the optical properties of a paint, printing ink or plastic into which they are incorporated. This section considers the way in which pigments affect three optical properties: colour, opacity and gloss. A pigment owes its optical properties to a combination of two effects which result from the interaction of the pigment particles with visible light: light absorption and light scattering. Colour is determined mainly by light absorption, while opacity generally depends principally on light scattering.

Colour

An object appears coloured if certain wavelengths of incident light are selectively absorbed. A painted surface will appear red, for example, if the pigment used selectively absorbs most of the wavelengths of incident light and reflects the red wavelengths. The colour of a pigment is, as with dyes, primarily dependent on its chemical constitution. The chemical structures of the most important pigments, and the colours which they give rise to, are discussed later in this chapter. However, two further factors which influence the colour of pigments are a result of the fact that they are used as solid crystalline particles. These are the crystal structure, i.e. the way in which the molecules pack in their crystal lattice, and the particle size.

An example of the effect of crystal lattice structure on the colour of a pigment is provided by the most important blue pigment, copper phthalocyanine (CuPc). This pigment is manufactured in two crystal modifications, referred to as the α-form and the β-form. These two forms have identical molecular constitutions but differ in the way in which the CuPc molecules are arranged in the crystal lattice, as illustrated in Fig. 6.1 This phenomenon is known as polymorphism and is a common feature of the crystal structures of many organic and inorganic pigments. As shown in Fig. 6.1, in the α-form, the planar CuPc molecules are arranged parallel to one another in stacks, while in the β-form, they adopt a 'herring-bone' arrangement. Both

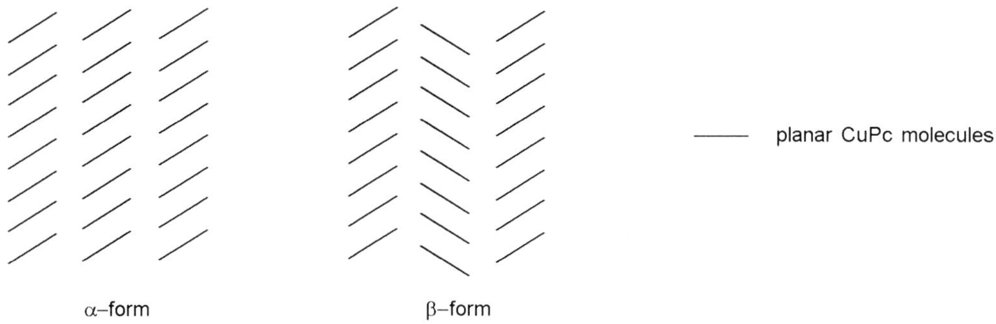

Fig. 6.1 The crystal lattice arrangements of molecules in α- and β-copper phthalocyanine. An example of polymorphism.

forms of CuPc are blue, but the colours are different. The α-form is a reddish shade while the β-form is greenish. An even more striking example of the influence of polymorphism on the colour of pigments is given by quinacridone, which is red in its γ-form but is violet in its β-form.

A further factor which is of critical importance in determining the colour of pigments, especially in influencing the strength or intensity of colour, is the particle size of the pigment. The theoretical physical basis underlying the influence of particle size on colour strength is complex and beyond the scope of this book. It is based on *Mie Theory* which may be used to provide a mathematical interpretation of the way in which pigment particles interact with light. The approach makes certain assumptions to simplify the mathematics. It considers interaction of monochromatic light (i.e. light of a single wavelength) with individual spherical particles in highly dilute systems. In addition, the treatment assumes that the particles are isotropic, i.e., that their optical properties are independent of the direction in which light travels through the particles. These are clearly assumptions which do not fully apply to real pigment systems. Establishing experimentally the influence of particle size on the colour strength of pigments also presents difficulties. The problems include the difficulty in measuring the tiny particles involved, which may in many cases be less than 0.1 μm in size. A number of techniques, including transmission electron microscopy and the disc centrifuge may be used for measuring particles of this order of magnitude, but each of these techniques has its limitations. A further complication is that pigment samples will contain particles showing a distribution of sizes and shapes, in varying degrees of aggregation. In spite of the difficulties, a generalisation, which agrees reasonably well both with the theoretical treatment and qualitative experimental observation, is that organic pigments show an increase in colour strength as the particle size is reduced, while with many inorganic pigments there is an optimum particle size at which the colour strength reaches a maximum, as illustrated in Fig. 6.2. For maximum colour strength, therefore, organic pigments are commonly manufactured in as fine a particle size as possible ($< 0.1\ \mu$m) while many

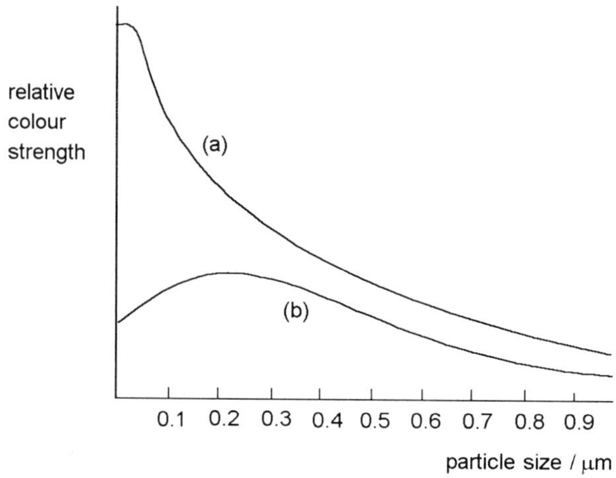

Fig. 6.2 The dependence of colour strength on particle size for typical (a) organic pigments and (b) inorganic pigments.

coloured inorganic pigments are produced at the particle size at which the colour strength is optimised (commonly 0.2–0.4 μm).

Opacity

The ability of a pigment to impart opacity is of fundamental importance for many surface coatings which are required to hide the surface to which they are applied and for certain plastics applications where opacity is a desirable feature. Opacity is mainly dependent on the degree of light-scattering by the pigment particles which is in turn dependent on a property of the material known as the refractive index (n). Refraction involves the deflection of waves of light as they pass from one medium into another, for example from air into the polymeric matrix of a paint film and from there into a pigment particle, as illustrated in Fig. 6.3. The ratio of the sine of the angle of incidence to the sine of the angle of refraction is the refractive index. Inorganic pigments which generally have a densely-packed ionic crystal lattice structure are high refractive index materials and are therefore capable of providing high opacity. In particular, titanium dioxide owes its importance as the most important white pigment principally to its ability to provide opacity as a result of its high refractive index. Titanium dioxide, like copper phthalocyanine and quinacridone, exhibits the phenomenon of polymorphism, existing in two forms which are produced commercially, namely *rutile* and *anatase*. The rutile form has a higher refractive index (2.71) than the anatase form (2.55) and is therefore more opaque. Opacity, like colour strength, is dependent on the particle size of the pigment, titanium dioxide pigments showing maximum opacity at around 0.28 μm.

In complete contrast, for some pigment applications high transparency is required. This is especially true of the coloured printing inks used in

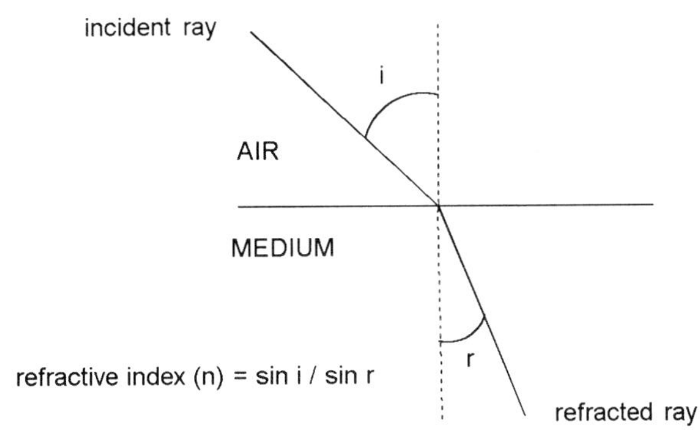

Fig. 6.3 The refraction of light.

multicolour printing processes. Generally such prints are produced from four inks comprising the three subtractive primary colours, yellow, magenta and cyan, and black. Transparency is required to ensure that the optical effect of the first colour printed is not obscured by subsequently printed colours (see Chapter 4). Organic pigments are generally molecular crystals with a low refractive index, a feature which ensures transparency, especially when they are manufactured in a fine particle size form and these are thus the pigments of choice for most printing ink applications.

Gloss

Gloss is the mirror-like reflection which is an important and desirable optical effect of many coatings. For example, the consumer expects that the paint on a car body should show extremely high gloss. The gloss shown by a painted or printed surface depends on a number of factors including the viewing angle and the nature of the film-forming polymer, its refractive index and the extent to which it forms a smooth film. However, the property may also be influenced by the pigment. In essence, gloss is determined by the degree of smoothness of the surface of the paint or ink film and will be reduced by the protrusion of pigment particles through the surface. In general, maximum gloss is achieved using pigments of small particle size at low concentrations, very finely dispersed in the coatings formulation.

6.1.2 The mechanical functions of pigments

In surface coatings and plastics, the mechanical properties are largely determined by the nature of the polymer used as the film-former (see Chapter 4). However, in some cases the pigment may be used to modify the mechanical properties of the polymer, either by playing a protective or reinforcing role.

Protective function

Some pigments may play a role in enhancing the protective properties of a paint film. For example, of particular importance for many exterior coatings applications is the ability of the pigment to enhance weather stability and corrosion prevention properties. An example of this is the use of micaceous iron oxide pigments in metal-protecting coatings. The flake-like particles of the pigment form a reflective surface thereby reducing degradation, mainly by UV radiation, of the film and in addition they form a barrier to moisture as an aid to corrosion protection.

Reinforcing function

Pigments may enhance the cohesiveness, elasticity, hardness and abrasion resistance of the polymers into which they are incorporated. A notable example of this is the ability of certain grades of carbon black pigments to reinforce rubber, particularly for demanding applications such as car tyres. When dispersed in rubber, the carbon pigment particles exhibit a high degree of structure involving chain-like clusters of particles and this provides reinforcing characteristics.

6.1.3 *Flow properties*

The flow properties, or rheology, of wet paints and printing inks are of fundamental importance in controlling the way these products are applied (see Chapter 4). For example, it is desirable that certain decorative paints, particularly for domestic use, should have non-drip qualities yet they must exhibit flow properties to provide an even surface on application to the substrate and, for example, to allow brushmarks to disappear. Flow properties are also of critical importance in application of printing inks by controlling the way the ink passes through the distribution system and is ultimately transferred on to the substrate during printing. While the flow properties of a paint or an ink are determined largely by the nature of the film-former and the solvent, especially its viscosity, pigments can play an important modifying role.

6.2 Requirements of pigments for application

The ability to produce the desired optical effect in the product into which it is incorporated is a prime requirement for a pigment. However, the pigments must in addition be capable of withstanding the effects of the environment in which they are placed, both in processing and in their anticipated useful lifetimes. A pigment will be selected for a particular application on the basis of its optical and technical performance but with due regard also to toxicological considerations and, inevitably, cost.

6.2.1 *Optical properties*

Colour

The brightest, most intense colours are in general provided by the use of organic pigments. The colours provided by inorganic pigments are as a rule weaker and duller. The ability to provide bright intense colours is of prime importance in printing inks. This is because most printed materials, for example publications and packaging, are generally highly coloured in order to provide the appropriate visual impact and because printing inks are usually applied as extremely thin films. Organic pigments are therefore usually preferred for printing ink applications.

Opacity/transparency

The ability of a pigment to impart opacity is a critical requirement for their application in paints which are generally required to hide the surface to which they are applied. Inorganic pigments, especially titanium dioxide, are used most commonly to produce opacity when it is required in paints, plastics or printing inks. Alternatively, and in complete contrast, high transparency may be essential. For example, transparency is required in printing inks used in multicolour printing processes. This is because the optical effect of the first colour printed must not be obscured by subsequently printed colours in order to ensure accurate colour reproduction. Organic pigments are inherently more transparent than inorganic pigments. This provides a further reason why organic pigments are preferred for printing ink applications. To produce a coloured transparent plastic article, an inherently transparent polymer is coloured either with dyes which dissolve in the polymer or with organic pigments.

6.2.2 *Fastness properties*

The fastness properties which are of most relevance to pigments in their applications are resistance to light, weathering, heat, and to solvents and other chemicals.

Lightfastness/weatherfastness

The properties of lightfastness and weatherfastness are clearly related although, on occasions, pigments which show good resistance to fading when exposed to light can perform less well under the combined attack of sunlight and moisture. These properties are determined principally by the chemical structure of the pigment but they may also depend to an extent on its concentration in the application medium and on the nature of the medium. Lighter shades, especially in combination with white pigments, tend to show poorer lightfastness than deeper shades. Lightfastness requirements will vary with the demands of particular applications, in particular the degree of external exposure which the product will be required to resist. For example, outstanding fastness to light is required for pigments used in automotive paints, such as those applied to cars and other vehicles, and in plastics used in the construction of the vehicle. On the other hand, it is considerably less important for low cost coloured printed materials, such as newspaper supplements.

Heat stability

The heat stability of a pigment refers to its resistance towards colour change at high temperatures. Colour changes resulting from inadequate heat stability, which can lead to off-shades or a failure to match shades, may well be due to decomposition of the pigment at high temperatures. Alternatively, they may be associated with the increasing solubility of a pigment as the temperature is raised or with pigment crystal phase changes. Heat stability is critically important in pigments for ceramics which are processed at temperatures in excess of 800°C and only certain inorganic pigments are capable of withstanding such temperatures. In the coloration of thermoplastics, heat stability is generally a critical feature in colorant selection. In this case, the degree of heat stability required of a pigment for a plastic application will depend not only on the processing temperature for the polymer in question, which can range from 150 to 350°C, but also on the time of exposure. Heat stability requirements in the coloration of thermosets, and paints and printing cured at elevated temperatures will also depend on the processing temperatures and times involved although these tend usually to be less severe than is the case with thermoplastics.

Solvent resistance

Since pigments in their application remain as discrete solid particles held mechanically within a polymer, they are required to resist dissolving in solvents which they may come into contact with. The extent to which solvent resistance is important for a particular application will depend on the nature of the solvent to which the pigment is exposed and the temperature of exposure since solubility increases with increasing temperature. Any tendency for the pigment to dissolve will lead to problems, for example in paints, of 'bleeding' of colour into overpainted layers of the paint. In plastics, a tendency to dissolve in the polymer may lead to problems of migration of colour. This may result in contact bleed, when a coloured plastic material causes staining of a dissimilarly coloured material with which it is in contact, plate-out, which refers to an accumulation of additive on the metal surfaces of plastics processing equipment, or solvent bleed which occurs when colour is leached from a plastic article when immersed in a solvent. 'Bloom' can also be observed when a powdery deposit of colorant appears on the surface of a paint or ink film or plastic product giving rise to poor rub-fastness. This generally results when a pigment of limited solubility dissolves in the application medium when hot and forms a supersaturated solution when cooled. In time, pigment molecules diffuse to the surface and crystallisation occurs.

Chemical resistance

Resistance towards changing colour when exposed to acids, alkalis or other chemicals is occasionally a feature of the selection of a pigment. For example, paints to be applied to fresh plaster require pigments which are resistant towards alkali. Acid-resistant pigments are generally used in the coloration of PVC which becomes acidic as it ages due to the formation of HCl, and also in the coloration of paints, inks or thermosetting plastics cured in the presence of acid catalysts. In plastic containers for reactive liquids, clearly resistance to the agent involved will be required.

6.2.3 Dispersibility

It is essential that a pigment is well dispersed into its application medium in order to achieve an even colour distribution. In addition, control of the degree of subdivision of the pigment in the medium may be important to optimise optical performance. For example, in the case of organic pigments, colour strength increases with a reduction in particle size (see Fig. 6.2) and a high degree of dispersion is therefore essential in order to achieve the maximum colour strength when it is incorporated, for example, into a paint or printing ink. Pigment powders which consist of clusters of aggregated particles must be subjected to conditions which will break down large aggregates into smaller aggregates and primary particles to the degree required to produce the optimum optical performance. The term 'dispersibility' refers to the ease with which the desired degree of dispersion is achieved. Over the years, pigment manufacturers have dramatically improved the dispersibility of the products produced commercially, thus minimising the energy requirements of the dispersion process. This is generally achieved by a surface treatment process in which the pigment particles are coated with surface active agents or organic resins. The result of these treatments is that the degree of particle aggregation is reduced and, in addition, such aggregates as are formed are of lower mechanical strength and hence easier to disrupt in the dispersion process. The principles of the surface treatment of pigments during their manufacturing process are discussed further in section 6.4. For a more detailed discussion of the nature of the pigment dispersion process when they are incorporated into paints, printing inks and plastics and a discussion of the manufacturing technology used, the reader is directed towards Chapter 10.

6.2.4 *Toxicological and environmental considerations*

Toxicological considerations must always be to the fore in colorant selection from the point of view of hazards presented in the workplace and exposure of the general public to the product. Some of the more important environmental issues which the colour industry are addressing are discussed in Chapter 12 of this book. Toxicological considerations are of critical importance for applications where ingestion is a possibility, such as in paints, printing inks and plastics which may come in contact with food and in toys and graphic instruments such as pens and pencils. In general, it is reassuring that most pigments may be considered as relatively non-toxic materials. In fact a number, including titanium dioxide, are acceptable as food additives. It may be argued that the insolubility of pigments means that they are capable of passing through the digestive system without absorption into the bloodstream and so present little hazard. There is, however, long-standing legislation limiting the use of certain pigments, notably those containing the toxic heavy metals lead and cadmium, in such applications. As a result of growing public concern on the more general impact of these pigments on the environment, extension of the legislation limiting their use further has taken place in recent years and this process seems set to continue into the future.

6.3 The chemical structures and properties of pigments

As described in Chapter 2, which presents an introduction to dye and pigment molecules, pigments are conveniently classified as either inorganic or organic. These two broad classes of pigments are of roughly comparable importance industrially. A number of generalisations may be made concerning the performance of inorganic and organic pigments in their applications, and these are summarised in Table 6.1, although it should be noted that exceptions to the general rules given in the table are found. Inorganic pigments generally are capable of providing excellent resistance to heat, light, weathering, solvents and chemicals and in those respects they can offer technical advantage over most organic pigments. In addition, inorganic pigments are generally significantly lower in cost than organics. On the other hand, they frequently suffer from the disadvantage of considerably inferior intensity and brightness of colour compared with organic pigments. Organic pigments are characterised in general by high colour strength and brightness although they are somewhat variable in the range of fastness properties which they offer. There is, however, a range of high performance organic pigments offering excellent fastness while retaining their excellent colour properties but these tend to be rather expensive. The ability to provide opacity or to ensure transparency provides a further contrast between inorganic and organic pigments. Inorganic pigments are, in general, high refractive index materials which are capable of giving high

Table 6.1 A comparison, in general terms, of the performance characteristics of inorganic and organic pigments.

Characteristic	Inorganic pigments	Organic pigments
Colour strength	low	high
Colour brightness	low	high
Opacity	high	low
Transparency	low	high
Lightfastness	excellent	moderate–excellent
Heat stability	excellent	moderate–good
Solvent resistance	excellent	moderate–excellent
Chemical fastness	good–excellent	moderate–excellent
Cost	low–moderate	moderate–high

opacity while organic pigments are low refractive index materials and consequently are transparent.

6.3.1 Inorganic pigments

Inorganic pigments are mainly fairly simple ionic species although there are a few structurally more complex materials. For the purposes of this section which provides a brief survey of the most important inorganic pigments in use commercially, they are classified further as non-coloured (achromatic) pigments, including hiding white pigments, non-hiding white pigments, black pigments and metallic pigments, and coloured (chromatic) pigments.

Hiding white pigments

By far the most important white opaque pigment is titanium dioxide (TiO_2). It owes its dominant position to its ability to provide a high degree of opacity and whiteness (maximum light scattering with minimum light absorption) and to its excellent durability and non-toxicity. The pigment is manufactured in two polymorphic forms, the rutile and the anatase form, the former being far more important commercially. The rutile form has a higher refractive index than the anatase form and is therefore more opaque. In addition, it offers better weathering properties. Zinc oxide (ZnO) and zinc sulphide (ZnS) are white hiding pigments which find some specialist applications but their lower refractive indices mean that they are less efficient than TiO_2 in producing opacity. White lead (basic lead carbonate), formerly the traditional white hiding pigment, has become virtually obsolete on the grounds of both inferior technical performance and toxicity.

Non-hiding white pigments

Non-hiding white pigments, sometimes referred to as extenders or fillers, are inexpensive white powders used in large quantities particularly by the plastics industry. They are of relatively low refractive index and thus they are capable of playing only a minor role in providing opacity. They are, however, used in a variety of other ways. For example, they may be used to modify the flow properties of paints and inks and to modify the mechanical properties and lower the cost of plastics. Commonly used non-hiding white pigments include calcium carbonate ($CaCO_3$), barium sulphate ($BaSO_4$), talc (hydrated magnesium silicate), china clay (hydrated aluminium silicate) and silica.

Black pigments

Carbon black is by far the most important black pigment and finds widespread use in paints, printing inks and plastics. Although there is a convincing argument that it ought to be regarded chemically as an organic pigment, it is generally classified amongst the inorganics. Carbon black pigments dominate because they provide intense black colours and an outstanding range of fastness properties at relatively low cost.

Metallic and pearlescent pigments

The most important metallic pigments are those based on finely-divided aluminium flakes which find their main use in automotive paints to produce a metallic effect, sometimes used in combination with transparent coloured pigments. Bronze pigments may be used similarly to give a gold or copper effect. Pearlescent pigments give rise to a white pearl effect often accompanied by a coloured iridescence. The most important of these pigments consist of platelets of mica coated with titanium dioxide.

Coloured inorganic pigments

The most important coloured inorganic pigments are iron oxides which are very low cost products providing shades ranging from yellow and red to brown and black. They are used extensively in paints, plastics and in building materials such as cement and concrete. Red iron oxide pigments consist principally of anhydrous Fe_2O_3 in its α-crystal modification. Yellow iron oxides, although often described as hydrated oxides, are more correctly formulated as the oxidehydroxide, FeOOH. Iron oxide pigments are characterised, in general, by excellent durability, high opacity, low toxicity and low cost. However, the yellow pigments show somewhat lower heat stability because of their tendency to lose water at elevated temperatures, in the process turning redder. The main deficiency of iron oxide pigments is that the colours lack brightness and intensity. The only other 'simple' oxide pigment of major significance is the chromium oxide, Cr_2O_3, C.I. Pigment Green 17. This is a tinctorial weak dull green pigment but with outstanding durability, including thermal stability to 1000°C. The mixed phase oxides are a group of coloured inorganic pigments of a range of colours which were developed originally for use in ceramics but which have subsequently found widespread application in plastics particularly because of their outstanding heat stability and weathering characteristics combined with moderate colour strength and brightness. Important commercial examples include cobalt aluminate blue, C.I. Pigment Blue 28, and nickel antimony titanium yellow, C.I. Pigment Yellow 53. Cadmium sulphide and sulphoselenide pigments provide a range of colours from yellows to reds and maroons which are rather brighter and stronger than most other classes of inorganic pigments. Cadmium sulphide itself (α-CdS) is yellow. The shade range of the pigments is extended by ion replacement. Partial replacement of Cd^{2+} by Zn^{2+} ions in the crystal lattice produces progressively greener shades of yellow, while replacement of sulphur by selenium gives rise to the orange, red and maroon cadmium sulphoselenides. The use of cadmium sulphide pigments is limited severely on the grounds of their potential toxicity and in most of the industrialised world there is legislation which restricts their use in applications where they are not seen to be essential. Accordingly their use is currently based on necessity, related to their excellent technical performance and in particular heat resistance, unmatched in their shade area by any organic pigment, to 600°C. Their principal use tends therefore to be in engineering plastics requiring high processing temperatures such as polyamides, acetals and PTFE. Lead chromate pigments provide a range of hues from greenish-yellow to yellowish-red. The mid-shade yellows are essentially pure $PbCrO_4$. The greener lemon chromes are solid solutions containing $PbSO_4$, while incorporation of molybdate ions into the lattice gives the orange and light red molybdate chromes. The use of these pigments, which provide bright colours with good fastness properties, is limited by the potential toxicity hazards associated in this case with the presence of lead and hexavalent chromium. Ultramarine blue, C.I. Pigment Blue 29, exhibits generally good durability and especially excellent fastness to light and heat, although rather poor resistance towards acids. It has a complex sodium aluminosilicate structure containing trapped S_3^- radical anions which are responsible for its colour. It has only a fraction of the tinctorial strength of copper phthalocyanine, the most important organic blue pigment. Prussian blue, or iron blue, C.I. Pigment Blue 27, a hydrated iron (III) hexacyanoferrate $Fe_4(Fe(CN)_6)_3.nH_2O$, is the longest established of all the synthetic colouring materials in current use. It is a low cost blue pigment with good lightfastness but relatively poor resistance to heat and alkalis.

6.3.2 Organic pigments

A useful classification of organic pigments is as either classical or high performance pigments. Classical organic pigments are the long established, relatively inexpensive commodity products used extensively in a wide range of printing ink, paint and plastics applications. This group includes mainly azo pigments and phthalocyanines. High performance pigments are capable of providing superior technical performance, generally at higher cost, and are more diverse chemically. Alternatively, organic pigments may be grouped into chemical classes, the most important being azo, metal complex, carbonyl and azomethine pigments. The description of the chemical structures of some of the more important chemical types of organic pigments contained in the following section complements the discussion in Chapter 2 of this book on how, for selected cases, organic pigment molecules are designed to meet certain application requirements.

Azo pigments

Azo pigments, both numerically and in terms of tonnage produced, dominate the yellow, orange and red shade areas in the range of commercial organic pigments. Fig. 6.4 illustrates the chemical structures of some important classical azo pigments. Simple classical monoazo pigments such as Hansa Yellow G (**1**), C.I. Pigment Yellow 1, and Toluidine Red (**2**), C.I. Pigment Red 3, are products which show good lightfastness, but rather poor solvent resistance because of their small molecular size. Their use is largely restricted to decorative paints. The pigments resist dissolving in the solvents used in these paints (either water or aliphatic hydrocarbons) at the low temperatures involved, but have a tendency to dissolve in more powerful solvents, especially if higher temperatures are involved. The most important yellow azo pigments, particularly for printing inks, but also for a range of paint and plastics applications are the disazoacetoacetanilides (Diarylide Yellows), which include C.I. Pigments Yellow 12 (**3a**), and 13 (**3b**). These products exhibit good colour strength and solvent resistance although rather poorer lightfastness than the Hansa Yellows. The Pyrazolone Oranges, such as C.I. Pigment Orange 34 (**4**), are similar both in chemical structure and in properties to the Diarylide Yellows and are used in similar applications.

The most important red azo pigments are metal salts, such as compounds **5a–5c**. These are products of high colour strength and brightness, high transparency, and good solvent resistance. They are especially important in printing ink applications, C.I. Pigment Red 57:1 (**5a**) for example being the principal pigment used in magenta printing inks. They are also suitable for use in many paint and thermoplastic applications. Metal salt pigments, commonly referred to as 'toners', have evolved from 'lake' pigments, now largely obsolete, which were essentially anionic azo dyestuffs precipitated on to inorganic substrates such as alumina and barium sulphate.

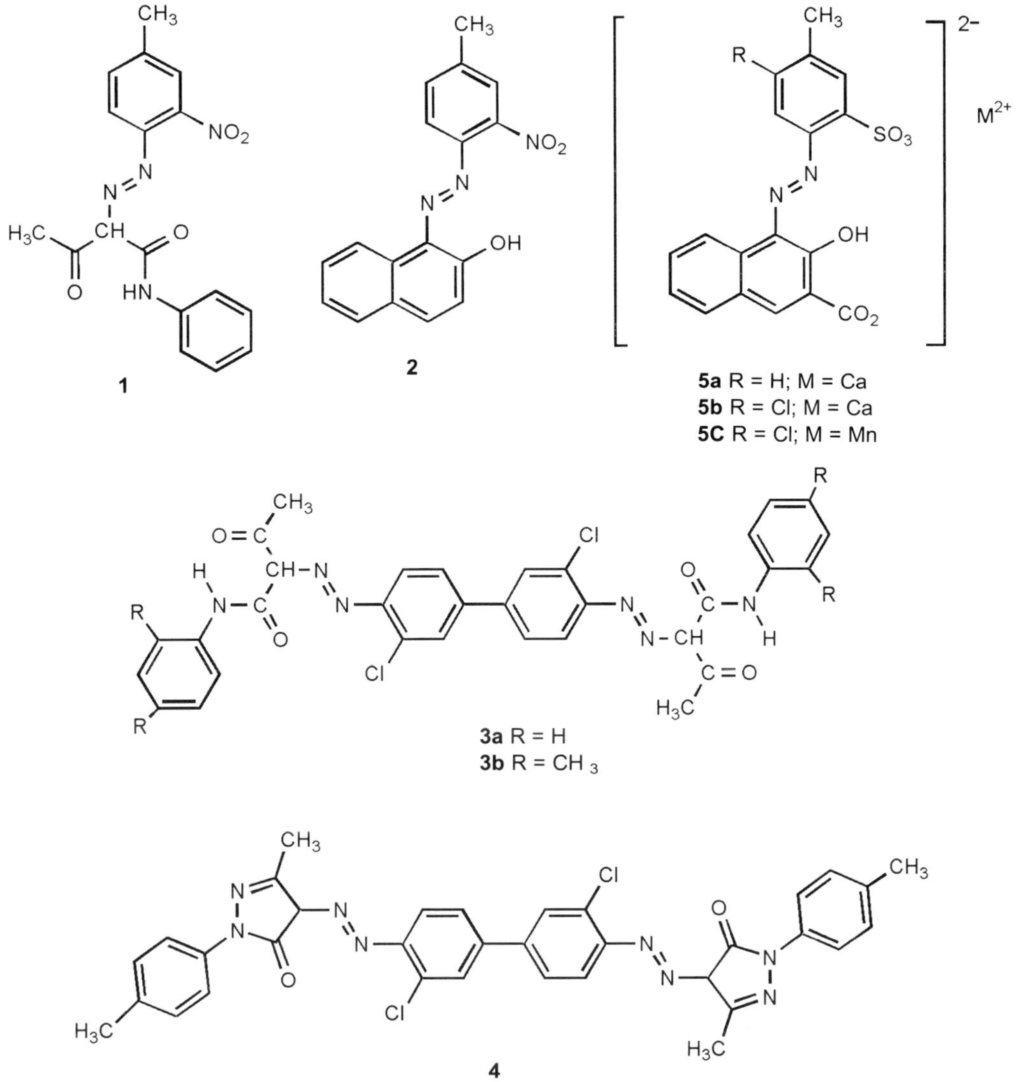

Fig. 6.4 The chemical structures of some important classical azo pigments. (**1**) Hansa Yellow G, C.I. Pigment Yellow 1; (**2**) Toluidine Red, C.I. Pigment Red 3; (**3a**) C.I. Pigment Yellow 12; (**3b**) C.I. Pigment Yellow 13; (**4**) C.I. Pigment Orange 34; (**5a**) C.I. Pigment Red 57:1; (**5b**) C.I. Pigment Red 48:2; (**5c**) C.I. Pigment Red 48:4

Copper phthalocyanines

Copper phthalocyanines provide by far the most important of all blue and green pigments. Copper phthalocyanine (**6**), (CuPc) C.I. Pigment Blue 15, is arguably the single most important organic pigment. Copper phthalocyanine finds widespread use in most pigment applications because of its brilliant blue colour and its excellent resistance towards light, heat,

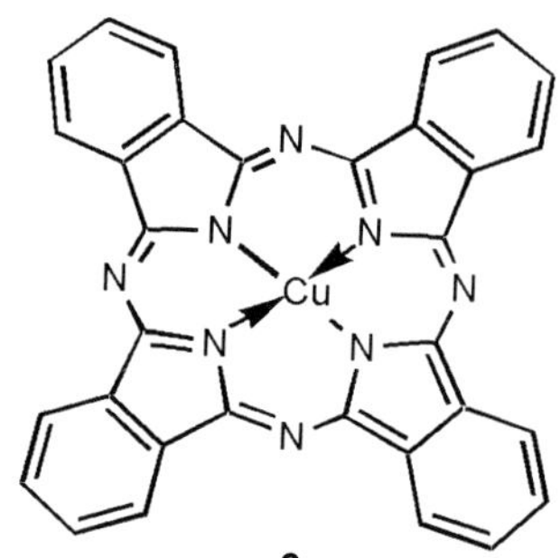

6

Fig. 6.5 The chemical structure of copper phthalocyanine.

solvents, acids and alkalis. In addition, in spite of its structural complexity, copper phthalocyanine is a relatively inexpensive pigment as it is manufactured in high yield from low cost starting materials. The β-form of the pigment is almost always the pigment of choice for cyan printing inks, and it is suitable also for use in most paint and plastics applications. The most important green organic pigments are copper phthalocyanines in which most or all of the ring hydrogen atoms are replaced by halogens. Polychloro, polybromo and polybromochloro copper phthalocyanines are produced, the shade of the pigments giving rise to progressively yellower shades of green with increasing bromine content. These copper phthalocyanine greens, which exhibit properties comparable to the blues, are also excellent pigments.

High-performance organic pigments

Copper phthalocyanines, although generally regarded as classical organic pigments, exhibit outstanding technical performance and so could equally well be described as high-performance organic pigments. This section contains a brief survey of a range of organic pigments encompassing a wide variety of structural types which have been developed in an attempt to match the properties of copper phthalocyanines in the yellow, orange, red and violet shades. High performance organic pigments are particularly suited to applications which require bright, intense colours and which also place severe demands on the technical performance of pigments, such as the coatings applied to car bodies, referred to as automotive paints. They provide excellent durability, especially light and weather fastness, combined with high colour strength and brightness. They do tend, however, to be rather expensive.

There are two classes of high performance azo pigments, the chemical structure of a representative example of each of which is given in Fig. 6.6. Disazo condensation pigments, developed by Ciba-Geigy, are a range of

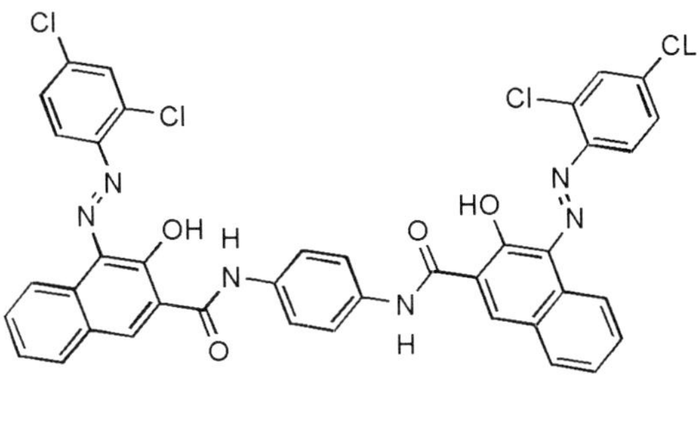

7

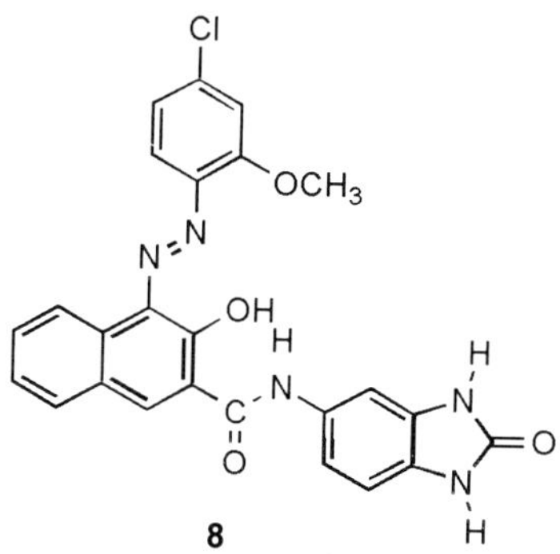

8

Fig. 6.6 The chemical structures of some high performance azo pigments. (**7**) C.I. Pigment Red 166, a disazo condensation pigment; (**8**) C.I. Pigment Red 183, a benzimidazolone azo pigment.

durable yellow, red, violet and brown products with structures such as C.I. Pigment Red 166 (**7**). These pigments derive their name, and also their relatively high cost, from the rather elaborate synthetic procedures involved in their manufacture which involves a condensation reaction. Hoechst have developed a range of benzimidazolone azo pigments, e.g. C.I. Pigment Red 183 (**8**), which range in shades from yellow to bluish-red and browns and with excellent fastness properties including heat stability to 330°C. Their good stability to light and heat and their insolubility is attributed to extensive intermolecular association as a result of hydrogen-bonding and dipolar forces in the crystal lattice. A similar type of intermolecular association accounts for the excellent range of fastness properties of the red and violet quinacridone pigments, such as compound (**9**). A diagrammatic representation of this intermolecular association is illustrated in Fig. 2.14 (Chapter 2).

A number of vat dyes developed originally for textile applications are suitable, after conversion to an appropriate pigmentary physical form, for use in many plastics applications. Examples of these so-called vat pigments include the anthraquinones, Indanthrone Blue (**10**), C.I. Pigment Blue 60, and Flavanthrone Yellow (**11**), C.I. Pigment Yellow 24 and the perinone (**12**), C.I. Pigment Orange 43. Other high performance pigments include the perylenes, for example C.I. Pigment Red 179 (**13**), the tetra-chloroisoindolinones, such as C.I. Pigment Yellow 110 (**14**), the dioxazine (**15**), C.I. Pigment Violet 23, and isoindolines, such as C.I. Pigment Yellow 139 (**16**). Arguably the most significant development in organic pigments in recent years has been the discovery and development of diketopyrrolo-pyrrole (DPP) pigments, such as C.I. Pigment Red 254 (**17**). These pigments are capable of providing brilliant saturated red shades with outstanding durability, including excellent lightfastness and thermal stability.

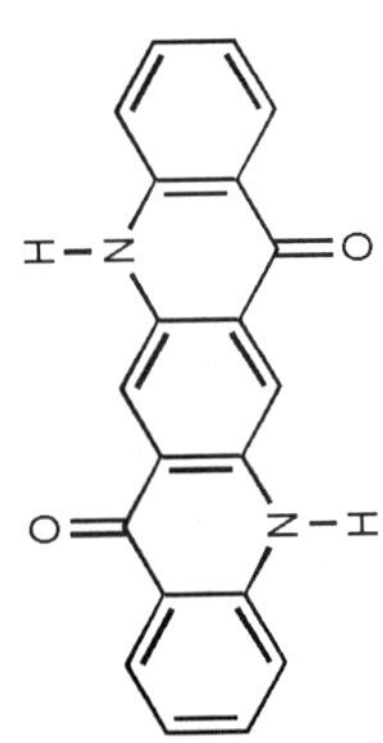

9

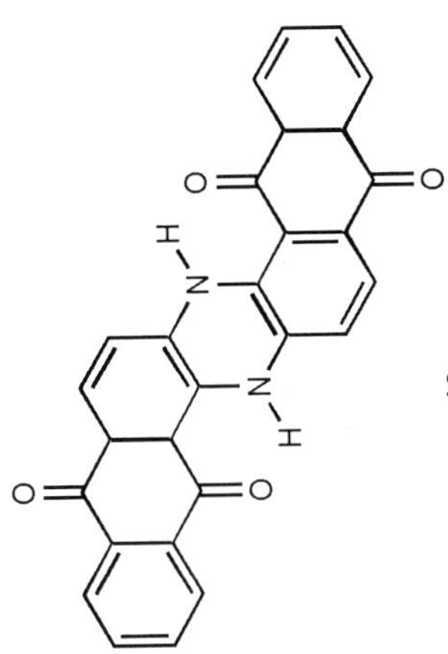

10

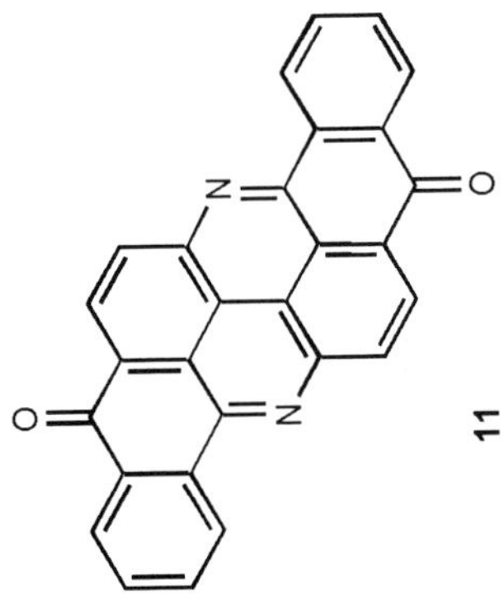

11

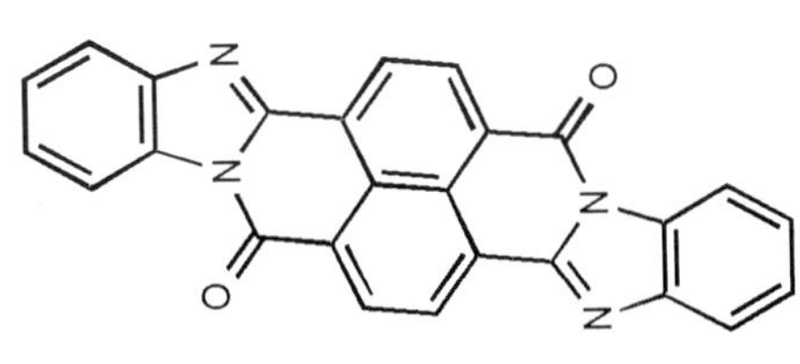

12

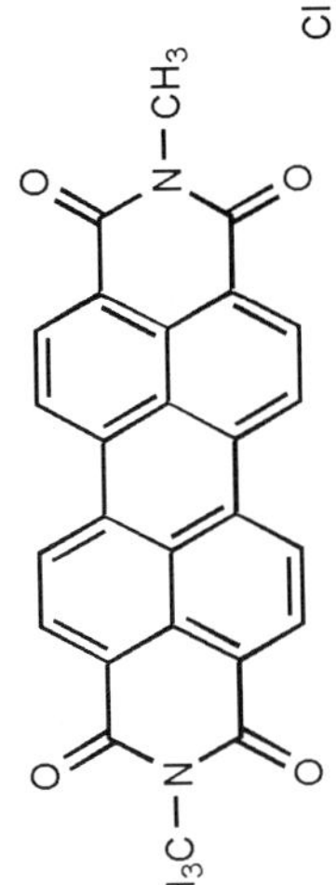

13

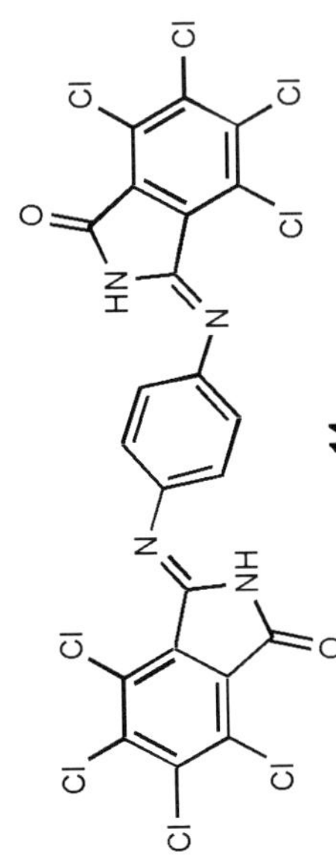

14

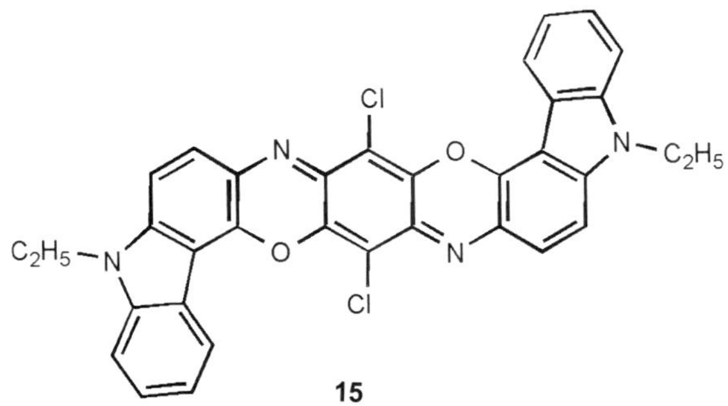

15

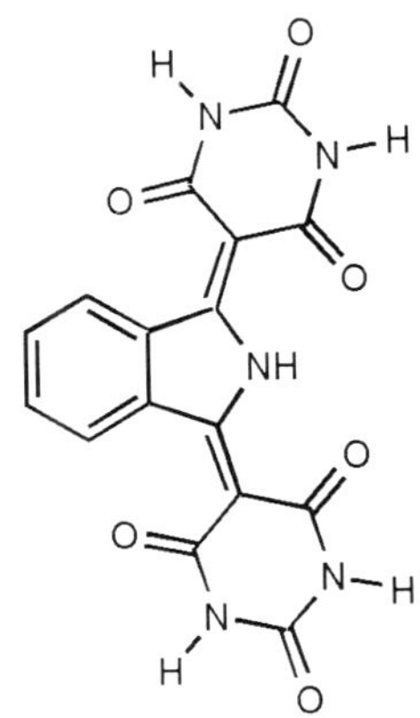

16

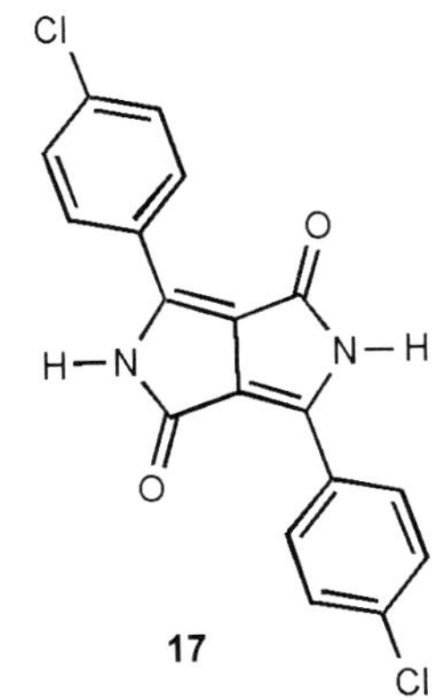

17

Fig. 6.7 The chemical structures of a range of high performance pigments. (**9**) Quinacridone, C.I. Pigment Violet 19; (**10**) Indanthrone Blue, C.I. Pigment Blue 60; (**11**) Flavanthrone Yellow, C.I. Pigment Yellow 24; (**12**) perinone, C.I. Pigment Orange 43; (**13**) perylene, C.I. Pigment Red 179; (**14**) tetrachloroisoindolinone, C.I. Pigment Yellow 110; (**15**) Dioxazine Violet, C.I. Pigment Violet 23; (**16**) isoindoline, C.I. Pigment Yellow 139; (**17**) DPP pigment, C.I. Pigment 254.

6.4 The manufacture of pigments

The process by which a pigment is manufactured may be considered as involving a number of stages. The first of these is the *synthesis* in which the sequence of chemical reactions which form the pigment is carried out. However, simply carrying out this chemistry does not necessarily give the product in a form which is optimised for a particular application. In a second phase of the process, commonly referred to as *conditioning*, the product is developed into an appropriate crystalline form, with a controlled particle size and shape distribution, and the surfaces of the particles may be altered by suitable treatments. Depending on the particular pigment, these two phases may be either distinctly separate or combined into a single process. Finally, the pigment is subjected to *finishing* in which it is converted into a form which is acceptable to the user.

Since this particular text is focussed on the application of colour, the chemistry of the synthesis of dyes and pigments is not considered in depth, but the reader will find many accounts of this topic in other reference works cited in the bibliography. The methods of synthesis of organic and inorganic pigments are quite diverse and dependent on the particular chemical type of pigment in question. Azo pigments and many inorganic pigments, including some iron oxides, cadmium sulphides, Prussian blue and lead chromates are manufactured by water-based precipitation reactions. Copper phthalocyanines and many high-performance organic pigments are commonly synthesised using elevated temperature organic solvent-based reactions. Some inorganic pigments, such as mixed phase oxides and ultramarines are manufactured using very high temperature solid state reactions, while the two largest tonnage pigments, titanium dioxide and carbon black may be manufactured using high temperature continuous gas phase reactions.

The conditions of synthesis of those organic and inorganic pigments which are prepared by precipitation from reaction in water may commonly be controlled to ensure that the pigments are formed directly at an appropriate particle size. Commonly they will precipitate from this aqueous medium in a poorly-defined crystal form, and so they are subsequently treated at elevated temperatures, either in the presence or absence of solvents (water or organic solvents), in order to develop high crystallinity and ensure that the products show optimum optical properties. Many inorganic pigments are subjected to very high temperatures towards the end of their manufacture in a process referred to as *calcination* (temperatures which organic pigments would not survive), in order to remove tightly bound water and to develop the appropriate crystal form. Organic pigments such as copper phthalocyanines, quinacridones and DPPs, which are synthesised at high temperatures by reaction in organic solvents, are produced initially in a large particle size form, which must be reduced to a fine particle size to make them useful pigmentary products. One way in which this may be achieved is by mechanical grinding, sometimes in the presence of abrasive materials such as inorganic salts. Alternatively, the pigment may be precipitated in a fine particle size by drowning out solutions of the crude pigments, for example in concentrated sulphuric acid, into water.

An extremely important feature of pigments is the nature of the surfaces of the individual particles, since it is here that the pigment comes into direct contact with the application medium. The surfaces of pigment particles which form directly in their synthesis are often highly polar, or hydrophilic, in nature which means that they show poor compatibity with many of the non-polar, or hydrophobic, media with which they will be required to interact when used in paints, printing inks and plastics. The pigments are therefore often treated with materials such as surfactants or resins, either during or after the synthesis phase of their manufacture in order to change the nature of the surfaces of the particles. The molecules of these surface-active agents contain a mix of hydrophilic and hydrophobic groups. Many such compounds may be considered as consisting of a polar 'head', often a sulphonic or carboxylic acid group or an amino group, and a 'tail', usually a long alkyl chain.

When the pigment is treated with the surface-active agent, the polar group attaches itself to the polar pigment surface, as a result of dipolar or ion-dipole forces of attraction. As a result, the alkyl chains provide the pigment with a new surface which is non–polar or hydrophilic and hence more compatible with an organic application medium. In addition, because of the steric barrier presented by these alkyl chains, the pigment particles are kept apart and are thus less susceptible to aggregation. Also, those aggregates which are formed are more easily broken down by the dispersion process. Natural resinous materials, such as wood rosin, the principal component of which is abietic acid (Fig. 6.8) are also used in the surface treatment of pigments. Another way in which the properties of pigments may be improved by surface treatment is the coating of inorganic pigments, such as titanium dioxide and lead chromates, with silica which stabilises the particles and leads to a marked improvement in lightfastness.

Inspection of the technical literature of pigment manufucturers reveals that they frequently will produce a range of grades of the same pigment for use in different applications. This is normally achieved by the use of specific

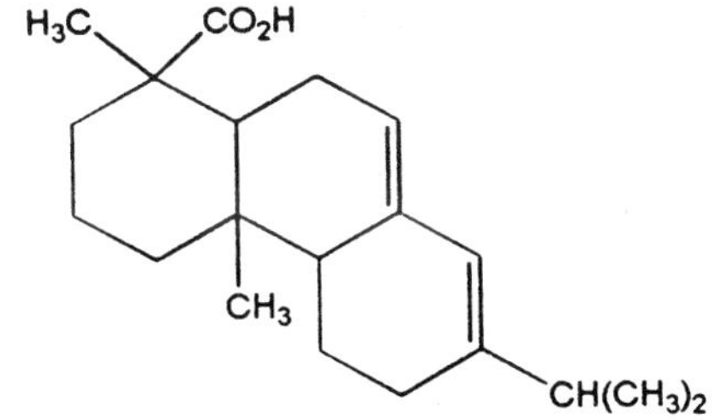

Fig. 6.8 Chemical structures of abietic acid, the principal component of rosin.

surface treatments which are tailored to ensure that the product provides the optimum performance in a particular application. Pigments are supplied to the customer in a variety of forms. Traditionally, for convenience most pigments have been supplied as dry powders. To achieve this, organic pigments which are isolated by filtration from their aqueous medium are dried at elevated temperatures. A problem with this stage of the manufacturing process, particularly with respect to certain organic pigments, is that drying can lead to hydrophilic aggregation and 'sintering' of the particles. The resulting clusters of particles may be hard in texture and thus difficult to disperse. This problem is minimised by the surface treatments described previously in this section, in which the surfactant or resin layer at the pigment surface presents a steric barrier which keeps the pigment particles apart as the water evaporates, and also by careful control of the drying conditions. There has been increasing concern in some industries with the inevitable level of dust which is generated in the handling of dry pigment powders, and pigment manufacturers have responded to this by the development of low-dusting, non-dusting and various types of granular forms of the dry pigments. Alternatively, pigments may be supplied in the form of pastes in which the problems associated with drying have been avoided. For applications of pigments in water-based paint and ink applications, it is common to supply the pigment as an aqueous paste or 'presscake'. For some applications, especially for paste ink system, organic pigments may be supplied as *flushed* colours. Flushing involves the treatment of an aqueous paste of the pigment with a viscous printing ink vehicle or a material compatible with such a vehicle, in either case the materials being organic in nature and immiscible with water. The flushing process makes use of mixing equipment capable of developing high shear conditions. Under these conditions, the organic pigment migrates from the aqueous phase into the viscous organic phase, and the pigmented organic phase may be separated from the water and incorporated directly into a printing ink. Flushed pigments are emerging as the dominant technology for use in paste printing inks, especially for use in the lithographic printing process (see Chapter 4). Many manufacturers of plastic articles find it convenient to make use of pre-dispersed concentrates or masterbatches of colorant in a liquid additive such as a plasticiser or in a compatible resin. Such concentrates are then easily incorporated by a fairly simple mixing procedure into the final polymer composition at an appropriate stage of the processing sequence (see Chapter 3).

Bibliography

Most of the texts cited in the bibliography at the end of Chapter 2 contain chapters which discuss aspects of the chemistry of pigments as well as dyes. In addition, the following texts deal specifically with pigments as colouring materials.

Treatise on Coatings, Vol. 3, Pigments: part 1, R.R. Myers (ed.), Dekker, New York (1975).
Pigment Handbook, Vol. 1: Properties and Economics, P.A. Lewis (ed.), John Wiley, Chichester (1988).
Pigments: Structures and Synthetic Procedures, R.M. Christie, Oil and Colour Chemists' Association, London (1993).
Industrial Organic Pigments: Production, Properties and Applications, W. Herbst and K. Hunger, VCH, Cambridge (1993).
Pigments for Inkmakers, J.D. Saunders, SITA Technology, London (1989).
Physical-chemical Aspects of Pigment Applications, J.T. Guthrie and l. Lin, Oil and Colour Chemists' Association, London (1994).

Chapter 7

Textile Printing

7.1 Introduction

Of the various methods of producing multi-coloured effects on textiles, such as embroidery or tufting, printing is the easiest and the cheapest. The printing of fabrics is long established, with evidence of it being carried out in India as long ago as the 4th Century BC. The worldwide annual output of printed textiles has increased year-on-year over the last 10 years to its present level of over 20 billion linear metres, which is enough to circle the world 450 times. Printed cotton accounts for about half of this production. The most important end-use for printed textiles is women's outer wear, with bedding also being significant. Printed cloth for shirts is also an important end-use. Other, less important end-uses, are curtains and upholstery.

A wide variety of techniques and methods exist for printing, all of which basically involve the application of the dye or pigment as a thick paste (the print paste) to localised areas of the textile fabric. Significant technological developments have occurred in the process during the last fifty years as the industry has responded to the specific requirements of the new synthetic fibres and to the changing demands of the market. Despite these developments and the increased understanding of the background chemistry, printing is still considered by some as an industrial art, rather than a scientific technology, an opinion which on occasion is not without some foundation!

Textile printing can be carried out using either dyes, or pigments. The decision on which to use is made on the basis of the particular design to be printed, the nature of the fabric and the intended end-use. Printing with pigments is equally as important as with dyes, but pigments are insoluble in water and have no natural affinity for textile fibres. In contrast to dyes which diffuse into the fibres, pigment printing is essentially a surface effect, so there is no diffusion into the fibre and the pigment molecules have to be held on the fibre surface by chemicals called binders. Pigment printing has advantages of technological ease and lower costs, but there are also situations where the resistance of pigment prints to abrasion and to laundering are not good. Pigment printing is described in greater detail in section 7.9.

In principle, the mechanisms involved in printing with dyes are similar to those in dyeing. After the print paste has been applied to the fabric surface, the fabric is dried to remove the moisture present in the print paste. The colour is then fixed either by dry heating or, more usually, by steaming. During steaming, the dye molecules adsorb on to the fibre surface and then diffuse to the interior of the fibre. However in a print, the dye molecules must first diffuse through the film of thickener in order to reach the fibre surface, so the time for steaming must be long enough to allow adequate fixation and diffusion to occur.

In selecting the dyes to be used, the first consideration is the nature of the fabric being printed. The dyes are selected on the basis of the application classes used to classify dyes in the conventional dyeing processes, described in Chapter 5. Consideration must also be given to fastness requirements for the intended end-use of the fabric. Next, the solubility of the dyes has the greatest importance. Only a small quantity of water is available in a print paste and if heavy depths are to be printed the limited solubility of some dyes could prevent their use. Also, during the steaming process, the dyes have to dissolve in a small volume of condensed steam to transport them into the fibre. Other important considerations are the diffusion characteristics of the dyes. The preference is for dyes of small relative molecular size, which are able to diffuse quickly through the print paste and after adsorption on to the fibre surface to diffuse quickly into the fibre matrix. Dyes which possess good diffusion characteristics require shorter steaming times, which in turn minimises damage to the fabric. Also such dyes tend to have lower affinities for the fibre and therefore ease of washing off when cross-staining occurs is easier.

7.2 Printing pastes

The print paste is the medium by which the dye or pigment is applied to the textile fabric. A typical print paste contains the dye, a solvent or solution assistant to aid the solubility of the dye, any special chemicals required to aid the fixation of the dye to the fibre and a thickening agent.

The application class of dye used must be one which is suitable for the type of fabric to be printed and the fastness requirements of the final product. Water-soluble dyes, such as acid, direct, basic, reactive, vat (leuco form) may be used, as well as the partially soluble disperse dyes and the totally insoluble pigments. It is often necessary to use a concentrated solution of a water soluble dye, so solution assistants may be required. For example thiodiethylene glycol (or thiodiglycol) is added to print pastes to aid the solution of acid milling dyes.

Other chemicals are incorporated in the print paste according to the particular class of dye being printed. For example ammonium tartrate or ammonium sulphate are added to generate the acidic pH necessary for the fixation of acid dyes. To provide the alkaline conditions required for the fixation of reactive dyes such as the vinyl sulphone type, sodium carbonate or sodium bicarbonate may be added. For the printing of vat dyes, a strong reducing agent and alkali are both required and typical print pastes in this case contain sodium dithionite, also known as sodium hydrosulphite ('hydros'), and sodium hydroxide. In addition, an anti-foaming agent and a wetting agent may be added to the paste. Some print pastes also contain a humectant, for example in the case of printing wool, when urea or glycerol are added, which have the effect of swelling the fibre and therefore aiding penetration of the dye into the fibre during the steaming process. Another consideration in the printing of wool is that the wool fibre can have a reducing action on azo dyes which destroys their colour. To counteract this effect, an oxidising agent such as sodium chlorate may be incorporated into the paste.

An extremely important component of all of print pastes is the thickening agent, which is necessary to ensure that the paste has satisfactory visco-elastic properties. These properties are vitally important because, as well as being viscous enough to hold the colour in place on the fabric, the viscosity must reduce enough to allow the paste to pass through the print screen under the influence of the shear forces created during application. These shear forces are created by the movement of the angled squeegee across the screen and cause the print paste to pass through the holes in the screen to the fabric underneath. On reaching the fabric surface, the viscosity must again increase, to prevent 'bleeding' of colour away from the local area to which it was applied. A wide variety of substances are available as thickeners and these may be derived from natural sources or may be synthetic. The particular thickening agent which is selected for a print paste depends on the application class of dye involved, the nature of the fabric to be printed and the process by which it is to be printed. The important requirements of the thickener are that it must be:

(a) compatible with all the other components of the paste;
(b) easy to dissolve in water;
(c) stable during the fixation stage;
(d) capable of being removed in the final washing-off process.

Clearly no one thickening agent can possibly meet the requirements of all printing processes and it is necessary therefore to select one which will function satisfactorily in the process to be used. Natural thickening agents are all polymeric and belong to a class of compounds called polysaccharides, which are based on simple sugars. Examples of such agents are starch ethers, gum tragacanth, locust bean gum, guar gum and alginates. All types can be modified chemically to alter their relative molecular mass, so that for any one of the types listed above, a range of products is available. As a general rule, the higher the molar mass, the less of the material that is required to be dissolved to obtain a certain viscosity. However the degree of chain branching in the molecule also influences viscosity. Polysaccharide molecules which are highly branched are more soluble and are required in greater concentration to produce a certain viscosity than simple linear polymers of the same relative molar mass. The printer has to decide which product to use, bearing in mind the amount of thickener that will be needed, the ease with which it can be dissolved and the overall visco-elastic properties of the final print paste required.

The nature of the dye being applied also has a bearing on the type of thickening agent which can be employed. For example, starch, which has a very similar chemical structure to cotton cellulose, is unsuitable for use in pastes for printing reactive dyes on cotton for obvious reasons and instead alginate thickeners must be used. In the printing of azoic dyes, alginate (or starch) thickeners should be avoided because they react with the diazo compound. Instead a locust bean ether is usually used. In the printing of vat dyes the thickener used must be able to withstand the strong alkaline conditions employed. For this purpose, a wide range of modified starches and gums is available.

A range of synthetic thickening agents based on polyacrylates has been developed, mainly for pigment printing. Whereas the thickening effect of the traditional polysaccharide types is due to their interaction with the water in the paste through the formation of hydrogen bonds, polyacrylates act by a repulsion effect of the carboxylic acid groups present. The success of the polyacrylates in pigment printing is due to their ability to give prints which are brighter and possess better fastness and handle properties than can be obtained with the polysaccharide type of thickeners. The use of polyacrylates has also been extended to the printing of polyester with disperse dyes. Unfortunately polyacrylate thickeners are susceptible to electrolytes, which act to neutralise the charge repulsion effect of the carboxylic acid groups on the molecule. Although disperse dyes are themselves nonionic, they contain in their commercial formulations anionic dispersing agents and residual electrolyte, and a reduction in the viscosity of the paste occurs. In response to this problem, the dye manufacturers have produced special 'low electrolyte' disperse dye ranges specifically for printing purposes. The print paste composition is very simple, since no acids or bases are required (the polyacrylate itself produces the optimum pH for the fixation stage), nor are any oxidising or reducing agents necessary. Despite the availability of the polyacrylates, thickeners from gums or alginates are still widely used for disperse dye printing. With the latter types of thickeners, the special 'low electrolyte' disperse dyes are not needed, the only requirement of the dyes being a very fine particle size distribution so that the pores in the print screen do not clog up. Also, because the dyes of the other application classes are themselves ionic, polyacrylates are unsuitable as thickeners for print pastes which contain these dye types.

An interesting area of innovation in printing processes has been the use of foam as the means of transferring the colour, instead of the conventional print paste. A foam, which is an unstable colloidal dispersion of air in a liquid, involves the use of only a small amount of liquid. Consequently, the drying processes are much quicker and involve the use of less energy. There are also advantages in increased dye penetration and migration so that foams may be used for printing carpets as well as lightweight fabrics. The resource efficient nature of printing with foams is likely to result in an increased use in the future.

7.3 Printing styles

There are three different printing 'styles' used to produce patterned effects on textiles, these being termed direct, discharge and resist. Each of these will be described in turn.

7.3.1 Direct printing

This method involves the direct application of the colour design to the fabric and is the most common method of textile printing. The dyes used for direct printing are those which would normally be used for a conventional dyeing of the fabric type concerned. The particular requirements of the process and the print pastes for the various fibre types are described below.

Direct printing of cotton fabrics

Prior to printing, cotton fabrics must be carefully prepared. Firstly it is necessary to singe the fabric by passing it rapidly, at full width, through a gas flame to burn off any protruding fibres or fragments of seed husk, thereby giving it a cleaner surface. Next, the fabric must be thoroughly desized. Sizes are chemicals, such as starches (flours, water-soluble gums), fats such as tallow or waxes, which are applied to the cotton yarn prior to weaving, to protect the yarn against fraying in the loom. These chemicals may contain reducing end groups which will cause the reduction of azo groups in the dye molecules, thereby destroying their colour. The desizing process involves treating the cloth with preparations which contain enzymes which break the sizes down to simple water-soluble products. Oxidative desizing, using hydrogen peroxide, is also increasing in popularity because of the whitening effect it imparts to the cloth.

Scouring, which involves treating the cloth with hot alkali, usually sodium hydroxide, is another important preparative process for cotton fabrics. This process causes the saponification of the fatty esters present in cotton into fatty acid soaps which are water soluble. At the high temperatures used for the process, any fatty alcohols and waxes melt and are emulsified by the fatty acid soaps which have been formed. The resulting cotton fabric possesses much improved wettability and absorbency characteristics, which are necessary for printing.

Better colour yields and brighter shades are obtained if the cotton has been bleached and mercerised. Bleaching, usually carried out with hydrogen peroxide, destroys any traces of natural colouring matters present in cotton fibres and results in cloth of superior base whiteness. Mercerising is a process which involves the treatment of cotton yarns under tension in very strong solutions of caustic soda and brings about a change in the crystal structure of the cellulose. Mercerised fibres have greater strength, and more importantly, greater lustre. Dyes show a greater substantivity for mercerised cotton so overall greater depths of shade are achievable.

Of all the possible application classes of dye which can be applied to cotton, it is only the reactive and vat classes which have any importance in printing. Nearly 50% of cotton is printed with reactive dyes, about 10% with vat dyes and the remainder is printed with pigments.

Reactive dyes are popular because dyes of relatively low molecular mass can be used, which possess good solubility, are faster diffusing and because of their lower affinity for the fibre, hydrolysed dye is more easily removed in the final washing off process. There is less tendency for them to stain adjacent white areas in the pattern. Reactive dyes as a class produce bright shades and cover a wide range of the colour gamut. A print paste for a reactive dye will typically comprise:

(1) reactive dye, as required by the colour recipe
(2) water
(3) thickener (as 3–12% solution)
(4) urea (to increase dye solubility and swell the fibre)
(5) sodium bicarbonate (to provide alkalinity for fibre reaction)
(6) sodium *m*-nitrobenzenesulphonate

The sodium bicarbonate is added last, when the paste has cooled, to prevent hydrolysis of the dye prior to printing. This so-called all-in-one method accounts for about 90% of cellulosic printing with reactive dyes. The remainder is processed using two stages in which the fabric is printed with the paste without the alkali. The fabric is then treated separately in alkali to bring about fixation. Two stage methods also provide a compatible means of printing reactive and azoic dyes in patterns.

Direct printing of wool fabrics

In comparison with cotton the quantity of wool printed is extremely small, probably because fabric cost, sampling costs and shorter run lengths all combine to make printed wool an expensive product. The careful pre-treatment of wool is more important than with any other fibre type and is essential if brightness of shade and full colour yields are to be obtained. Some wool fabrics require to be relaxed and stabilised, which is generally achieved by a process called *crabbing*. This process imparts a permanent set into the cloth by treating it with boiling water. Efficient scouring, to remove spinning oils or warp size, is also required. Spinning oils are added at levels between 2% for worsted yarns to 15% on carpet yarns, to reduce frictional effects in weaving. In the main, fine wool worsted fabric is printed and the levels of spinning oils present on this type of fabric are at the lower end of this range. The scouring process traditionally involved treating the fabric in a solution of soap (2–3%) and sodium carbonate at a pH of about 9 and a temperature of no higher than 55°C. Nowadays it is common practice to use non-ionic synthetic detergents. Indeed, some spinning oils used are the self-scouring types.

Another important pre-treatment for wool fabrics is acid chlorination, an oxidative process which produces an increased number of strong acid groups, for example —SO_3 and causes cleavage of disulphide and peptide cross-links. The resulting fibres have increased polarity and are therefore more hydrophilic. Also they have greater accessibility and are more readily swollen in water.

The dye classes most widely used for printing wool are the acid milling, metal complex and reactive types. Reactive dyes are often preferred because they are more soluble in water and the use of dye solvents is not required. Shorter steaming times for fixation are required since it is only necessary to form a dye–fibre bond to achieve the required wash fastness, so the process is less costly. A disadvantage of reactive dyes is that they tend to have poor migration properties which can lead to unlevelness in large blotches.

A typical recipe for printing wool is:

(1) dye (acid milling, metal complex, or reactive)
(2) water
(3) thickener (usually guar gum derivatives, sometimes mixed with starch ethers)
(4) urea (to aid dye solution and promote fibre swelling)
(5) thiodiethylene glycol (dye solvent – particularly necessary if acid milling dyes are used)
(6) citric acid, or acid donor such as ammonium sulphate (to provide the acidic conditions required for dye affinity)
(7) sodium chlorate (oxidising agent to counteract the reducing action of wool on the dye)
(8) wetting agent
(9) antifoam

Direct printing of polyester fabrics

The fabrics should be scoured prior to printing to remove any sizes or spin finishes. It is also preferable to heat set fabrics after scouring to confer suitable stability for printing. It is not usually necessary to bleach polyester fabrics unless especially white grounds are required. In such cases sometimes a fluorescent whitener is added after bleaching to enhance the whiteness.

Disperse dyes are used for printing polyester and the dye manufacturers produce special printing brands in which the dye particles possess much smaller average particle size distributions than those for application by dyeing. Particle size is an important consideration, because if the particles are too large they will not pass through the fine holes in the printing screens. The dyes used must have a high sublimation fastness. Fixation after application of the print paste normally involves treatment in high temperature steam at about 180°C and under such conditions sublimation of some dyes on to adjacent white areas may occur. As a rough guide, dyes of higher molecular mass usually have the better fastness to sublimation. The fact that polyester has a high glass transition temperature means that prints have excellent wash fastness as the dye effectively becomes locked into the fibre matrix once the steaming stage is finished.

A typical print paste formulation is:

(1) disperse dyes
(2) water
(3) thickener (usually alginate or locust bean ether type of low solids content or polyacrylate if low electrolytes dyes are used)
(4) fixation accelerator or carrier (to improve dye adsorption and diffusion)
(5) acid donor (sodium dihydrogen phosphate, to prevent formation of alkaline conditions which would degrade the dye)
(6) mild oxidising agent (sodium *m*-nitrobenzenesulphonate)

Direct printing of acrylic fibres

Very little printing of acrylics is carried out today. The only pre-treatment necessary for acrylic fibres is scouring to remove any sizes or knitting lubricants. Bleaching is not usually necessary since the natural whites are satisfactory for most purposes.

Basic dyes are used, which give prints of excellent brightness and good fastness to light and washing. The same characteristics apply to printing as in dyeing, in that the dyes have very high affinity for the fibres but below the T_g of the fibre (about 75–80°C), the rate of diffusion of the dye into the fibre is very slow. In formulating the print paste, no anions which are capable of forming an insoluble complex with the dye cation, must be added. A typical print paste formulation is:

(1) basic dye
(2) water
(3) thickener
(4) dye solvent (thiodiethylene glycol)
(5) acid (acetic acid)
(6) acid donor (citric acid or ammonium sulphate, to maintain acid conditions at all stages)
(7) fixation accelerator (dicyanoethyl formamide)

Direct printing of polyamide fabrics

The normal preparation of the fabric involves a scouring process followed by heat setting to impart stability. Of the dye classes normally used for polyamide fibres, the acid dye type is most commonly used for printing. Solubility is a crucial factor in dye selection though solubility can be increased by the presence of the dye solvent thiodiethylene glycol, which additionally swells the fibres, thus aiding diffusion.

An acid pH is required to confer dye affinity for the fibre so citric acid or ammonium sulphate is added to the paste. The thickener used must be stable to the acid pH so usually locust bean ether derivatives are used. A typical print paste recipe is:

(1) acid dye
(2) water
(3) thickener (locust bean ether derivative)
(4) thiodiethylene glycol (dye solvent)
(5) thiourea (fibre swelling agent)
(6) ammonium sulphate or ammonium tartrate (acid donor)

7.3.2 Discharge printing

In this method the fabric is pre-dyed to a solid shade by a traditional dyeing process and the colour is then destroyed locally, by chemicals incorporated in the print paste especially for that purpose. The result is a white patterned discharge on a coloured ground.

The type of chemical used to discharge the colour is usually a reducing agent, the most common being sodium formaldehyde sulphoxylate (C.I. Reducing Agent 2). The discharging action is based on the reduction of the azo group in dyes:

$$ArN{=}NAr' + 4H \rightarrow ArNH_2 + Ar'NH_2$$

The amine compounds formed are washed out at the end of the process. The actual choice of reducing agent used depends on the dyes being used and the fibre being printed. For example, for the discharge printing of wool, zinc formaldehyde sulphoxylate (C.I. Reducing Agent 6) is the preferred agent because it functions in the weakly acidic pH range and thus causes less fibre damage than the sodium compound, which requires an alkaline pH. As a general rule, heavier depths of shade are more difficult to discharge than lighter depths. In cases where dyed grounds are difficult to discharge, alkali in the form of soda ash may be added to the print paste, reinforced by perhaps a dispersing agent and a penetrating agent. Particular care is necessary in formulating the paste containing the discharging agent since 'bleeding' occurs more readily than with pastes containing dyes, which results in a halo effect.

The quality of the white discharge produced depends on the ease with which the dye can be reduced. The whiteness can be improved by incorporating a fluorescent whitening agent, which is unaffected by the reducing agent, into the print paste. However if the ground shade cannot be removed completely, fluorescent whitening agents may not improve the white and in such situations finely ground zinc oxide may be added to the print paste instead.

An imitation discharge effect can be produced by printing the ground shade with a print paste containing a heavy loading of zinc oxide or titanium dioxide and a resin binder. This approach avoids the need for a reducing agent because the effect is produced by masking the dyed ground shade.

Another effect can be produced by incorporating a dye which is resistant to the reducing agent into the print paste. The dye is simultaneously applied to the fabric as the ground shade is destroyed and a coloured discharge (called an 'illuminated discharge') effect is produced. The dyes used for this purpose are called 'illuminating colours', though unfortunately there are relatively few dyes available which are resistant to the reducing agents used. However, it is possible to produce, especially on wool, intricate patterns on strongly contrasting grounds. The clarity and sharpness of the prints is greater than can be achieved by direct printing.

Discharge printing is easiest to carry out on cellulosic, wool and silk fabrics. The synthetic fibres, such as cellulose acetates, polyester and nylon are hydrophobic and adequate penetration of the fibre by the reducing agent is difficult to achieve. In the case of polyester it is particularly difficult to remove the dye, because at temperatures below the glass transition temperature of the polymer, the dye is effectively locked within the fibre. Much work has been carried out in the development of a process to discharge disperse dyes on polyester, especially the new micro-fibres with their silk-like handle. An approach to this problem is to pad-dye the fabric with dischargeable disperse dyes then dry the fibres at a temperature of no greater than 80°C. The fabric is then printed with a paste containing zinc formosul and discharge resistant disperse dyes. Finally the fabric is steamed and given a thermofixation treatment, to produce an illuminated discharge.

The developments in recent years, especially of rotary screen printing, have reduced the need for discharge printing. This style is also relatively expensive, since a dyeing process is required, together with the printing of the discharge paste. However it is a style which is still used because it enables intricate patterns to be produced of a quality that is not possible by other methods. Also if the nature of the design is such that there is a large area of solid ground shade present, this is best achieved by a dyeing process. The higher costs of discharge printing are usually offset by the fact that it is a style most often used for high value products where the selling price is high enough to justify its use.

7.3.3 Resist printing

In this method of printing the fabric is first printed with a substance called a 'resist' which will prevent the dye from being taken up in a subsequent dyeing process. The resist functions by either mechanically preventing the dye from reaching local areas of the fabric or by chemically reacting with the dye or the fibre, to prevent adsorption.

A traditional form of resist printing is the 'Batik' style, particularly famous in Java, in which paraffin wax is used to prevent the dye reaching the fibre. It is necessary for the wax to be applied to both sides of the cloth and it must be applied carefully so that it does not actually penetrate into the cloth, otherwise the final pattern will not be clearly defined. The dyes which can be used in the subsequent dyeing process are limited to those which can be applied from relatively low temperatures, otherwise the wax resist may soften (or melt) and become detached from the cloth. It is preferable therefore that the dyes have a high affinity for the cloth at the low dyeing temperatures necessary.

Vat dyes (indigo is most commonly used) and solubilised vat dyes, with their high substantivities are very suitable for resist printing. Vat dyes of course must be applied in their leuco form and oxidised to the parent form after dyeing. Also useful are the cold-dyeing reactive dyes, though most have only a modest substantivity for cellulose, so the repeated dipping of the cloth into the dye liquor is necessary if heavier depths of shade are required. Azoic dyes are better in this respect because they give deep shades easily, producing prints with a very good rubbing fastness.

After dyeing, the wax must be removed. This is accomplished by the repeated treatment of the cloth in boiling water, until as much of the wax as possible has been removed. The wax floats on the water and can be skimmed off and used again. Finally the cloth is treated in a hot solution of sodium carbonate which emulsifies any remaining wax. This wax cannot be recovered, so the emulsification treatment is left to the end.

There are a number of ingenious methods for producing resist effects by chemical means. For example a resist for reactive dyes can be produced by pre-printing the fabric with a print paste containing a weak organic acid such as citric or tartaric acid. Since reactive dyes require an alkaline pH for fixation, a resist effect will be obtained. The production of a resist on wool can be achieved by the use of sulphamic acid, NH_2SO_3H. This is effective with reactive dyes because it reacts with the basic amino groups and hydroxyl groups in the wool which reactive dyes require for fixation. As well as blocking these groups the sulphamic acid also introduces an anionic group which serves to repel the anionic reactive dyes.

Resist printing gives results which are as good as those achieved by discharge printing, but has the advantage that it has a greater versatility in terms of use. Most application classes of dyes can be resisted but relatively few dyes can be discharged. Both styles are used when a design requires a white or coloured ('illuminated') pattern on a coloured background because they are capable of giving prints of very high quality in terms of sharpness of definition and fastness properties. Such prints very often cannot be obtained by the other printing styles.

7.4 Printing processes

There are five main methods of printing a fabric, these being the block, roller, screen, heat transfer and ink-jet methods. The heat transfer method differs from the others in that it involves the transfer of colour from the design printed on paper through the vapour phase into the fibres of the fabric. With the other methods the dye or pigment is applied to the fabric surface through a print paste medium. The ink jet printing process however is a comparatively recent innovation and is referred to as a 'non-impact' method, because the print paste is fired on to the textile from a jet which is not actually in contact with the fabric.

7.4.1 Block printing

The blocks are usually made of wood and the design is hand carved, so that it stands out in relief against the background surface. The print paste is applied to the design surface on the block and the block then pressed against the fabric. The process is repeated with different designs and colours until the pattern is complete.

Block printing is a slow, laborious process and is not suitable for high volume commercial use. It is a method still practised in the oriental countries where markets exist for the types of printed fabrics produced.

7.4.2 Roller printing

Roller printing has traditionally been preferred for long production runs because of the very high speeds possible. It is also a versatile technique since up to a dozen different colours can be printed simultaneously. The basic roller printing equipment, shown in Fig. 7.1, consists of a number of copper faced rollers in which the design is etched. There is a separate printing roller for each colour being printed. Each of the rollers rotates over the fabric under pressure against an iron pressure roller. A blanket and backing cloth rotate over the pressure roller under the fabric and provide a flexible support for the fabric being printed. A colour doctor blade removes paste or fibres adhering to the roller after contact with the fabric. After the impression stage the fabric passes to the drying and steaming stages.

In order to obtain sharply defined prints, high quality engraved copper rollers are required. It is essential that each colour is printed in register, that is, without any overlapping of the colours of the different patterns, unless this effect (fall-on effect) is specifically required. The pressure of each print roller against the cloth can be controlled independently by either hydraulic or pneumatic systems, so that adjustments can be made to ensure that each print roller gives a uniform pressure across the width of the fabric. These control systems also allow the register of the prints to be controlled and provide adjustments to enable fabrics of different thicknesses to be printed.

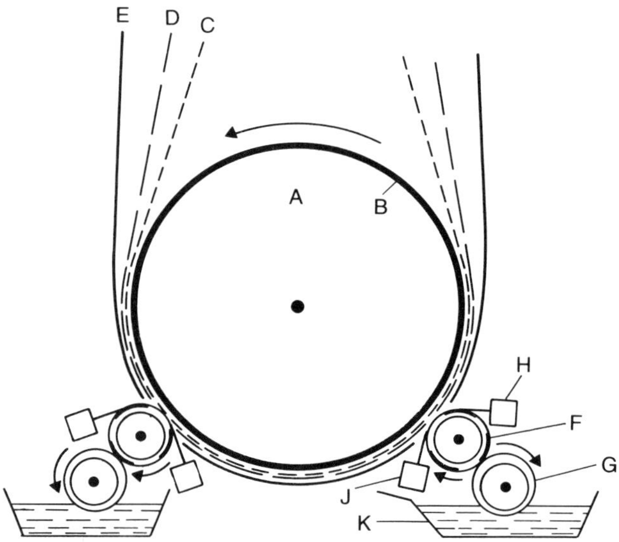

Fig. 7.1 Diagram of basic two-colour roller printing machine. A, Iron pressure roller (impression cylinder); B, thick layer of textile backing (lapping); C, endless printing blanket; D, back grey; E, fabric; F, engraved printing roller; G, furnishing rollers; H, lint doctor; J, colour doctor; K, colour box containing printing paste.

It is usual to print the paler depths of a pattern first, to reduce the danger of contamination of the following print rollers. Another difficulty in printing multi-colour patterns is that the first colours to be printed become crushed as the print rollers following (applying other colours) press against the fabric. This crushing effect impresses the colour further into the fabric and the final visual result is a lightening of the shade. This is a problem which only applies to the first colours to be applied and can be compensated for by adjusting the strength of the colour in the colour boxes.

One of the problems in controlling the register of the print in multi-colour machines is the time delay between adjusting the pressure of the relevant print rollers and being able to assess visually the quality of the resultant pattern. Systems for monitoring the print register as the fabric leaves the print roller, coupled with the automatic control of the pressure of the appropriate print rollers, have been developed.

Considerable care is required in producing the engraved copper rollers, which coupled with their high manufacturing costs, means that long production runs are required to make the process economic. The textile market currently requires relatively short runs, probably often repeated and for this type of demand rotary screen printing is far more suitable. In consequence, roller printing is declining rapidly (from about 50% of output in 1974 to about 10% in 1995) and it is likely it will be used no longer by the year 2000.

7.4.3 Screen printing

This type of printing has increased enormously in its use in recent years because of its versatility and the development of rotary screen printing machines which are capable of very high rates of production. An additional significant advantage is that heavy depths of shade can be produced by screen printing, a feature which has always been a limitation of roller printing because of the restriction to the amount of print paste which can be held in the shallow depth of the engraving on the print roller. Worldwide, some 61% of all printed textile fabric is produced by the rotary screen method and 23% by flat screen printing.

There are two basic types of screen printing process, the flat screen and the rotary screen methods. Flat screen processes may be operated by hand or be automated, though the former method is not fast enough to be commercially feasible. The screens, sometimes referred to as 'silk' screens because silk was once used as the screen fabric, must be stable and maintain their tension across the frame especially when wet and be resistant to chemical attack by the components of the print paste. The screen fabric is usually woven polyester and is drawn very tightly over a metal frame. The screen fabrics are available in a range of meshes, the choice of which depends on the nature of the fabric (for example coarse or fine) and the type of the design (for example large areas or small detail) to be printed.

A separate screen is required for each colour in the design to be printed. The pattern for a particular colour is produced by coating the screen with a viscous polymer solution which is sensitive to light. The polymer used is usually either gelatine or poly(vinyl alcohol), sensitised with ammonium dichromate, so that on exposure to ultraviolet radiation cross-linking occurs, giving a water insoluble product. Then by exposing the coated screen through the positive for the colour and washing with water to dissolve out the unexposed polymer, the screen fabric will contain the pattern through which the print paste can pass. The cross-linked polymer in the exposed areas of the screen presents a barrier to the passage of the paste.

During the printing process the screen is placed on the fabric, against registration stops to ensure accurate positioning. The viscous print paste is applied against the inside edge of the frame and then forced through the open pores of the screen by a squeegee, which is a rubber blade mounted in a wooden or metal support. The squeegee is drawn at an angle across the screen, maintaining as constant a pressure as possible. The screen is then lifted and moved to the next part of the fabric and the process repeated. Often the screen is actually moved to the next but one position on the fabric to allow time for some drying of the paste to occur, thus preventing pick-up of colour, from a previous pass, by the frame.

Printing is carried out on a long flat table (some 20–60 metres long), covered with a layer of felt and a washable blanket. It is important that the fabric does not move or shrink during printing and for this reason it is stuck to the blanket with a water soluble adhesive.

The flat screen process has been semi-automated in systems which move the screens mechanically and use magnetic squeegees which operate at constant angle, rate and pressure. However productivity is still low because the cloth has to be stationary during the drawing of the squeegee across the screen. Production rates are faster with machines in which all the screens needed for the various colours of the design are positioned accurately over the printing table, the gaps between the screens being an integer of the design repeat. The fabric is stuck to the under blanket at the entry end of the table and fed along intermittently, one screen distance at a time. The screens are lowered on to the fabric and the printing of each colour carried out simultaneously, whilst the fabric is stationary. The screens are then lifted and the fabric (and under blanket) moved to the next position.

Rotary screen printing machines allow a much faster rate of production than is possible with flat screen methods, with speeds of up to 60 metres per min or more being possible. The screens are cylindrical and rotate in contact with the fabric, giving truly continuous production. The squeegee remains stationary inside the screen (Fig. 7.2) and it is the rotation of the screen against the squeegee which forces the printing paste through the mesh of the screen on to the fabric moving past. As with flat screen printing a separate screen is required for each colour to be printed. On average about 7 colours per design are printed, but the trend is to an increasing number and rotary printing machines are available now which are capable of printing up to 26 different colours per design.

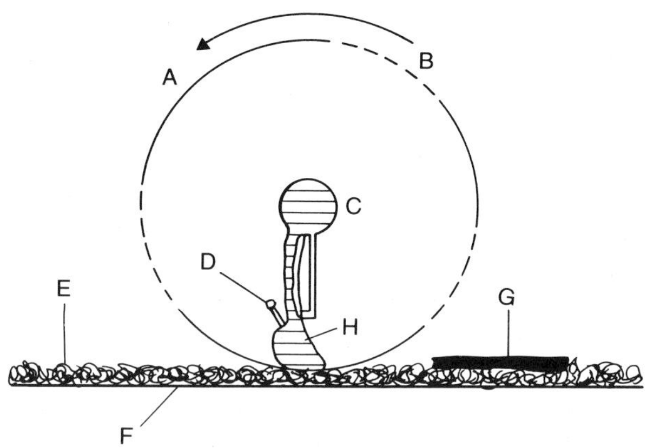

Fig. 7.2 Cross-section of a rotary screen. A, Screen; B, open (design area); C, print paste feed; D, level control; E, fabric; F, blanket; G, printed area; H, squeegee blade.

The key to the development of rotary screen printing was the invention of the seamless rotary screen, made of nickel foil and possessing good durability. More recently, phosphor–bronze wire mesh screens with a polyester coating have been used. Early problems in the development of rotary screen printing were caused by the tendency of the screen to deform along their length. However screens possessing the necessary rigidity can now be produced and indeed a comparatively recent major development has been the ability to print wide widths of fabric. Over the last 20 years, flat and rotary screen printing machines up to three metres wide have been developed, often with the ability to print big repeat patterns. The demand for wider printing widths has come from the manufacturers of bedding, curtains and carpets. Systems have been devised to raise and lower the rotary screens intermittently, with a single colour of the pattern perhaps being printed in sections by two or even three screens, to cope with large repeats.

The rotary and flat screen methods have become widely adopted for carpet printing, for which the development of the wide-width machines have been especially useful. It is important in carpet printing that enough print paste is applied to penetrate the pile and for this purpose a variety of refinements to the method have been adopted, such as machines which apply a vacuum to the reverse side of the fabric to draw the paste through the screen into the pile. Another interesting development is that of machines with double magnetic roller squeegees. In these machines the print paste can be fed under pressure into the small space between the two squeegees, so that the paste is forced through the screen down into the carpet pile.

7.4.4 Transfer printing

Transfer printing techniques involve the transfer of a design from one medium to another. The most common form used is heat transfer printing in which the design is printed initially on to a special paper, using conventional printing machinery. The paper is then placed in close contact with the fabric and heated, when the dyes sublime and transfer to the fabric through the vapour phase.

The heat transfer technique therefore requires dyes which sublime within a fairly narrow range of temperature and which are substantive to the fibre. Since ionic dyes will not sublime easily or without degradation, the type of dyes used is restricted to the non-ionic disperse type. Consequently the process is only really suitable for printing polyester, nylon and acetate fibre types. The actual disperse dyes used represent only a small fraction of all those available, because the rate of sublimation and the temperature at which sublimation occurs depends on the molecular weight and polarity of the dye molecule. The different coloured dyes in the print must be compatible in terms of their sublimation properties.

A most important stage in the process is the printing of the paper. Usually gravure methods are used, but flexography is also suitable. Care is necessary in controlling the register and the quantity of the printed paper, which requires great skill on behalf of the paper printer. In contrast, the actual heat transfer process to the textile fabric can be carried out by people who have no particular printing skills. The equipment required, a simple calender, has a low capital cost and occupies little space. The process

(Fig. 7.3), which involves pressing the fabric and paper together and heating at 180–210°C for about 15–20 seconds, is non-aqueous, causes no effluent problems and requires no after-treatments. By applying a partial vacuum to the underside of the fabric, the transfer of the dye from the paper is speeded up, enabling higher molecular weight or more polar dyes, with better fastness properties to be used. The vacuum also increases the extent of penetration of dye into the fabric so that heavier fabrics and even carpets can be printed successfully.

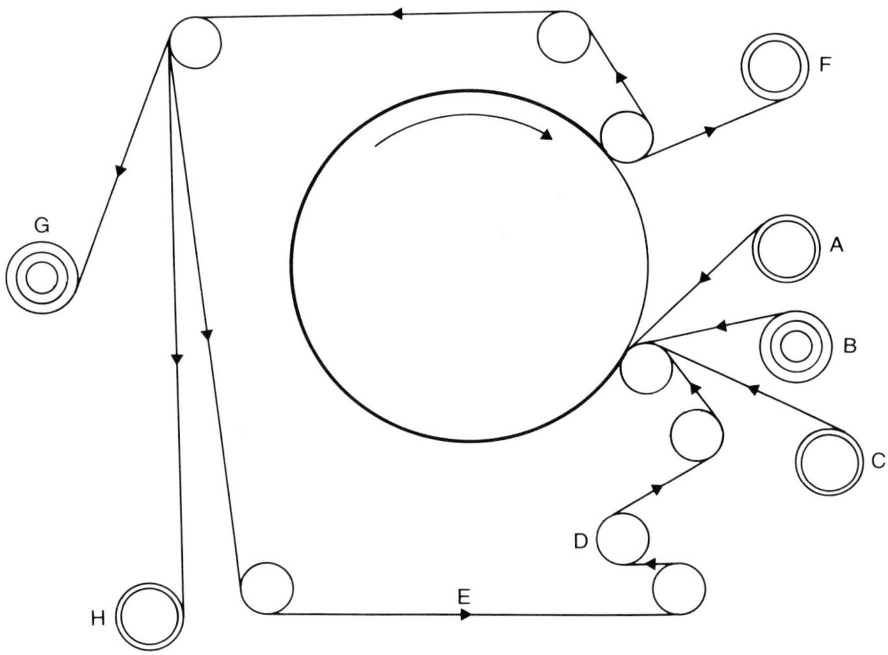

Fig. 7.3 Diagram of continuous transfer printing machine. A, Transfer paper; B, fabric; C, backing paper supply; D, tension rollers; E, endless blanket; F, used transfer paper; G, printed fabric; H, backing paper.

A great deal of effort has been devoted to adapt the technique of transfer printing to fibres other than polyester, such as the natural types – wool and cotton. This is difficult however because the affinity of disperse dyes for natural fibres is very low and the ionic dyes which are normally applied to these fibre types do not sublime. Some ingenious alternative transfer systems have been developed for these fibres, of which the most commercially important is the wet or migration method for wool. In this method the design is printed with α-bromoacrylamino reactive dyes for wool on to a special paper in a similar way to the heat transfer papers. The printed paper is brought into contact with the fabric which has been pre-padded with a thickened and acidified liquor. The fabric and paper are held in contact under pressure for 30–60 seconds at about 100°C, after which the fabric is given a final washing off treatment to remove any unfixed dye. This method however is not suitable for cotton.

Other methods have been devised which are based on modifying the fibre to improve its substantivity for disperse dyes. For example, by pre-treating cotton with benzoyl chloride, or wool with benzoic anhydride, which confer hydrophobic character, some substantivity to these fibres for disperse dyes is imparted. However these processes involve chemical pre-treatments of the fabrics and have therefore not achieved any commercial success.

A system which has achieved some measure of success is the melt transfer system. In this process the colorant (usually a pigment) and a binder are printed on to a paper. When the paper is placed in contact with the fabric and heat applied, the binder melts and the design is 'marked-off' on to the fabric and solidifies on cooling. This process has become widely used in garment printing and has found a particular niche in printing slogans on leisure garments such as T-shirts.

7.4.5 Ink-jet printing

There has been considerable interest in the technology surrounding non-impact printing, mainly for the graphic market, but the potential benefits of reductions in the time scale from original design to final production has led to much activity in developing this technology for textile and carpet printing processes. The types of machines developed fall into two classes, drop-on-demand (DOD) and continuous stream (CS), the principles of which are described in Chapter 4.

The first commercially available system was introduced in 1991 and was a continuous stream type which used the Procion P (monochlorotriazinyl) reactive dyes in the ink formulations for printing cellulosic fabrics. The technology of the continuous stream system requires a very exacting specification of conductivity, so the dyes used in the inks must be of exceptionally high purity. The reactive dyes require hot alkaline conditions for the formation of the covalent bond with the fibre, and the necessary chemicals are present in the print paste of conventional printing methods. However these chemicals cannot be included in the ink applied by ink-jet printing, so it is necessary to pad these on to the fabric prior to the jet printing stage. Special pre-treatment chemicals have been developed for application at the initial padding stage which also enhance the colour development of the dye. This is a significant bonus because the amount of dye which can be applied by an ink-jet printer is only a fraction of that contained in a print paste for application by conventional printing methods and the visual depths of shade achieved therefore not as high. A chemical pre-treatment has also been developed which imparts a hydrophilic coating on polyester fibres, which acts as a receiving layer for reactive dyes. Designs can therefore be ink-jet printed on to cotton/polyester blends and even on to 100% polyester fabrics.

The most significant advantages of jet printing textiles are:

(1) Ease of transfer from design stage to sampling or production. Designs produced on a CAD system can be digitised and fed directly to the printing head. The design can be printed quickly in various colourways so the customer can make informed decisions on preferred designs.
(2) Small samples of exclusive designs such as might be produced on ties or scarves can be produced in a very short time, as can one-off sample prints.
(3) Screens are not used, so screen preparation and storage is not required.
(4) Design flexibility is enhanced because the production of designs with large repeats is just as easy as those with smaller repeats. The size of the repeat can be limited with conventional screen processes.

Jet printing offers much potential for the future in textile and carpet printing. In section 7.10 the way in which the process is developing, from design to sampling and then to production, is described.

7.5 Carpet printing

The printing of carpets only really achieved importance after the introduction of tufted carpets in the late 1950s. Until then the market was dominated by the woven Wilton carpets and Axminster designs were well established, but by the 1980s tufted carpet production accounted for some 80% (by area) of UK production. Much of this carpet production was printed because the range of patterns possible to produce using tufting machines was limited and there was a desire to produce a greater flexibility of design for these types of carpet.

Initially printing was achieved using especially robust flat screens on machines fitted with a vacuum facility to aid penetration of the printing ink into the pile of the carpet. The prints produced had poor wear fastness however and were aimed at the cheaper end of the market. The quality improved especially with the introduction of nylon and also with the introduction of shallow loop carpet types. An alternative method which produced speckled designs on tufted carpets was that of warp printing the yarns with random bands of colour. On forming the tufted carpet the speckled effect was obtained, a design effect which has always remained popular.

A wide variety of machines have been developed for the direct printing of carpets, a type typically used being the Stalwart machine. In this machine, the carpet pile is pressed against the print roller, of which there can be up to four, mounted one above the other at an angle of 45° to the vertical. Each print roller applies a different colour and the design on its surface is cut into the rubber surface and covered with foam rubber.

In recent years rotary-screen printing has become the most widely used method of carpet printing, as it has done for textile printing. It is likely that this trend will continue because of the high production speeds possible, low associated labour costs and flexibility of patterning possible. The main difference from rotary screen printing of textiles is that the squeegees are much larger, usually circular and positively driven. In some types a hydrostatic pressure is used to force the print paste into the carpet pile. Fig. 7.4 illustrates some of the principles involved.

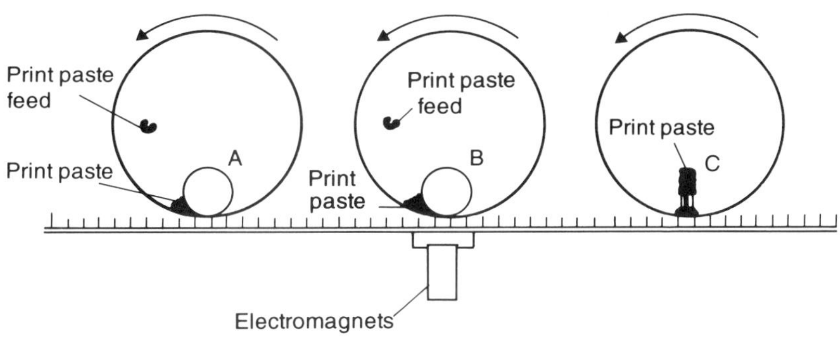

Fig. 7.4 Rotary screen printing of carpets. A, Driven squeegee; B, magnetic squeegee; C, hydrostatic pressure system.

Multicolour effects on carpets have also been produced by spray and jet printing machines. One such example is the Kusters system in which the machine was designed to apply a random shower of droplets of dye liquor on to the carpet which had been pre-padded to a solid ground shade. By varying the viscosity of the dye liquor, the pick-up of the padding liquor and the steaming conditions, the diffusion of the droplets into the carpet could be controlled. In this way, the patterns produced could be made to be sharply defined against the solid shade background or less so if desired. Commercial machines have been developed for the jet printing of carpets. Generally, machines for the ink-jet printing of carpets differ from those for textiles in that the former tend to have a coarser resolution, about 40 lines per inch compared with about 300 lines per inch with the DOD and CS types used for textiles. As indicated in section 7.10 this technique has many advantages, not least of which are the ease with which the design information can be transferred to the print head and the high definition of patterns possible. One of the limiting factors is the slower speed of operation, compared with those of rotary screen methods.

7.6 Drying and fixation

After the print paste has been applied by roller or screen methods, drying is necessary to produce prints of clear definition. The moisture must be removed from the print paste at a rate fast enough to ensure that any 'bleeding' of colour is prevented.

Similarly the fixation stage, usually carried out by steaming, is critical and needs to be well controlled. During steaming the printed areas absorb moisture, thus allowing the dye to diffuse into the fibres. Steaming times of about one hour are required for dyes to diffuse adequately, but in the case of fibre reactive dyes, much shorter times are adequate, since it is only necessary to bring about the reaction between the dye and the fibre. Depending on the chemical reactivity of the particular reactive dyes being applied, steaming times of as short as 15 seconds may be all that is required. The condition of the steam used is another important consideration, in terms of it being too dry or too wet, the balance depending on the type of dye being applied.

7.7 The flash-ageing technique

Flash-age printing was developed to overcome the process difficulties caused by the very low stability of highly reactive dyes. In this method the fabric is first printed with the dye and thickener. Immediately after passing through a padding liquor containing a concentrated solution of caustic soda and sodium silicate, it is steamed in a chamber at 120–130°C for 45–60 seconds. Unfortunately this method suffers from the disadvantage that large quantities of caustic soda and sodium silicate are required, which produces large quantities of alkaline effluent. On the plus side however the process uses much less urea.

This has become a common method for producing high quality prints of vat dyes and is basically a two-stage process. The print paste, consisting of the vat dye (in its insoluble 'parent' form) dispersed in a thickening agent, is applied to the fabric and dried. The fabric is then padded in an alkaline bath of sodium dithionite to reduce the dye in situ, and steamed. The steaming time required is only about 20–40 seconds, so the steam consumption is economical and the machinery comparatively simple. Another advantage is that since the dye is applied without the presence of a reducing agent, the dried print is stable and can be stored if necessary before steaming. With a careful selection of vat dyes, prints of excellent definition and possessing a high depth of colour can be obtained.

7.8 Washing off

The final process in textile printing is the removal of the thickener and other auxiliary chemicals in the print paste, together with any unfixed dye.

The fabric is usually rinsed well with cold water and then soaped in a hot solution of detergent. The severity of the soaping treatment depends on the dyes used. Those dyes which possess high substantivity for the fabric are difficult to remove and require the longest soaping treatment. In the case of reactive dyes, it is preferable to use dyes of low substantivity for printing so the soaping treatment is less vigorous. There is also less chance of adjacent white areas being stained.

7.9 Pigment printing

Pigment printing merits a separate treatment because of its extensive use and technically different characteristics from printing dyes. About half of the 20 billion metres of cloth which are printed annually worldwide are pigment printed, though there are significant variations in different countries. For example, about 70% is pigment printed in the USA, but only about 10% in Russia.

Pigments are insoluble in water and have no substantivity for textile fibres, so it is necessary to fix them by the use of a binder. The binders used are polymeric film-forming materials which enclose the pigment particles and provide a permanent bond between them and the fibre. The film when formed is several microns thick, so inevitably it has an influence on the handle characteristics of the fabric, especially if the fabric is thin. Also the rubbing, washing fastness and dry cleaning fastness properties are all directly those of the binder and the desired performances are not always possible to achieve with the monomers available.

The technique of pigment printing is however a relatively simple one, consisting basically of printing the paste (by conventional printing methods), drying and fixing. Since no wet after-treatments are required, the whole process is a very economical method of printing. Further, the simplicity of the process is such that it can be applied to all fibre types, even those which are difficult to print by traditional methods, such as the commercially important polyester/cotton blends, and even glass fibres. However it is generally less suitable for wool or silk fabrics because the handle of woven and knitted fabrics is quite badly affected and the low fixation temperatures which have to be used inhibit the development of full fastness properties.

The significant advantage of pigment printing is the extensive range of colours which are available. The requirement of pigments with high purity and brilliance can restrict the choice somewhat, but pigment manufacturers have been able to enhance these properties of pigments by modification of the crystal form. An interesting innovation in recent years has been to deposit especially brilliant soluble dyes on colourless 'carriers', or to entrap them in these carriers. The 'carrier' materials have been improved to the extent that brilliant pigments having excellent fastness to light are now available. The high tinctorial strength of pigments enables certain special effects to be produced which cannot be achieved by the use of dyes, for example full-cover overprints.

The binder is a particularly important component in pigment printing since fastness properties, such as rubbing, are those of the binder itself. The binder film has to be colourless, smooth, flexible and able to adhere to the fibres without being tacky. The film is produced by the polymerisation of appropriate monomers in the print paste. The most widely used monomers are ester derivatives of acrylic acid, butadiene or vinyl acetate. These are water-insoluble and are used in the form of 0.1–0.5 μm droplets in water. In order to obtain the desired characteristics described above, it is necessary to co-polymerise different monomers. In addition, monomers are added which contain reactive groups, so that during the heat treatment following film formation, a certain amount of cross-linking occurs, thereby enhancing the fastness properties. However, very careful selection of the monomers, their proportions and the polymerisation conditions used are necessary to avoid detrimental effects on the handle of the fabric.

As in printing with dyes, it is necessary to incorporate a thickener into the print paste in order to achieve sharply defined prints. The water soluble thickeners used in printing dyes are unsuitable for pigment printing however because they become bound within the binder film, where they are capable of swelling, even after fixation. Consequently the wet fastness of the prints will be low. The types of thickeners used are called emulsion thickeners, which comprise two immiscible liquids (usually water and white spirit) which have been mixed together by a high-speed stirrer in the presence of an emulsifier. One of the liquids becomes dispersed in the form of fine droplets (the inner phase) within the other (the outer phase). The viscosity increases if the volume of the inner phase is higher than that of the outer phase, so that the particles of the inner phase come into contact with one another, thereby restricting their mobility. If the inner phase consists of white spirit, the emulsion is referred to as an oil-in-water (O/W) type; if the inner phase is water, the emulsion is a water-in-oil (W/O) type. Both types are used for pigment printing and because both phases are volatile, these thickeners do not influence the final handle or fastness properties of the printed fabric.

In recent years the release of organic compounds into the atmosphere has become an environmental issue and indeed in some countries it is now prohibited by law. As a result white spirit emulsions are being phased out and replaced by totally aqueous systems. These systems are based on polyanionic compounds which are capable of swelling to produce solvated particles of similar dimensions to the droplets in the white spirit based emulsions.

Print paste formulations for pigment printing can be quite complex, a consequence of the technical and commercial demands made on the process. Typically, the formulation will include:

(1) water
(2) pigments

(3) binder, probably a blend of different types
(4) white spirit, if O/W or W/O emulsions are used, or synthetic thickening agent
(5) water-retaining agent
(6) cross-linking agent
(7) adhesion promoter
(8) catalyst
(9) handle modifier
(10) antifoam agent

All these components are not always required, it very much depends on the conditions under which printing is being carried out. In order to simplify the preparation of pastes, the manufacturers of auxiliaries market specially formulated mixtures of the individual components.

Inevitably the coverage of large areas of a fabric with a pigment print has an effect on the handle of the fabric and for this reason it is more common for pigments to be printed on designs in which the patterns have only a low coverage. Pigment printing has the advantages of process simplicity and high production rates. With no washing off being required, it is also very economical. Pigments can be applied to any fibre type and the prints have excellent fastness to light, though abrasion resistance and resistance to frequent washing, especially of blends containing synthetic fibres can be poor. The most useful outlets for pigment printed fabrics are in areas such as soft furnishings, bed linen and cotton and viscose apparel.

7.10 The future of textile printing

The market demands of the textile industry have changed over recent years. Designs now change more frequently, with fashion and textile houses presenting new collections up to six times per year, compared with the former practice of producing one summer and one winter collection. Also much less fabric is produced, in anticipation of being sold, instead only the amount of a particular design that has been ordered is actually produced. As a consequence manufacturers hold very little stock and rely on being able to respond quickly to repeat orders. Average print runs vary considerably between different geographical regions. In Europe the average is about 1000 metres, in North America 6000 metres, but in Africa it is 10 000 metres. The particular need in Europe to supply repeat, relatively short runs of the same quality means that reproducibility, flexibility and speed of production have to be important aspects of modern printing processes. In the UK, print runs of as small as 300 metres are not unusual. A new market for printed textiles is the automotive industry, where car interiors are increasingly seen as having potential for innovative styles.

The developments taking place in printing practices are directed at achieving the above demands and to this end integrated computerised systems are being introduced. With a computer aided design (CAD) system a designer can transfer the design to a computer-aided engraving unit, where a laser engraver produces the screens for the various colours involved. The design and colour information can be stored digitally on computer file for future reference. The computer can be used to predict the recipes for the dyes in the print paste for the each of the various colours in the design. The development of automated monitoring and control systems, for example for the supply of print paste to the screen during printing, or the colour measurement of the final product is also taking place.

Increasing pressures are being placed upon textile printers by national and local government authorities, to limit discharges into air, water and, via landfill, to the ground. A large amount of printing is carried out using reactive dyes, of which between 10–40% remains unfixed and has to be removed from the fabric to ensure satisfactory wash fastness. None of the methods currently available for reducing colour from waste water streams, such as chemical destruction (oxidation/reduction), physical removal (ultra-filtration), physico-chemical methods (sedimentation/flocculation), or adsorption (activated charcoal) are entirely satisfactory. This is an issue which will have to be given further consideration in the future.

Although expensive and sometimes difficult to obtain, alginates are widely used thickeners. Alginates are major contributors to biological oxygen demand (BOD) and chemical oxygen demand (COD) in the effluent from printing processes and at present no recovery system is available. Urea is another widely used auxiliary in print paste formulations, but since strict limitations are now placed on the nitrogen content of aqueous effluents, considerable effort is being expended in seeking methods for

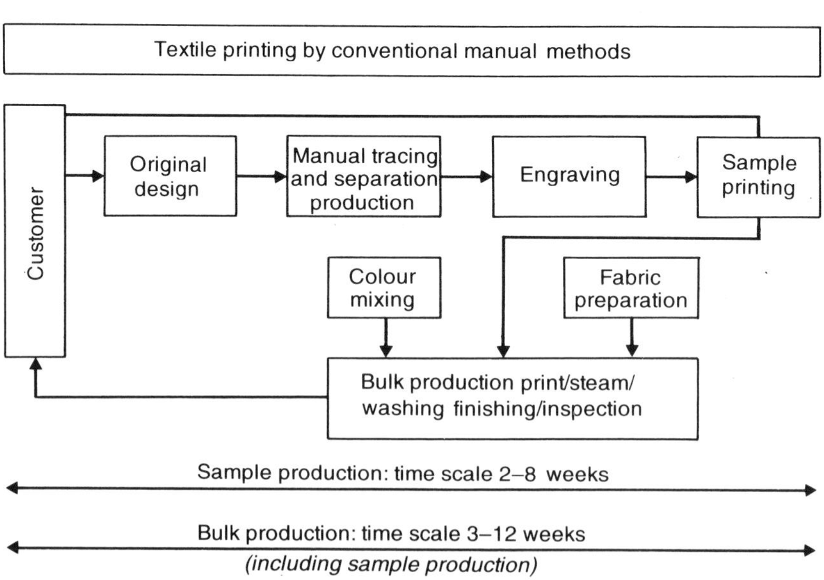

Fig. 7.5 The stages of conventional textile printing.

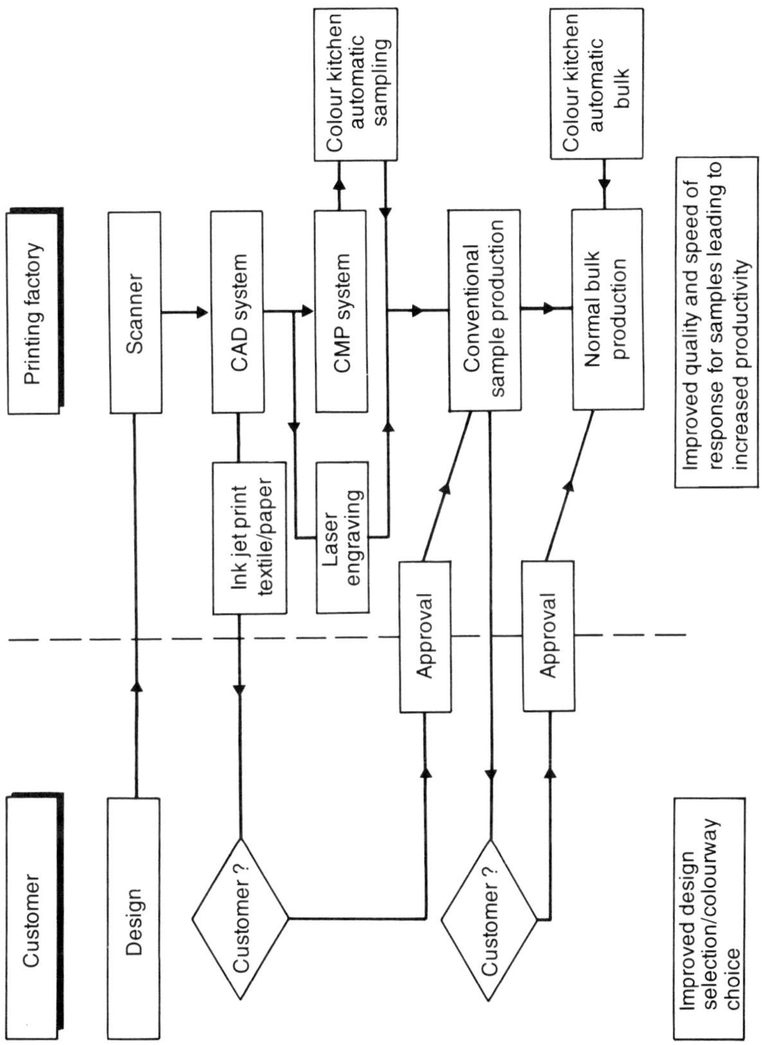

Fig. 7.6 The future of textile printing showing the interactions between printer and customer.

eliminating or at least reducing significantly urea in print pastes. Urea is an effective swelling agent for cellulose, it disaggregates and solubilises dyestuffs, improves colour yield and gives better levelness and smoothness of the print. A large number of alternatives to urea have been tried without success and it seems unlikely that a chemical will be found that is technically as good and as cost-effective.

An interesting area of development is that of ink-jet printing, a technology that has seen considerable advances in recent years for paper printing (see Chapter 4). The ink-jet process is a clean and environmentally friendly printing process for textiles. In combination with CAD systems, considerable possibilities arise for textile designers, whereby small realistic samples can be produced in a short time, without the need to engrave screens. The customer can then make informed decisions about the design, the colours and the repeat, and only when completely satisfied will it be necessary for screens to be engraved and a conventional sample printed for final approval prior to bulk production. Fig. 7.5 shows the stages involved in textile printing by conventional methods. In the future, the scheme shown in Fig. 7.6 may well become the norm. This type of printing will therefore allow designers to be much more responsive to their customers' needs.

Bibliography

Textile Printing, Edited by Leslie W.C. Miles, 2nd edition, Society of Dyers and Colourists, Bradford, 1994.

An Introduction to Textile Printing, W. Clarke, 4th edition, 1980, Butterworths.

Chapter 8

Introduction to the Physical Basis and Measurement of Colour

8.1 Introduction

The measurement and specification of colour are essential processes in both the colour manufacturing and colour using industries. The manufacturers of dyes and pigments require to ensure that the colorants they sell possess the correct colour. Colorants are manufactured by a wide variety of highly complex chemical processes and even minor changes in processing conditions from one batch to another can result in variations in the colour of the final product. The manufacturers need to take great care to ensure that the dyes and pigments they sell are of consistent quality, in terms of their colour, strength and application properties. In order to ascertain that this is the case, detailed analyses need to be carried out to check these properties, all of which involve the measurement of colour.

In the same way, the users of the colorants need to ensure that the colours of the products they are manufacturing are exactly as required and are consistent from one item to another. In the case of textiles, it may be for example, that the sleeve of a shirt is cut from a different batch of cloth from the front panel, so it is essential to have accurate colour control of the different batches of cloth to avoid complaints from the consumer. Colour is an important aspect in most consumer products and instrumental colour measurement and control find application over a wide variety of industries, from textiles, to foods, plastic articles, paper products, toiletries and paints.

The quality of coloured items can be assessed by conducting analyses either in solution, where the transmission properties are measured, or on the solid material, where the reflectance characteristics are used. Typically, analyses in solution are used to ascertain the quality, in terms of hue and strength, of dyes and are conducted mainly by the manufacturers of dyes. Also the users of dyes may also carry out such tests, as a check, prior to using them. The principles of these tests are described below in sections 8.4 and 8.5. The colour measurement of opaque materials, which represent most consumer items, is based on reflectance measurements from which colour co-ordinates are computed. These co-ordinates define a colour and therefore allow decisions to be made as to whether batches of items are on-shade. The theoretical details of this subject are dealt with in sections 8.7 and 8.8.

Instrumental colour measurement is now a routinely used tool and probably one of its most useful applications is the basis it has provided to enable the computation of the amounts of dyes (or pigments) to apply to a substrate to produce a desired colour. This technology is widely used, not only in the textile dyeing and printing industries, but also in those involving the use of pigments, such as the paint and the paper printing industries. Some stores which retail paint have such recipe prediction systems so that a customer can have a paint formulated precisely to match any colour. Instrumental colour measurement is also widely used to compute the size of the colour difference between pairs of colours, a facility which aids colourists to decide whether batches of materials are acceptable colour matches to the standard colour. It can also be used to assess the degree of metamerism between colours and to obtain an objective evaluation of the quality of whites. This latter use is of particular relevance in the retailing of white paint and white textiles.

The measurement of colour has now become highly automated and the technology has developed to the extent whereby the measurements can be used to provide feedback to a coloration process, thus providing on-line control. This technology is well developed for the control of textile dyeing and printing processes. A particular advantage of this type of control is that it can be used to control the rate of adsorption of dyes during a dyeing process, in order to facilitate the attainment of uniformly dyed goods.

Computer-aided design (CAD) systems have exploited the principles of colour physics and by linking them to colour printers, a design produced on screen can be printed directly to paper for inspection. The designer then has the opportunity to view the design as it would appear as a woven, printed or knitted textile fabric (in preference to viewing the image on screen, which is emitted light). This procedure provides the flexibility of producing a range of designs for inspection without actually having to weave, knit or print physical samples, which is a time-consuming operation. The systems can also provide the information about the colours used in the design so that dye or pigment recipes can be computed for production.

This chapter provides the theoretical background to light and colour and explains how the principles of colour specification are applied in industry.

8.2 Electromagnetic radiation

Two models are used to describe the properties of electromagnetic radiation – these are the wave model and the particle model – neither of which alone explains satisfactorily all of the properties of radiation. Whilst some behaviour, such as reflection, refraction, interference and diffraction effects can be accounted for by the wave model, other properties, such as the photoelectric effect are best described by the particle model.

On the basis of the wave model, electromagnetic radiation can be regarded as a wave motion, (Fig. 8.1), consisting of electric and magnetic fields, moving at right angles to each other and to the direction of propagation. This model of radiation is illustrated in Fig. 8.2.

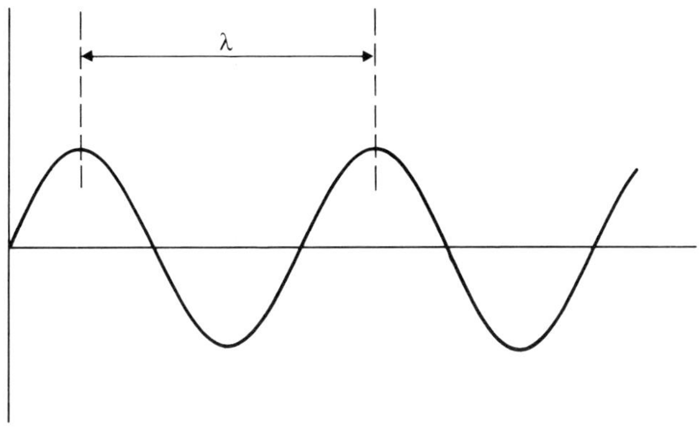

Fig. 8.1 Electromagnetic radiation waveform.

Electromagnetic radiation may be defined by wavelength λ and has characteristic properties which depend upon the wavelength. Thus different regions of the electromagnetic spectrum can be classified in terms of wavelength ranges, as shown in Table 8.1. The different areas of the electromagnetic spectrum have different properties and so interact with matter in different ways. Rays of very short wavelength such as cosmic and X-rays possess high energy and are very penetrating. Ultraviolet radiation has higher energy than visible light, and is an important constituent of sunlight, especially for photosynthesis reactions. Infrared radiation has lower energy than visible light but gives a heating effect when incident on objects. The rays of very long wavelength have well known applications in microwave ovens, radar, television and radio.

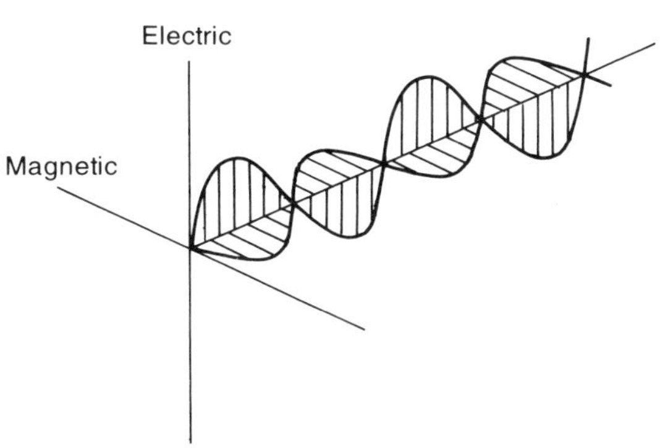

Fig. 8.2 Propagation of alternating electric and magnetic fields.

Table 8.1 Classification of regions of electromagnetic spectrum.

Type of radiation	Wavelength λ/m	Frequency ν/Hz
Cosmic rays	10^{-17}	10^{22}
	10^{-16}	10^{21}
	10^{-15}	10^{20}
	10^{-14}	10^{19}
	10^{-13}	10^{18}
Gamma rays	10^{-12}	10^{17}
	10^{-11}	10^{16}
X-rays	10^{-10}	10^{15}
	10^{-9}	10^{14}
Ultraviolet	10^{-8}	10^{13}
Visible	10^{-7}	10^{12}
	10^{-6}	10^{11}
Infrared	10^{-5}	10^{10}
Far infrared	10^{-4}	10^{9}
Microwave	10^{-3}	10^{8}
Radar	10^{-2}	10^{7}
Television	10^{0}	10^{5}
VHF radio	10	10^{4}
Medium wave radio	10^{2}	10^{3}
Long wave radio	10^{3}	10^{2}

There are no distinct boundaries between the types of radiation shown in Table 8.1. A gradual change in the property of the radiation occurs as the wavelength changes. A common feature of all types of radiation is that they have the constant velocity, c, of $2.9979 \times 10^{8}\,\mathrm{m\,s^{-1}}$, regardless of wavelength. The frequency, ν, of radiation, which is the number of waves passing a fixed point every second, is obtained by dividing the velocity by the wavelength:

$$\nu = \frac{c}{\lambda} \tag{8.1}$$

Equation 8.1 shows that the higher the frequency, the shorter the wavelength, since the speed of radiation is constant.

The visible region of the electromagnetic spectrum, as the name implies, is the region of wavelengths to which the eye is sensitive and radiation in this region is referred to as light. This region constitutes only a very small part of the total electromagnetic spectrum and extends in wavelength from approximately 3.6×10^{-7} to 7.8×10^{-7} m. To avoid the use of powers of ten, the wavelength of light is usually expressed in nanometres (nm), where 1 nm $= 10^{-9}$ m, so on this scale the visible region extends from approximately 360 to 780 nm. The sensitivity of the eye to radiation is very low at each of these extremes and in practice the visible spectrum may be taken as 380 to 710 nm. In fact, some instruments used for the measurement of colour (see section 8.9), only cover the range 400 to 700 nm.

The colours which are perceived when the various wavelengths of spectral light are viewed are shown in Table 8.2. Ordinary white light, such as sunlight, contains all of these wavelengths, though not in equal intensities (see section 8.7.2.1). White light can be separated into its component wavelengths by a process called dispersion. If the light passes from one transparent medium (such as air) into another (such as glass), refraction occurs and the directions of the various wavelengths are altered to different extents. This effect occurs when light enters raindrops – the dispersion of light caused by the internal reflection and refraction processes results in the familiar sight of the rainbow. The effect can also be produced by passing white light (such as sunlight) through a glass prism.

Table 8.2 Perceived colours of visible spectral light.

λ (nm)	Colour
380–400	violet
400–435	indigo
435–480	blue
480–490	greenish blue
490–500	bluish green
500–560	green
560–580	yellow green
580–595	yellow
595–605	orange
605–740	red

The particle model considers radiation as the flow of particles, which can be regarded as bundles (or quanta) of energy. Electromagnetic radiation is a form of energy and the energy of radiation of frequency ν is given by Planck's equation:

$$E = \mathrm{h}\nu \tag{8.2}$$

where h is called Planck's constant and has a value of 6.626×10^{-34} J s. Planck's equation is important in that it unites the particle and wave models of radiation. The particles, or quanta of energy, have a magnitude of energy hν and are referred to as photons. This means that a wave of frequency ν can be considered as a flow of photons, each having energy $h\nu$. An ordinary particle in motion has a momentum mv, which is the product of its mass and velocity and similarly a photon has momentum also. However a photon has no mass and if it comes to rest by being absorbed by an atom or a molecule, its energy is converted into another form. The photon then ceases to exist.

One photon of radiation can be absorbed by one atom, ion or molecule, so the molar energy of radiation is given by

$$E = \mathrm{N}h\nu \tag{8.3}$$

where N is Avogadro's number, $6.023 \times 10^{23}\,\mathrm{mol^{-1}}$.

Substitution of Eqn 8.1 yields Eqn 8.4:

$$E = \frac{\mathrm{Nhc}}{\lambda} \tag{8.4}$$

and this equation can be used to determine the molar energy of radiation of a given wavelength. For example, radiation in the ultraviolet region at 250 nm, will have a molar energy given by:

$$E = \frac{6.023 \times 10^{23} \times 6.626 \times 10^{-34} \times 2.998 \times 10^{8}}{250 \times 10^{-9}}$$

$$E = \frac{0.1196}{250 \times 10^{-9}} \qquad = 4.786 \times 10^{5}\,\mathrm{J\,mol^{-1}}$$

$$\text{or } 478.6\,\mathrm{kJ\,mol^{-1}}$$

Similarly, green light at 550 nm will have an energy of $0.1196/550 \times 10^{-9} = 217.5\,\mathrm{kJ\,mol^{-1}}$. Infrared radiation of say $2500\,\mathrm{cm^{-1}}$ will have a wavelength of 4×10^{-4} cm, or 4×10^{-6} m. Consequently its molar energy will be $0.1196/4 \times 10^{-6} = 29.9\,\mathrm{kJ\,mol^{-1}}$.

A comparison of these energies shows that as the wavelength of radiation increases, from the ultraviolet, through the visible, to the infrared regions, the energy decreases. It is interesting to note that the strengths of most covalent bonds are about $300\,\mathrm{kJ\,mol^{-1}}$, so ultraviolet radiation has sufficient energy to rupture bonds in many molecules and cause photochemical reactions to take place. This is an important feature in relation to the behaviour of the retina to light and the perception of colour. Because of its high energy ultraviolet radiation should never be viewed with the naked eye, otherwise permanent damage of the retina in the eye may occur.

8.3 Absorption of radiation

All substances absorb radiation in one or more of the ultraviolet, visible or infrared regions of the spectrum. Since the eye is sensitive to light, any substance which absorbs in the visible region appears coloured. Conversely, those substances which absorb radiation only in regions outwith the visible region will be colourless. In order to understand why substances absorb radiation it is necessary to consider the various types of energy they possess.

A molecule in space possesses three types of energy. Firstly it possesses rotational energy, which arises from the rotation of the whole molecule about its centre of gravity. Secondly it possesses vibrational energy which is due to the stretching or bending of chemical bonds. Thirdly it possesses electronic energy since the electrons which occupy the various molecular orbitals around the nuclei are in continuous motion. These energies are illustrated in Fig. 8.3.

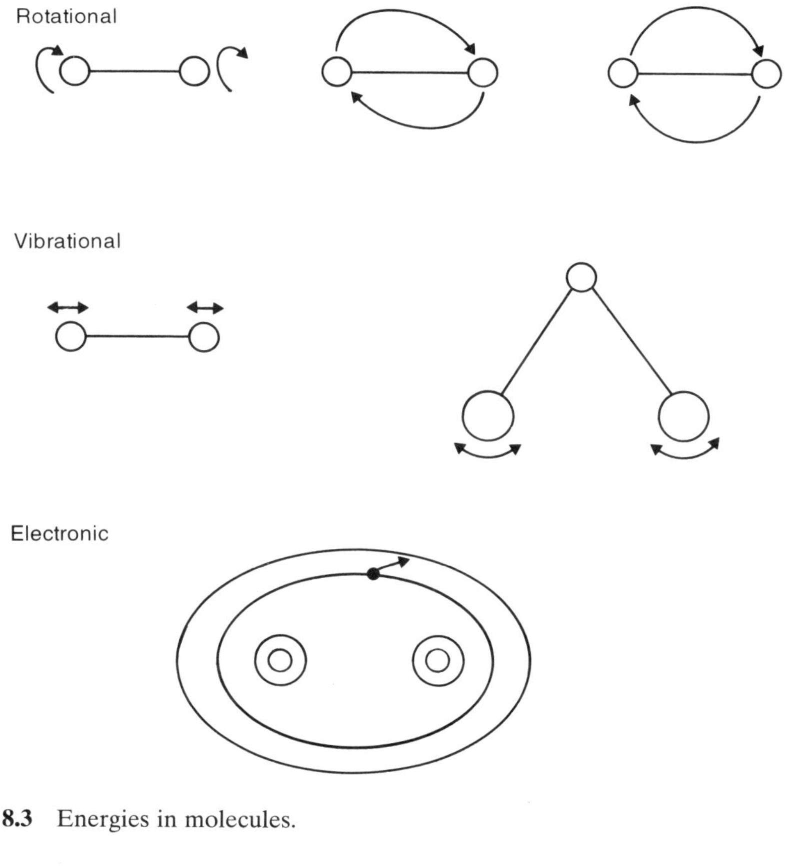

Fig. 8.3 Energies in molecules.

When radiation with a wide range of frequencies is incident on a molecule, only those frequencies whose energies correspond to the differences between the various energy levels in the molecules can be absorbed. For example, when radiation in the ultraviolet or visible region is absorbed, an electronic transition will occur, when an electron is promoted from the orbital it occupies in the ground state to an orbital of higher energy. The energy required for this transition to the excited state is provided by radiation of a specific frequency, this frequency being given by Planck's equation (Eqn 8.2). Excited states are not stable and very quickly a molecule will return to the ground state, usually dissipating the absorbed energy as heat as it does so.

Electronic transitions require the relatively high amounts of energy possessed by ultraviolet and visible radiation. Infrared radiation has much lower energy and absorption of this type of radiation is insufficient to cause electronic excitation. However these energies do correspond to those required to cause vibrational transitions in molecules and infrared absorption spectroscopy is an extremely useful technique in the qualitative analysis of compounds because the precise frequency of radiation absorbed depends on the particular type of the chemical bond involved. The third type of transition which may occur in molecules, the rotational transition, requires energies which correspond to radiation in the microwave region. The various transitions which may occur are represented in Fig. 8.4.

When a molecule absorbs radiation in the ultraviolet or visible regions of the spectrum, the large energies involved may cause simultaneous electronic, vibrational and rotational transitions to occur. It is the electronic transition however which is primarily responsible for absorption of radiation in the visible region and it is this which is responsible for the colour of dyes and pigments. These substances absorb light in specific regions of the visible spectrum and it is this selective absorption of radiation which gives rise to their individual colour. The colour seen depends on the wavelengths of light not absorbed by the dye or pigment, that is, the light reflected or transmitted to the eye. A useful guide to determining the colours of light absorbed by dyes is the 'opponent colours' diagram (Fig. 8.5), in which blues and yellows are opposites, as are reds and greens. Hence for example, a yellow dye absorbs light in the blue region of the spectrum, and vice-versa.

8.4 Quantitative aspects of light absorption

Coloured substances absorb light in the visible region of the spectrum. The amount of monochromatic light absorbed by a dye in solution is related to the concentration of the solution and to the path length of solution through which the light passes. These relationships are defined by Beer's and Lambert's laws respectively. Such measurements are made on the solution contained in a glass cell (Fig. 8.6) having a known path length (usually 1.0 cm) which is the length of the path through which light is actually in contact with the solution.

According to Beer's law, an increase in the concentration of the dye in solution, will give a proportional increase in the amount of light absorbed, if the cell length remains constant. Similarly, Lambert's law applies to a proportionate increase in the amount of light absorbed, with increase in cell length, if the concentration remains constant. These two laws can be combined to give the expression:

$$I_t = I_0\ 10^{-\varepsilon cl} \tag{8.5}$$

where ε is the molar absorptivity (or molar absorption coefficient) and is a constant for a given substance at a given wavelength. This equation can be written in the form:

$$\log\left(\frac{I_0}{I_t}\right) = \varepsilon cl \tag{8.6}$$

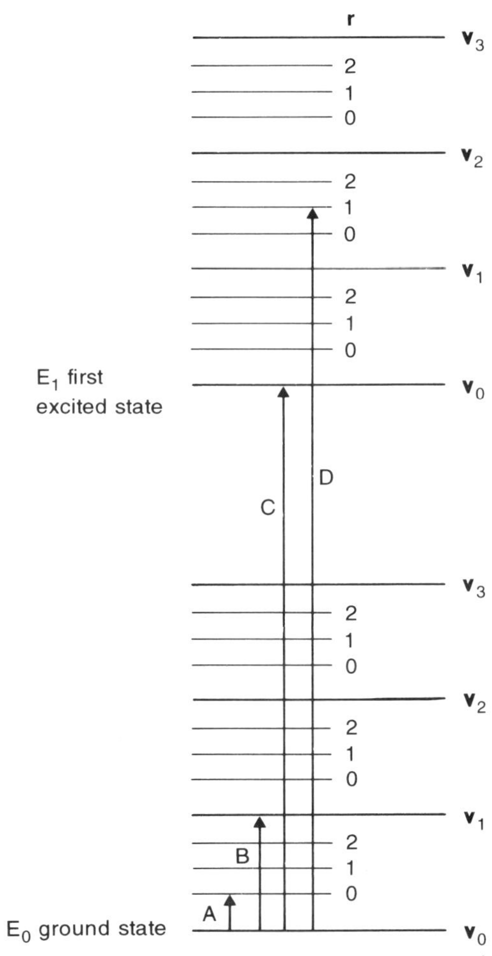

Fig. 8.4 Schematic diagram of molecular energy levels and energy changes. A, Rotational transition (far IR/microwave regions), $\Delta E \sim 10\ \text{J mol}^{-1}$; B, vibrational transition (near IR), $\Delta E \sim 1\ \text{kJ mol}^{-1}$; C, electronic transition (UV/visible regions), $\Delta E \sim 100\ \text{kJ mol}^{-1}$; D, simultaneous electronic, vibrational and rotational transition; V, vibrational quantum levels; **r** = rotational quantum levels.

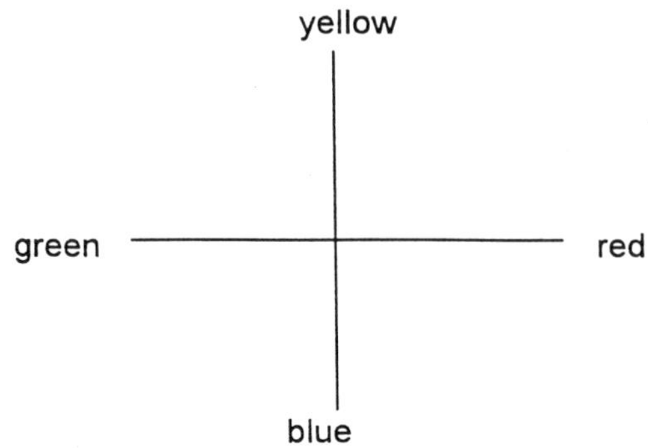

Fig. 8.5 'Opponent colours' diagram.

This relationship can be simplified by considering the situation in which the intensity of the incident radiation, I_0, is 100 arbitrary units. The intensity of the transmitted light, I_t, is then the percentage transmission, T, so we can write:

$$\log\left(\frac{100}{T}\right) = \varepsilon cl \tag{8.7}$$

The term log(100/T) is called the absorbance, A, so the Beer–Lambert law becomes simply:

$$A = \varepsilon cl \tag{8.8}$$

Thus, according to this law, the absorbance of a solution at a given wavelength is directly proportional to its concentration (at constant cell length) and directly proportional to the cell length (at constant concentration). Since the absorbance A is the logarithm of the ratio of the intensities of incident and transmitted light, it has no units. If $T = 100\%$, then $A = 0$. At the other end of the scale, if $T = 0\%$, then $A = \infty$. In practice

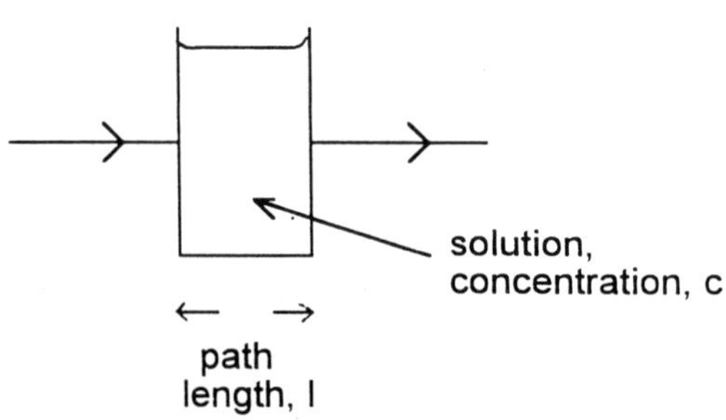

Fig. 8.6 Measurement of transmission of monochromatic light by solutions.

the normal range of transmission measured with the greatest accuracy by most instruments is 100% down to 1%T, which corresponds to an absorbance range of 0–2 respectively. Some high performance instruments will measure values of %T as low as 0.1% quite satisfactorily, which corresponds to an absorbance range of 0–3.

A graph of the absorbance of a dye solution plotted as a function of wavelength (through the visible region) is called the absorbance spectrum and is measured using a spectrophotometer. In Fig. 8.7 the absorption spectra of solutions of a red dye of different concentrations are plotted.

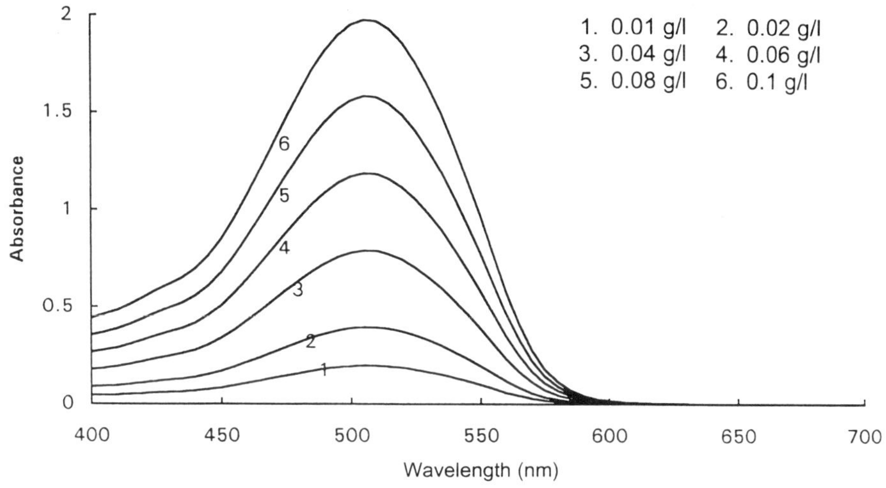

Fig. 8.7 Absorbance spectra of solutions of C.I. Acid Red 18. 1, 0.01 g/l; 2, 0.02 g/l; 3, 0.04 g/l; 4, 0.06 g/l; 5, 0.08 g/l; 6, 0.1 g/l.

According to the Beer's Law, if any one wavelength is selected and the absorbances of each solution at this wavelength plotted against its concentration, a straight line should be obtained (Fig. 8.8). It should be apparent that the absorbances vary more with concentration at some wavelengths than at others. It is found that the absorbance varies most with concentration (that is, the slope of the line in the graph of Fig. 8.8 is steepest) at the wavelengths where the absorption of light by the dye is high. In the case of the red dye in the example, these are the wavelengths in the middle of the visible spectrum.

The wavelength at which the slope of the line is the steepest (and the absorbance the highest) is called λ_{max}. In the case of the red dye C.I. Acid Red 18 referred to in Fig. 8.7, this wavelength is 505 nm. For the quantitative analysis of dyes, λ_{max} is the wavelength which should be used because it is the wavelength at which:

(a) absorbance varies most with concentration, so the analysis is at its most sensitive;

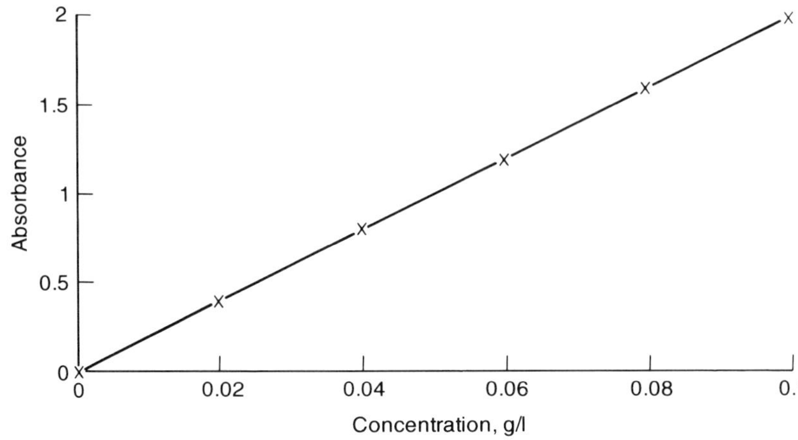

Fig. 8.8 Absorbance solutions of C.I. Acid Red 18 at 505 nm.

(b) any background absorbance due to impurities has minimal effect on the accuracy of the absorbance readings of the dye;
(c) the use of non-monochromatic radiation is least critical, because at the peak of the absorbance spectrum, the absorbance readings are fairly similar over a few nanometres on either side of λ_{max}.

8.5 Quantitative analysis of dyes in solution

8.5.1 One-component systems

A calibration graph is plotted of absorbance at λ_{max} against concentration, for a range of solutions of known concentration. The slope of this line is the absorption coefficient. The absorbance of the dye solution of unknown concentration (for example, a sample from a dyebath liquor) is measured at the same wavelength. The concentration can then be obtained either by:

(a) reading the concentration from the calibration graph, or
(b) dividing the measured absorbance by the absorption coefficient.

8.5.2 Two-component systems

If the solution being analysed contains two dyes, X and Y, then the procedure is more complex. It is necessary to select two wavelengths at which to carry out the analysis, usually the λ_{max} wavelengths for each dye. It is best if these two wavelengths are far apart, and at which one dye absorbs strongly, the other weakly, and vice-versa.

It is then necessary to calculate the absorption coefficients, ε, for each dye at each of these two wavelengths, that is a total of four extinction coefficients. These are calculated from absorbance measurements made on the two dyes individually. Finally, the absorbances of the mixture of unknown concentrations are measured at each of these chosen wavelengths, and the following simultaneous equations solved for the concentrations c_X and c_Y:

$$\text{at } \lambda_1 \qquad A_{\lambda,1} = \varepsilon_{X,1}\, c_X + \varepsilon_{Y,1}\, c_Y$$

$$\text{at } \lambda_2 \qquad A_{\lambda,2} = \varepsilon_{X,2}\, c_X + \varepsilon_{Y,2}\, c_Y$$

(assuming all absorbance measurements are made in 1cm cells, so $l = 1$).

8.5.3 Three-component systems

To analyse a solution which consists of a mixture of three dyes, the same basic procedure is adopted as is used for two-component systems. In this case the method has to be extended, so that three wavelengths are selected (again, usually λ_{max} of each dye) and three simultaneous equations solved. The analysis of three-component systems can be less accurate, especially if the wavelengths chosen are not very far apart and there is significant overlap of the absorbance bands of the three dyes. This analysis should therefore be carried out with care.

The procedures explained above can be applied to substances which absorb radiation in other regions of the spectrum, such as the ultraviolet and infrared regions. However, the practical techniques employed are different, mainly because the solvents commonly used to dissolve the dye or pigment for the analysis, themselves strongly absorb radiation in these regions. Also ordinary soda-glass cells cannot be used to contain the solutions because soda-glass too absorbs infrared and ultraviolet radiation. For analytical work involving the ultraviolet region, cells made of special silica glass must be used; for analyses in the infrared region, the cells used must be constructed from sodium chloride plates.

8.6 Fluorescence and phosphorescence

Compounds which fluoresce are used in a variety of applications. Perhaps the most common is in fluorescent lamps which are used extensively for lighting offices and shops (see section 8.7.2). The inside of the tubes are coated with substances which fluoresce when ultraviolet light is incident on them, resulting in the emission of visible light. Some dyes and pigments are also fluorescent and use is made of the brightening effect they produce in the manufacture of garments where high visibility is essential, such as in safety clothing worn by the staff of the emergency services. Many textiles sold as 'whites' are treated with compounds called fluorescent whitening agents, which have the optical effect of improving their whiteness and brightness. Such whites give a blue glow when they are viewed under ultraviolet radiation.

In section 8.3, reference was made to the transitions which may occur when a molecule absorbs radiation. The various molecular orbitals are each associated with a definite energy and the electrons which occupy them themselves spin about their own axis. The spin angular momentum of an electron is either $+\frac{1}{2}$ or $-\frac{1}{2}$. A maximum of two electrons can occupy an orbital and only when they have opposite spins, so that the net spin angular momentum is zero. States arising from electronic configurations are described by what is termed their multiplicity, m, where:

$$m = 2S + 1$$

in which S is the total spin angular momentum. In any state in which all electrons are paired in orbitals, the net spin is zero, so the multiplicity is unity. These states are called *singlet* states and are labelled S. If, on the absorption of radiation, the molecule is raised from the ground state (S_0) to the first excited electronic state (S_1^*), the spin of the promoted electron may remain the same, in which case the excited state is still a singlet one and an $S_0 \rightarrow S_1^*$ transition has occurred. However if the spin of the electron changes sign when it is promoted, the net spin, S, will be unity and the value of the multiplicity three. The excited state is now referred to as a triplet state.

The energy transitions that occur in a molecule exhibiting fluorescence are illustrated in Fig. 8.9. When the molecule absorbs radiation, it is excited from the lowest vibrational level in its ground state (S_0) to a range of vibrational levels in the singlet first excited state (S_1^*). During the time the molecule spends in the excited state (only about 10^{-7} seconds), energy is dissipated from the higher vibrational levels, and the lowest vibrational level is attained. Fluorescence occurs if the molecule then emits radiation as visible light, as it reverts from this level to various vibrational levels in the ground state.

The molecule may alternatively dissipate its energy, as it reverts to the ground state, by a non-radiative process, such as collisional deactivation. As a result there will be a reduction in the fluorescence intensity and in many cases to such an extent that the fluorescence effect is lost completely. The intensity of light emission from a fluorescent dye is characterised by its quantum yield, which is the ratio between the emitted and absorbed quanta. In some strongly fluorescent materials this may approach unity. In the case of fluorescent dyes, the visual effect produced, no matter whether the dye is in solution or incorporated into a textile fabric, plastic material or surface coating, is due to the colour attributable to the selective absorption of light, supplemented by the colour of the fluorescence emission.

Because the molecule loses vibrational energy prior to the emission of light as it returns from the S_1^* (excited) to the S_0 (ground) state, the fluorescence emission occurs at longer wavelengths (lower energy) than those which were absorbed initially. The difference between the emission and absorption wavelength maxima is referred to as the *Stokes shift*. The

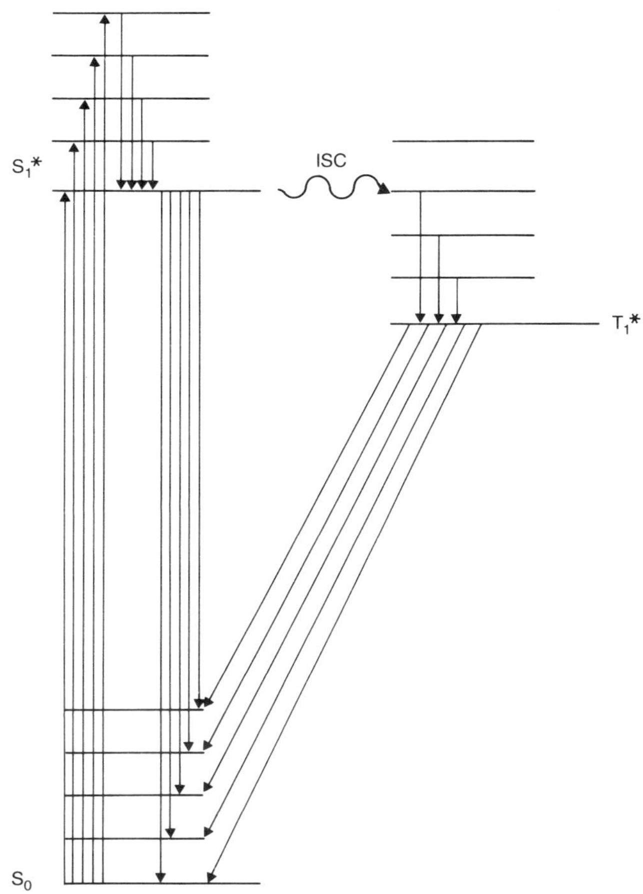

Fig. 8.9 A state diagram showing the electronic and vibrational absorption transitions and the relaxation processes leading to fluorescence and phosphorescence.

magnitude of the Stokes shift is an important parameter for certain applications of fluorescent dyes. For example, in dye lasers, solar collectors and certain biological stains (see Chapter 11), a large Stokes shift is desirable to minimise the overlap between the absorption and the emission spectra. In general the magnitude of the Stokes shift is about 50–70 nm, but in certain dyes the shift may be as high as 150–250 nm.

This process which leads to the effect of fluorescence can, in principle, occur anywhere in the electromagnetic spectrum where the absorption of the radiation causes electronic and vibrational excitation. For the production of coloured effects, it is the absorption and emission of radiation in the near ultraviolet and visible regions of the spectrum which is the most important. Fluorescent dyes absorb light in the visible region and also emit light (at longer wavelengths) in the visible region. It is this emission gives such dyes their characteristic brightness. The reflectance curve (see section

8.7.2) of a typical yellowish-green fluorescent coloured material (Fig. 8.10), shows that light is absorbed in the region 420–450 nm and the fluorescence emission occurs over the wavelength range 500–570 nm. Use of the fluorescence effect is also made with a class of compounds called fluorescent whitening agents and these are dealt with in section 8.10.5.

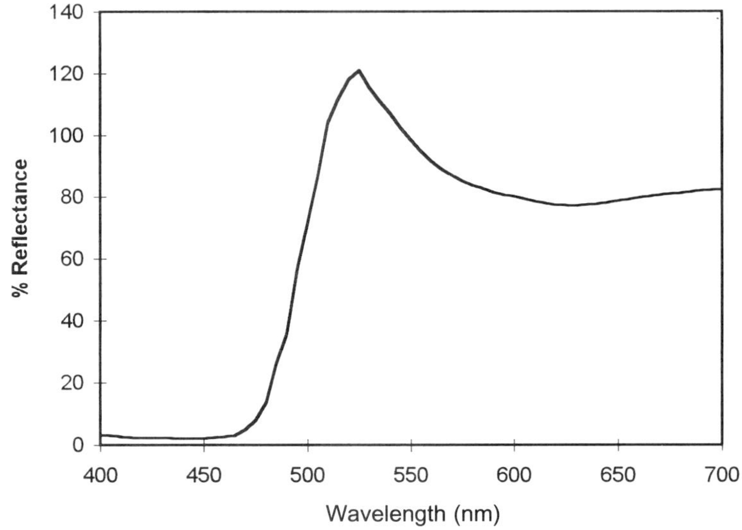

Fig. 8.10 Spectral reflectance curve of a typical yellowish-green fluorescent coloured material.

A set of vibrational levels in a T_1^* state can be seen in the right side of Fig. 8.9. This represents a *triplet* state and is assumed when spin inversion of the excited electron occurs, through a process called intersystem crossing (ISC). The triplet state always has lower energy than the corresponding singlet state. When a molecule reverts to the ground state from a triplet state, the emitted light is termed *phosphorescence*. In comparison with the processes which lead to fluorescence emission, those which lead to phosphorescence emission are slow and in some cases the latter can continue for up to 10 seconds after the irradiating light source has been switched off. In comparison fluorescence emission ceases instantly the irradiating source is switched off.

8.7 Specification of surface colours

The theoretical basis for colour specification has long been established – it was formalised by the C.I.E. in 1931 – but such was the mathematical complexity of the subject that its practical application in the colour manufacturing and colour using industries initially was hindered by a lack of computing facilities. The stage of development has now been reached whereby colour measuring instruments interfaced to their own dedicated microcomputers are the norm, the cost of which are within the budget of most companies.

Apart from the developments which have taken place in computers, developments have also been made in colour measuring instruments. In particular, improvements in performance, accuracy, reproducibility, ease of use and reduction of measurement time to a few seconds have been made, with the consequent benefit of convenience of use and reliability of measurements.

8.7.1 Additive and subtractive colour mixing

Fundamental to colour specification is an understanding of the laws of colour mixing. These laws are derived from observations of mixing colour on an additive and subtractive basis. Additive colour mixing refers to the mixing of coloured lights and the rules below follow from what is observed when the lights are shone together on to a white screen, which will reflect almost all of the incident light to the eye.

The three primary coloured lights are *red*, *green* and *blue* and are called *primary colours* because they cannot be produced by mixing together any other two colours. Also a wide range of colours can be produced by mixing these three primary colours in different proportions. A mixture of equal proportions of all three primaries produces white:

red + green + blue = white

If two primary colours are mixed in equal proportions a secondary colour is produced, for example:

red + blue = magenta
red + green = yellow
blue + green = cyan

thus magenta, yellow and cyan are secondary colours.

Any two colours, which when mixed in equal proportions produce white, are called complementary colours. From the above relationships, it can be seen that:

magenta + green = white
yellow + blue = white
cyan + red = white

and the pairs magenta/green, yellow/blue and cyan/red are examples of complementary colours. This behaviour leads to the concept of the colour space, which shows the relationships of colours to each other. Thus the colour circle can be described by spacing the three primary colours equally round a circle, as shown in Fig. 8.11(a). The secondary colours are placed on the circle between their two component primaries and white, which is an equal mixture of all three primaries, is placed in the centre of the circle. The colour circle in Fig. 8.11(a) is two-dimensional and shows two attributes of colour perception, namely hue and saturation (or colourfulness). The third attribute, lightness, requires a third dimension and is added at right angles to the plane of the colour circle and a colour solid is formed. In order to describe the relationship of colours to each other therefore, a three-dimensional space is required, as shown in Fig. 8.11(b).

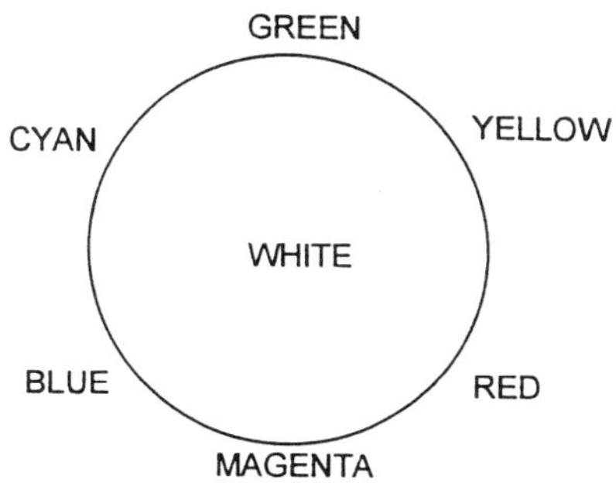

Fig. 8.11(a) The colour circle.

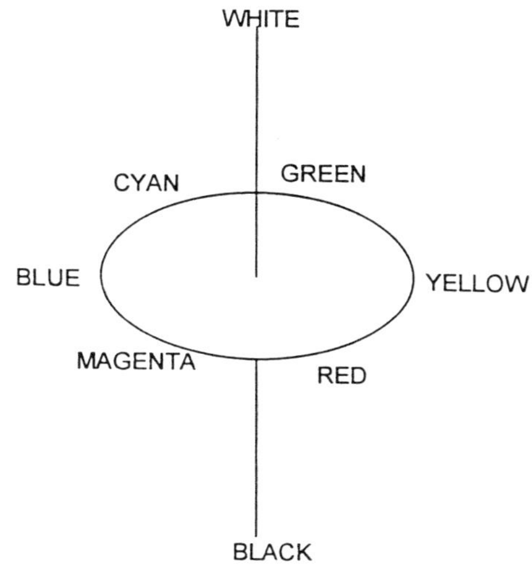

Fig. 8.11(b) The colour solid.

Subtractive colour mixing applies to the mixing of dyes and pigments and is the type of colour mixing with which we are most familiar. Dyes and pigments absorb light most strongly in a characteristic region of the visible spectrum and reflect the rest of the light. The colour of the light reflected is the colour seen. For example, a magenta coloured material absorbs light most strongly in the green region of the spectrum and reflects light in the blue and the red regions, which together appear magenta. Similarly, a yellow coloured material absorbs light in the blue region of the spectrum and reflects light in the green and red regions (as well as the yellow region).

Thus of the original incident white light, the various regions of the spectrum are reflected to different extents by coloured materials. The colour of the light absorbed is the complementary colour of the material. In the case of the two examples above, the magenta coloured material absorbs its complementary colour which is green, and the yellow coloured material absorbs its complementary colour which is blue.

8.7.2 *Factors affecting colour vision*

The factors which affect the perception of coloured materials are:

(a) the type of light source under which the materials are viewed;
(b) the light-reflecting properties of the material;
(c) the observer and the response of his eyes to the light entering them.

The viewing geometry can be represented by the diagram in Fig. 8.12.

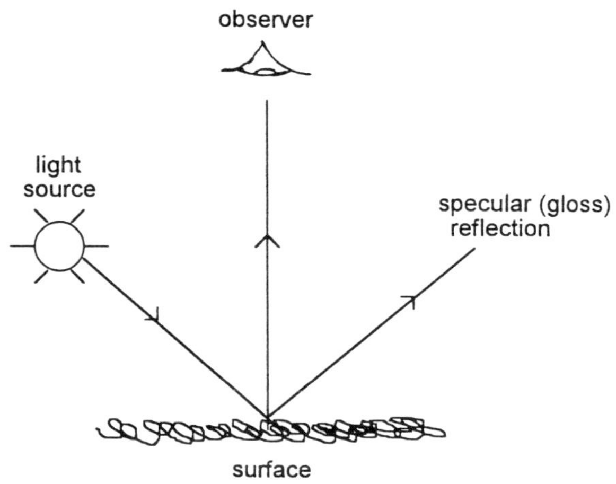

Fig. 8.12 The viewing geometry.

The ways in which the properties of each of these factors are specified will now be considered.

Light sources

The properties of light sources can be described in two ways:

(a) Spectral energy distribution
This is the amount of light energy emitted by the source at each wavelength through the visible region of the spectrum and is usually expressed relative to that emitted at 555 nm. Fig. 8.13 shows the distributions of a tungsten lamp (e.g. a household light bulb) and average daylight.

The sun and tungsten filament lamps are examples of incandescent sources of light. These are sources which emit light energy because of their temperature and their theoretical energy distribution across the spectrum can be calculated for any given temperature using Planck's radiation law. This law gives the spectral concentration of radiation from a source, M_e, as a function of wavelength, λ, and temperature T (Kelvin) of the source.

$$M_e = c_1\lambda^{-5}\left(e^{c_2/\lambda T} - 1\right)^{-1} \text{ W.m}^{-3} \quad (8.9)$$

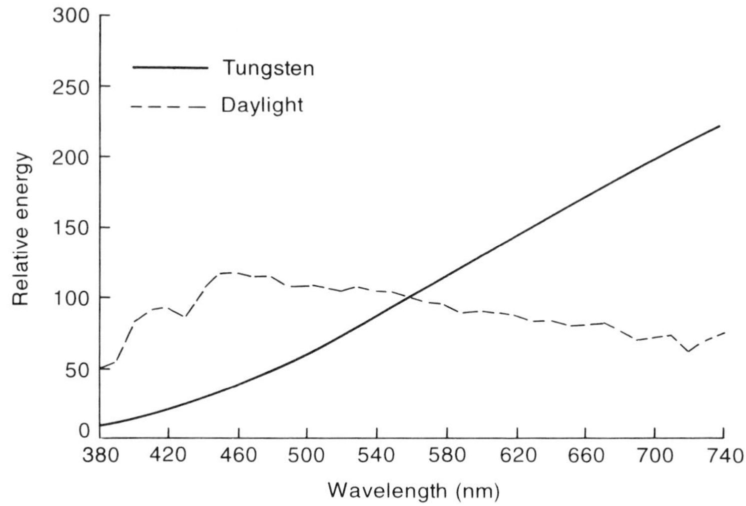

Fig. 8.13 Relative spectral energy distributions of average daylight and tungsten light.

where $c_1 = 3.74183 \times 10^{-16}$ W.m^2 and $c_2 = 1.4388 \times 10^{-2}$ m.K. The distributions obtained for different temperatures are illustrated in Fig. 8.14. The curves show that no light is emitted in the visible region until a temperature of about 2000 K is reached. At temperatures of 2000–3000 K the emission is very much higher in the red region than the blue region, but as the temperature increases, the amount of blue light increases relative to the red light.

Another common type of source is the fluorescent lamp, the emission of which is very different from that of either tungsten or daylight. Fluorescent sources consist of a mercury discharge tube the inside of which is lined with powders which fluoresce (see section 8.6). These powders are excited by the ultraviolet radiation from the mercury discharge and the light emitted from the tube consists of the intense emissions of the mercury in the violet, green and yellow regions, together with the fluorescence emission of the powders. Depending on the choice of powders in the tube, fluorescent sources of different hues can be obtained, for example cool white, or warm white effects. Even though the colour of these tubes can be made similar to daylight, the intense mercury line emissions cause the spectral energy distributions to be quite different.

Another type of fluorescent source now in common use, called the three-line type, has strong, narrow emission peaks in the blue, green and red regions, but emits very little light in the regions in-between. This type of source has a very high efficacy and has the effect of enhancing the brightness and richness of colour appearance. Fig. 8.15 shows the spectral energy distributions of normal and three-line type fluorescent sources.

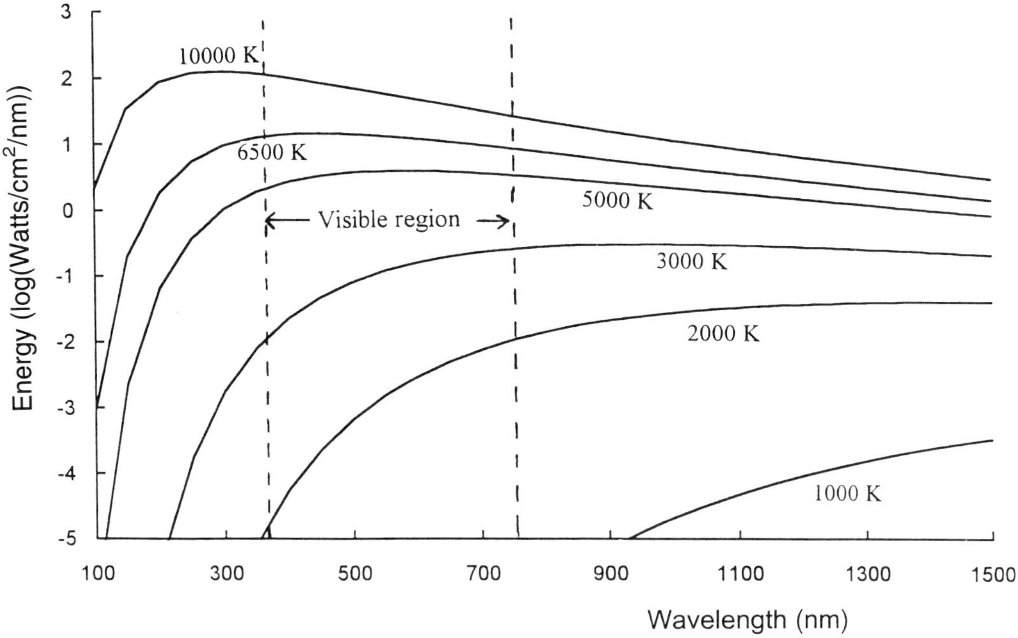

Fig. 8.14 Spectral energy distribution of incandescent light sources at various colour temperatures.

(b) Colour temperature

The physical concept of the ideal black body (Planckian) radiator can be used to describe the appearance of a source. Such a radiator is very difficult to achieve in practice, but its theoretical energy distribution in the visible region and hence its chromaticity (see section 8.8.5) can be calculated for a given temperature using Planck's radiation law, as described above. The colour temperature is defined as the temperature of the ideal black body radiator which has the same chromaticity as that of the source.

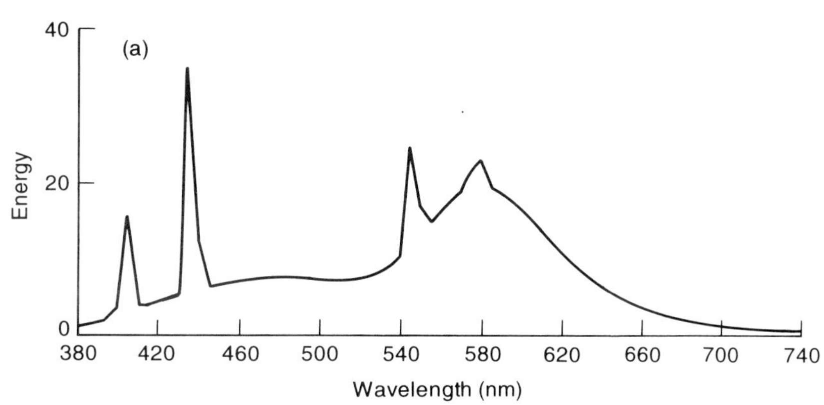

Fig. 8.15(a) Spectral energy distribution of typical fluorescent source.

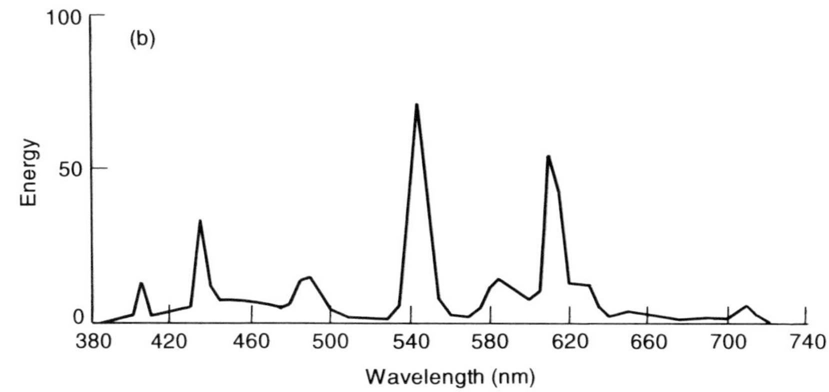

Fig. 8.15(b) Spectral energy distribution of typical three-line fluorescent source.

Strictly, the concept of colour temperature should only be applied to sources whose spectral energy distributions are similar to that of the ideal black body radiator, a condition which exists only in the case of incandescent sources. Daylight and light from fluorescent sources do not have a chromaticity which approximates to the ideal black body at any temperature. Such sources are defined by a correlated colour temperature, which is the temperature of the ideal black body radiator when its colour is closest to that of the source. Sources and illuminants are discussed again in section 8.8.3.

Surfaces

The colour of a surface depends on the proportion of the incident light that it reflects at each wavelength. The light reflected from a surface of a substrate is the net result of the various absorption, refraction and scattering processes which occur, as illustrated in Fig. 8.16. Reflectance values are measured by a spectrophotometer and the practice adopted is that the light reflected by a material is measured relative to that reflected by a reference white material, at the same wavelength. This procedure gives the reflectance R_λ, which is commonly expressed as a percentage. The graphs in Fig. 8.17 give typical examples of spectral reflectance curves for various colours.

8.7.3 The eye

The perception of colour is the result of the interpretation by the brain of nerve impulses stimulated in the retina of the eye by the action of light. Fig. 8.18 shows the important features of the human eye. The retina is a light-sensitive layer which lines most of the interior of the eyeball. There are two types of light-sensitive elements in the *retina* which are called rods and cones, because of their shapes.

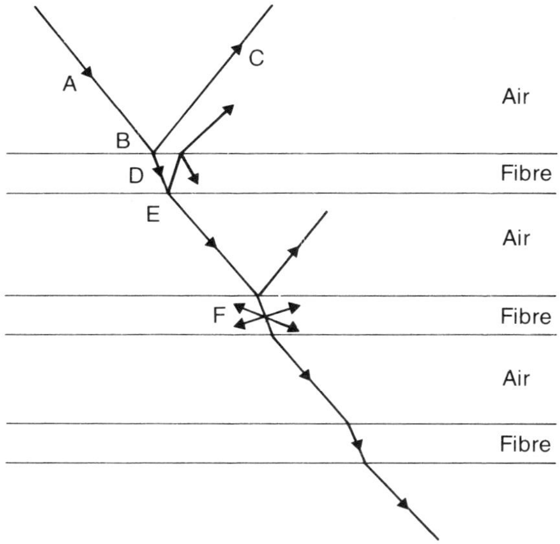

Fig. 8.16 Reflection and refraction effects at fibre surfaces. An incident ray of light A striking the fibre surface at point B will be either reflected from the surface (ray C) or refracted into the fibre (ray D). On refraction into the fibre the direction of the ray is bent towards the normal because the fibre is a denser medium than air. On emerging from the fibre at point E the direction of the ray is bent towards the normal because the fibre is a denser medium than air. On emerging from the fibre at point E the direction of the ray is bent away from the normal to a direction parallel with the incident ray. As the ray passes through the fibre its reduction in intensity is reduced by absorption due to dye molecules and scattering (F) due to pigment particles.

The *rods*, of which there are about one million in an eye, allow perception at very low levels of illumination, but do not impart colour vision. Many rods are connected to an individual nerve fibre and it is this feature which gives sensitivity to small amounts of light, though the arrangement prevents the perception of fine detail. The sensitivity of the rods to light is due to a pigment called rhodopsin. Under normal daylight viewing conditions nearly all the rhodopsin is bleached out. Under very dull lighting conditions, nutritive processes reform rhodopsin in the rods, a process which takes about half an hour to complete.

The *cones* are responsible for colour vision and daylight vision is due almost entirely to them. There are about seven million cones in an eye, each having an individual connection to a nerve fibre, a feature which allows the perception of fine detail. The *fovea* is a small area (about 1 mm^2) at the back of the retina and almost centred on the optical axis of the eye. It contains only cones, about 50 000 of them, but no rods, and is the region in which colour vision is most acute. There are three types of cone receptors, each sensitive to different regions of the spectrum, the short (blue), medium

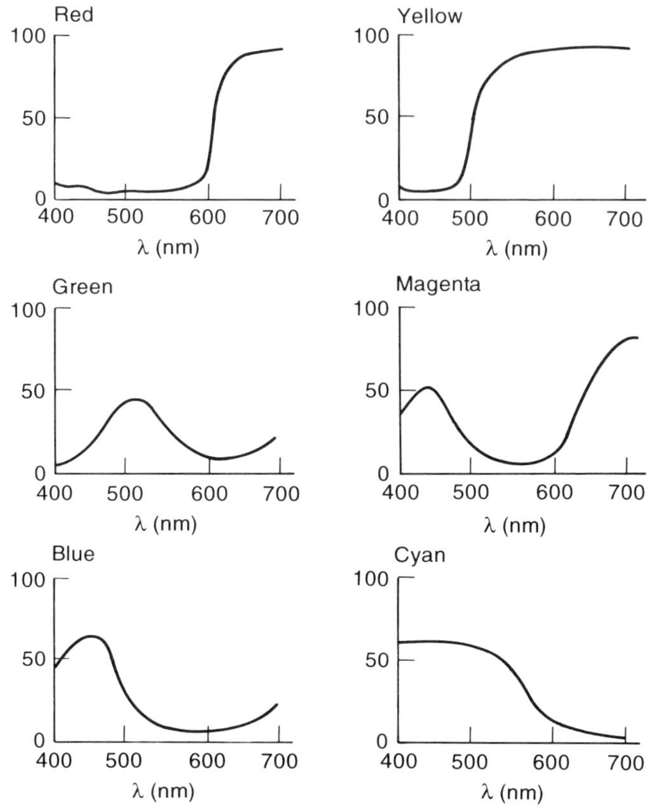

Fig. 8.17 Spectral reflectance curves of various colours.

(green) and long (red) wavelength regions. There are approximately equal numbers of red and green sensitive cones and about ten times fewer blue sensitive cones. On moving outwards from the fovea to the extremities of the retina, the concentration of cones decreases rapidly, whilst that of the rods simultaneously increases. All the nerves from the rods and cones exit from the eye through the optic nerve to the brain. The area at which the optic nerve enters the eye is called the 'blind spot', since it is the area at which no rods or cones are present.

8.8 The C.I.E. system of colour specification

The basis of all colour measurement work is the C.I.E. system of colour specification, agreed in 1931. The initials C.I.E. stand for Commission Internationale de l'Eclairage, which is an organisation devoted to international cooperation and exchange of information among its member countries on all matters relating to the art and science of lighting.

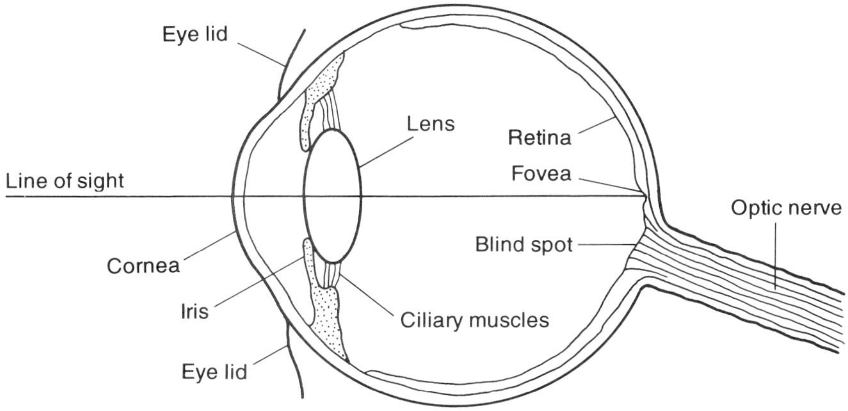

Fig. 8.18 Sketch of the cross-section of the human eye.

8.8.1 The C.I.E. standard observer

The C.I.E. system of colour specification is based on the properties of additive mixtures of coloured lights, determined by experiment. In experiments carried out using the arrangement illustrated in Fig. 8.19, it is found that a wide variety of coloured lamps can be matched by adjusting the proportions of the red, green and blue lights in the mixture.

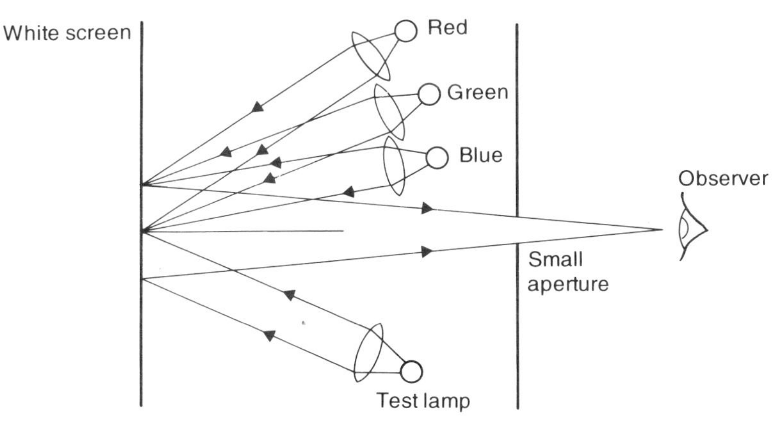

Fig. 8.19 Matching coloured lights.

A colour matching equation to represent this situation can be written:

$$C[\mathrm{C}] = R[\mathrm{R}] + G[\mathrm{G}] + B[\mathrm{B}]$$

which means that C units of colour [C] can be matched by R units of the red primary [R], G units of the green primary [G] and B units of the blue primary [B]. The primary colours are given the symbols [R], [G] and [B]. The amounts of these primaries required in the match, R, G and B, are called tristimulus values.

The gamut of colours that can be matched by a given set of primaries depends on the richness of the primary colours. It is found in this experiment that it is not possible to match all colours, even if the primaries are pure spectral primaries, which are the richest colours which exist. For example, a bright yellow light cannot be matched by any combination of the red and green spectral primaries, until its richness has been reduced by adding blue light first. In this case the colour matching equation is (where [C] is the yellow):

$$C[\mathrm{C}] = R[\mathrm{R}] + G[\mathrm{G}] - B[\mathrm{B}]$$

or

$$C[\mathrm{C}] + B[\mathrm{B}] = R[\mathrm{R}] + G[\mathrm{G}]$$

because the blue primary has to be added to the yellow, so that it can then be matched by a mixture of the red and green primaries. The tristimulus values R, G and B specify the appearance of a colour but the fact that for some colours, one of the tristimulus values could be negative, is inconvenient. The extent of this problem was illustrated by the results of colour matching experiments carried out separately by Wright and Guild. Using visual colorimeters of similar arrangement to that shown in Fig. 8.18, they asked a number of observers to match light of each wavelength through the visible spectrum. Although Wright used pure spectral primaries (red of 700.0 nm, green of 546.1 nm and blue of 436.8 nm) and Guild used filter primaries, the results when corrected for these differences were found to be in very good agreement.

The combined results were expressed as the tristimulus values required to match unit energy of light at each wavelength of the visible spectrum. The tristimulus values (also called distribution coefficients) are given the special symbols $\bar{r}_\lambda$, $\bar{g}_\lambda$, $\bar{b}_\lambda$. The values are plotted in Fig. 8.20, in which it can be seen that at many wavelengths, one of the values ($\bar{r}_\lambda$, $\bar{g}_\lambda$, $\bar{b}_\lambda$) is negative.

To overcome the problem of negative tristimulus values, the C.I.E. adopted three imaginary primaries [X], [Y], [Z], which may be considered to be very saturated red, green and blue primaries respectively. Although they cannot be made, they are defined mathematically, in terms of the real primaries [R], [G], [B] used by Wright, that is spectral lights at 700.0, 546.1 and 436.8 nm respectively. The imaginary primaries were so defined that white light from an equi-energy spectrum (that is, equal amounts of each wavelength) would be matched by equal amounts of them:

$$S_{\mathrm{E}} = 0.333[\mathrm{X}] + 0.333[\mathrm{Y}] + 0.333[\mathrm{Z}]$$

By the use of transformation equations, it was possible to calculate from the $\bar{r}_\lambda$, $\bar{g}_\lambda$, $\bar{b}_\lambda$ values, the amounts of the imaginary primaries $\bar{x}_\lambda$, $\bar{y}_\lambda$, $\bar{z}_\lambda$ required to match unit energy of light at each wavelength of the spectrum. These amounts are all positive and are called the C.I.E. 1931 Standard Observer colour matching functions. Again, they are tristimulus values and are represented in Fig. 8.21. The values of $\bar{x}_\lambda$, $\bar{y}_\lambda$, $\bar{z}_\lambda$ at each wavelength are

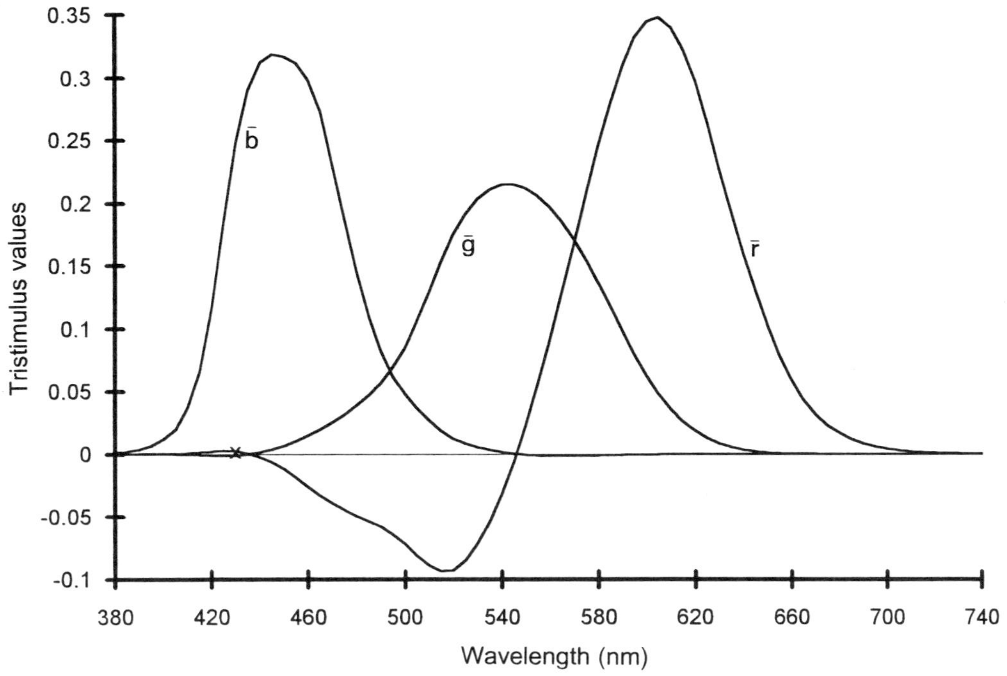

Fig. 8.20 Tristimulus values $\bar{r}_\lambda$, $\bar{g}_\lambda$, $\bar{b}_\lambda$ obtained from colour matching experiments.

published by the C.I.E. (C.I.E. Publication 15) and in most texts on colour measurement.

8.8.2 Calculation of tristimulus values of surface colours

The amounts of the imaginary primaries required to match unit energy of each wavelength of the spectrum are the standard observer functions $\bar{x}_\lambda$, $\bar{y}_\lambda$, $\bar{z}_\lambda$. If all of the wavelengths of white light are viewed simultaneously, the amounts of the imaginary primaries required will be given by their sums throughout the spectrum. These sums are called tristimulus values and are given the symbols X, Y, Z. The mathematical process for their computation is represented by the equations:

$$X = \int_{360}^{780} \bar{x}_\lambda \, d\lambda \qquad Y = \int_{360}^{780} \bar{y}_\lambda \, d\lambda \qquad Z = \int_{360}^{780} \bar{z}_\lambda \, d\lambda$$

The normal viewing situation is that in which the light from a source is reflected from a surface to the eye. In this case, account has to be taken of the fact that the source emits different amounts of light at each wavelength and also that the surface reflects only a proportion of the light incident on it. The above equations must therefore be modified and become:

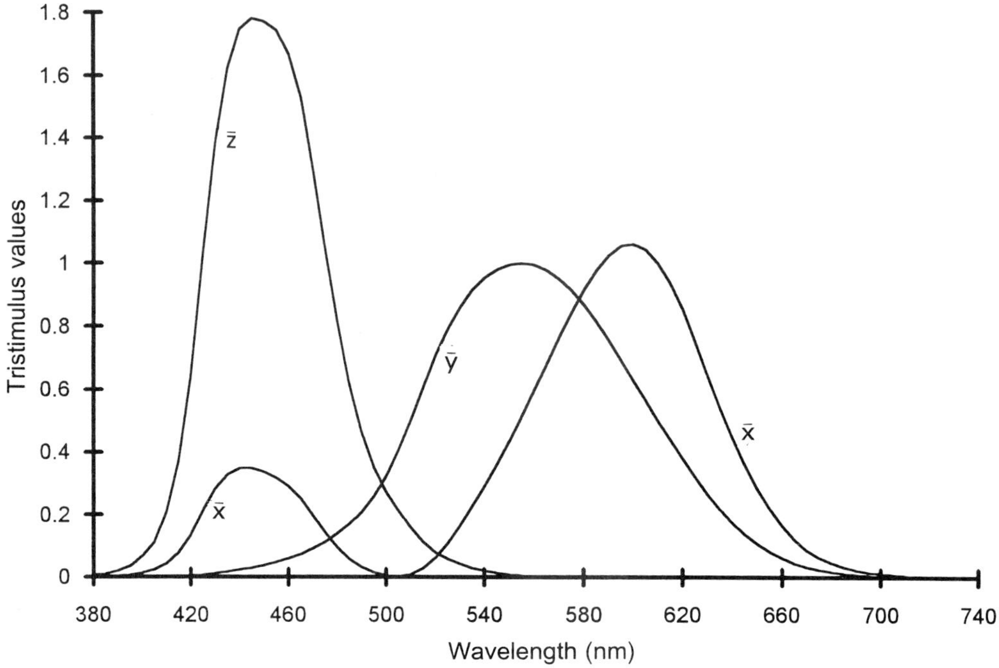

Fig. 8.21 C.I.E. 1931 colour matching values $\bar{x}_\lambda$, $\bar{y}_\lambda$, $\bar{z}_\lambda$.

$$X = k\int_{360}^{780} E_\lambda \bar{x}_\lambda R_\lambda \mathrm{d}\lambda$$

$$Y = k\int_{360}^{780} E_\lambda \bar{y}_\lambda R_\lambda \mathrm{d}\lambda$$

$$Z = k\int_{360}^{780} E_\lambda \bar{z}_\lambda R_\lambda \mathrm{d}\lambda$$

where $k = \dfrac{1}{\int_{360}^{780} E_\lambda \bar{y}_\lambda \mathrm{d}\lambda}$

(which makes $Y = 100$ for a perfect white surface;
for a perfect black surface $X = Y = Z = 0$)

E_λ = relative energy of the source;
R_λ = reflectance of the surface.

Thus to calculate the X, Y, Z values which define a colour, it is necessary to multiply the E_λ value, the standard observer value ($\bar{x}_\lambda$, $\bar{y}_\lambda$, $\bar{z}_\lambda$ respectively), the R_λ value and the wavelength interval together, at each wavelength, and then sum these products for each wavelength over the visible spectrum. This is a long process but is simplified in practice by using an approximation method called the weighted-ordinate method:

$$X = \sum E_\lambda \bar{x}_\lambda R_\lambda \qquad Y = \sum E_\lambda \bar{y}_\lambda R_\lambda \qquad Z = \sum E_\lambda \bar{z}_\lambda R_\lambda \tag{8.11}$$

Here the multiplications are carried out at regular wavelength intervals throughout the visible region, usually at 10 or 20 nm intervals.

These equations can be used to specify the appearance of a coloured material (of known R_λ values) under any source, without actually having to measure the reflectance of the material under that particular source. Provided that the colour is non-fluorescent, the R_λ values are independent of the type of light source used in the spectrophotometer. All that is required are the spectral energy (E_λ) values of the source and the $\bar{x}_\lambda$, $\bar{y}_\lambda$, $\bar{z}_\lambda$ values. In fact the values of the products, $E_\lambda \bar{x}_\lambda$, $E_\lambda \bar{y}_\lambda$, $E_\lambda \bar{z}_\lambda$ at each wavelength, which are called weighting functions, are published by the C.I.E. for a variety of illuminants and wavelength intervals (for example 5, 10 or 20 nm intervals). They are also available in textbooks on colour physics. As an example, Table 8.3 gives the values for Illuminant D_{65}, 10° observer, at 20 nm intervals.

This is an important feature of the C.I.E. system since it means that colour appearance can be specified using the standard illuminants specified by the C.I.E. (see below). The values of the weighting functions can be stored in a computer, so it is only necessary to measure the $\%R_\lambda$ of a sample and carry out the calculations described above, to obtain the tristimulus values of a coloured material.

8.8.3 Standard illuminants

There are many different sources of light under which coloured materials may be viewed, such as daylight (of which there are various phases), tungsten and the various fluorescent types. However, to check that a material has the correct colour, it is necessary to define carefully the light source under which it is viewed. It is important at this stage to distinguish between sources and illuminants. A source is a real emitter of light, such as a lamp or the sun, whereas an illuminant is defined in terms of a spectral energy distribution and is therefore a numerical specification. It may or may not be possible to make a source which represents an illuminant.

The C.I.E. originally defined three standard illuminants called A, B and C. Illuminant A represents a tungsten filament lamp operating at a colour temperature of 2856 K. The spectral energy distribution of Illuminant A is calculated from Planck's radiation law (Eqn 8.9) which applies to incandescent sources.

Illuminants B and C correspond to different phases of daylight and the C.I.E. has published details of how these illuminants could be produced as sources, by passing light from a tungsten lamp through liquid filters of defined composition. Source B was meant to correspond to noon sunlight and source C to average north sky light, but it has been shown that neither corresponds well to any common phase of daylight. In particular they are deficient in ultraviolet content when compared with daylight. Consequently the use of these Illuminant specifications has been discontinued.

In response to this problem a series of D Illuminants were defined, for the correlated colour temperatures of various forms of daylight. Their spectral energy distributions are shown in Fig. 8.22. The energy distribution for the correlated colour temperature of 6500 K used for colour specification; this Illuminant is called D_{6500}, or more commonly D_{65}.

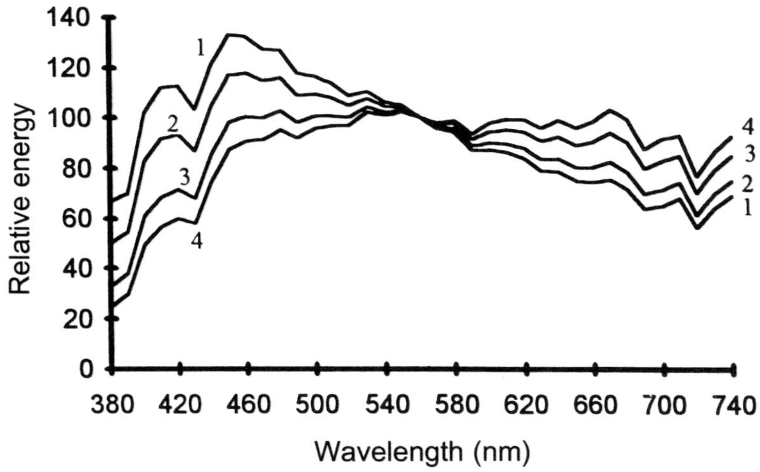

Fig. 8.22 Spectral energy distributions of C.I.E. Standard Illuminants. 1, D_{7500}; 2, D_{6500}; 3, D_{5500}; 4, D_{5000}.

The D series of Illuminants are a substantial improvement over B and C in representing daylight. However they are Illuminants (defined by spectral energy distributions published by the C.I.E.) and no manufacturer has yet managed to produce a source which emits light with precisely the same energy distribution as Illuminant D_{65}, though reasonably good approximations to the colour temperature of 6500 K are available.

8.8.4 1964 supplementary standard observer

The work of Wright and Guild, the results of which were used to define the C.I.E. 1931 standard observer, was carried out using a colorimeter constructed so that only a 2° field of view was observed. This is a much narrower field of view than is the normal viewing condition and so new colour matching experiments were carried out by Stiles and Birch in 1964. The experiments were similar to those of Wright and Guild but a much wider field of view was employed (10°). By a similar procedure to that described earlier, it was then possible to calculate 10° standard observer functions, denoted $\bar{x}_{10,\lambda}$, $\bar{y}_{10,\lambda}$, $\bar{z}_{10,\lambda}$. The subscript 10 distinguishes these data from the original 1931 2° values. Weighting values for $E_\lambda\bar{x}_{10,\lambda}$, $E_\lambda\bar{y}_{10,\lambda}$, $E_\lambda\bar{z}_{10,\lambda}$ at each wavelength for various illuminants (for example A, D_{65}, TL84, CWF) have also been published by the C.I.E. The values given in Table 8.3 are for Illuminant D_{65}, 10° observer. When using these weighted ordinates to calculate X, Y, Z tristimulus values, the reflectances must be used as a factor, that is, on a 0–1.0 scale, not as a percentage. Then, in the case of the perfect white, for which $R_\lambda = 1.0$ at all wavelengths, it can be seen that $X = 94.8272$, $Y = 100.0000$, $Z = 107.3906$. The perfect white would only have $X = 100$, $Y = 100$, $Z = 100$ if it were viewed under an equi-energy light source, for which $E_\lambda = 1.0$ at all wavelengths. An example of the way in which the weights in Table 8.3 are used to calculate the tristimulus values of a colour is given in Table 8.4. It will be seen that the weights in Table 8.3 (at each wavelength) are multiplied by the reflectance of the sample at that wavelength. The products are then summed, to give the tristimulus values for the sample.

Table 8.3 Weighting values for Illuminant D_{65}, 10° observer.

λ (nm)	$E_\lambda\bar{x}_{10,\lambda}$	$E_\lambda\bar{y}_{10,\lambda}$	$E_\lambda\bar{z}_{10,\lambda}$
400	0.2516	0.0236	1.0906
420	3.2317	0.3301	15.3824
440	6.6805	1.1069	34.3820
460	6.0964	2.6206	35.3562
480	1.7213	4.9378	15.8979
500	0.0589	8.6695	3.9972
520	2.1845	13.8473	1.0457
540	6.8093	17.3537	0.2373
560	12.8953	17.1539	0.0025
580	16.4686	14.1481	−0.0022
600	17.2340	10.1056	0
620	12.8953	6.0212	0
640	6.2267	2.5867	0
660	2.1113	0.8268	0
680	0.5736	0.2222	0
700	0.1209	0.0460	
Totals	94.8272	100.0000	107.3906

The X, Y, Z tristimulus values specify a colour viewed by the C.I.E. standard observer, under one of the Standard Illuminants. It must be remembered that an individual person may not have the same visual response characteristics as the C.I.E. standard observer and also that the light source under which the material is actually viewed may not correspond to a C.I.E. Standard Illuminant. Hence the X, Y, Z values computed for a colour may not relate precisely to the actual colour seen by an individual observer.

Table 8.4 Calculation of tristimulus value for a yellow coloured material (refer to the weights given in Table 8.3).

λ (nm)	%R	$E_\lambda \bar{x}_{10,\lambda} R_\lambda$	$E_\lambda \bar{y}_{10,\lambda} R_\lambda$	$E_\lambda \bar{z}_{10,\lambda} R_\lambda$
400	3.16	0.0079	0.0007	0.0345
420	3.18	0.1028	0.0105	0.4892
440	4.14	0.2766	0.0458	1.4234
460	6.94	0.4231	0.1819	2.4557
480	14.99	0.2580	0.7402	2.3831
500	25.72	0.0151	2.2298	1.0281
520	43.53	0.9509	6.0277	0.4552
540	59.18	4.0297	10.2699	0.1404
560	72.97	8.8750	12.5172	0.0018
580	78.63	12.9493	11.1247	–0.0017
600	80.90	13.9423	8.1754	0
620	81.91	8.5451	4.9320	0
640	82.61	5.1439	2.1369	0
660	83.08	1.7541	0.6869	0
680	83.84	0.4809	0.1863	0
700	84.23	0.1018	0.0387	0
Total		$X = 59.87$	$Y = 59.31$	$Z = 8.41$

8.8.5 Colour co-ordinates and colour appearance

One feature of the C.I.E. system of colour specification is that the Y tristimulus value is related to the lightness of a colour. The higher the Y value, the lighter the appearance of the colour. The other two attributes of colour appearance, hue and saturation can also be specified, in this case using the chromaticity co-ordinates x, y, z, of the colour, where

$$x = \frac{X}{X+Y+Z} \qquad y = \frac{Y}{X+Y+Z} \qquad z = \frac{Z}{X+Y+Z}$$

$$\text{and } x + y + z = 1$$

A plot of y vs x is called a chromaticity diagram and is shown in Fig. 8.23. The x,y values of each wavelength of light are plotted in this diagram and a smooth curve drawn through them, called the spectrum locus. Also shown in the diagram are the locations of Illuminants A (tungsten) and D_{65} (average daylight).

The hue of a colour is indicated by the dominant wavelength. For example, the yellow colour referred to in Table 8.4 is located at $x = 0.469$, $y = 0.465$. To obtain the dominant wavelength, a line is drawn from the location of Illuminant D_{65} ($x = 0.317$, $y = 0.329$) through the location of the colour, to the spectrum locus. The dominant wavelength is read at the point where the line intersects the spectrum locus, in the example (Fig. 8.23) this is 579 nm. This is approximately the wavelength of spectral light, which

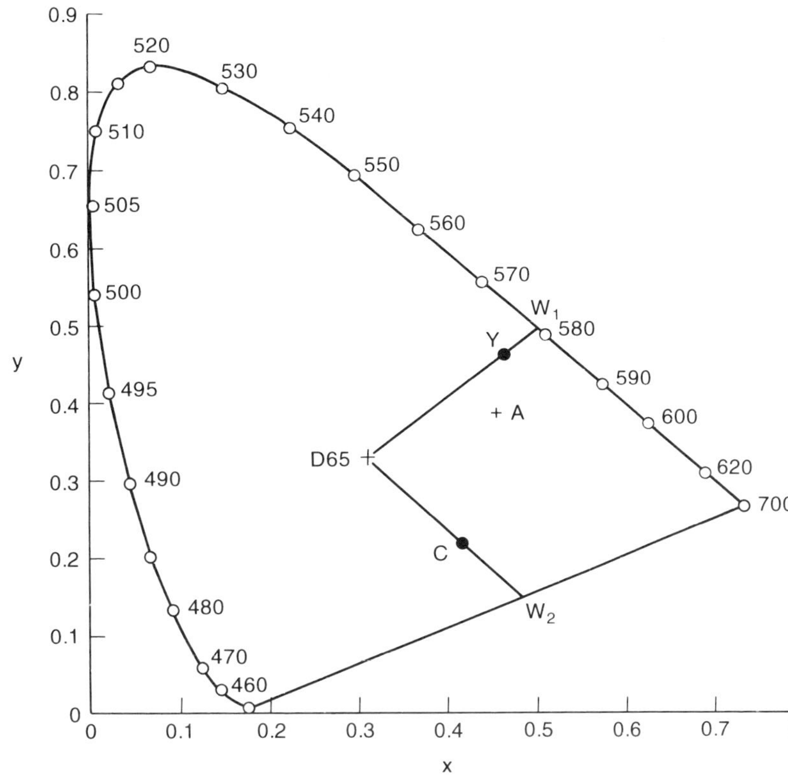

Fig. 8.23 C.I.E. chromaticity diagram showing locations of Illuminant D_{65} and Illuminant A. The point labelled Y indicates location of the yellow colour (see text).

when mixed with Illuminant D_{65} will match the hue of the colour. As a rough guide, it can be regarded as the wavelength of spectral light whose hue most closely matches the hue of the colour.

If the X, Y, Z tristimulus values, and hence x, y chromaticity co-ordinates had been determined for the yellow colour under another illuminant (for example Illuminant A), then the dominant wavelength would be determined by drawing the line from the location of Illuminant A through the colour to the spectrum locus. Under Illuminant A the chromaticity co-ordinates of the yellow colour are in fact $x=0.540$, $y=0.438$, giving a dominant wavelength of 587 nm. This wavelength has a slightly redder hue than that determined for Illuminant D_{65}, indicating that this yellow colour possesses a redder hue under tungsten light than it does under daylight. It can be seen that the C.I.E. system can predict the way in which the hue of a colour will change as the type of light under which it is viewed changes. However care must be taken in making such an interpretation since it ignores the ability of the human colour vision system to compensate for changes in the type of lighting conditions. A consequence of this property is that the colours of many objects remain reasonably constant to us, regardless of the nature of the light source under which they are viewed (see section 8.10.1).

The saturation of a colour is indicated by the excitation purity, which is the ratio of the lengths DY/DW_1 in the chromaticity diagram (Fig. 8.23). The richer the colour, the nearer towards the spectrum locus it will lie and the more the excitation purity will approach the limit of 1.0. In the case of colour *C*, the excitation purity is given by the ratio DC/DW_2. The colour *C* lies in the pink/purple region and it is not possible to determine a dominant wavelength since these are non-spectral hues. In this case the complementary dominant wavelength is determined, by extending a line in the opposite direction. The wavelength on the spectrum locus at which the line intersects in this case now corresponds approximately to the hue of light most strongly absorbed by the colour.

It is important to remember that colour is three-dimensional and that the chromaticity diagram is two-dimensional. The missing attribute, lightness, (indicated by the *Y* tristimulus value) can be regarded as being on an axis at right angles to the plane of the chromaticity diagram, as indicated in Fig. 8.24. In the case of the yellow colour mentioned above, under Illuminant D_{65} its *Y* value is 59.31 and under Illuminant A its *Y* value is 67.58. Hence the C.I.E. system indicates that the yellow colour will appear lighter under Illuminant A than under Illuminant D_{65}.

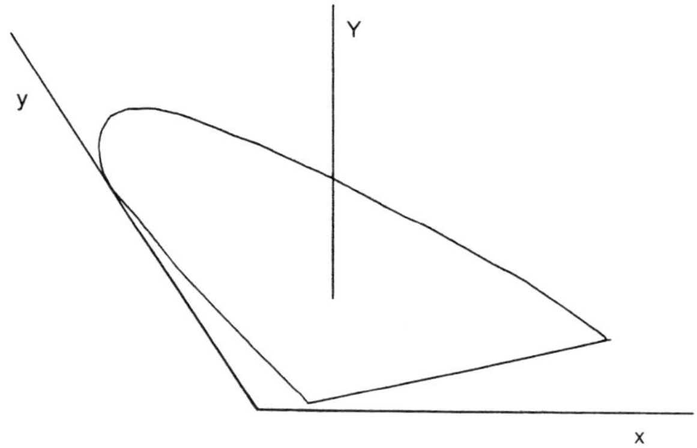

Fig. 8.24 Relationship of lightness to the chromaticity diagram.

8.9 Instruments for colour measurement

There are two basic types of instrument for colour measurement, spectrophotometers and photoelectric tristimulus colorimeters. The spectrophotometer is the most fundamental instrument for colour measurement, but it does not measure the colour of a material directly. It measures the light reflected (or transmitted) from the material, and these values are used to compute the tristimulus values, as described in section 8.8.

Colorimeters are designed to give the tristimulus values of materials directly. The light reflected by the material is passed separately through three coloured filters, specially designed so that their transmissions correspond to the $\bar{x}_\lambda$, $\bar{y}_\lambda$, $\bar{z}_\lambda$ colour matching functions of the C.I.E. standard observer. These instruments are able to give the tristimulus values of a colour directly, without the need for computation. However, many of the colorimeters on the market are interfaced with dedicated microcomputers, so that other functions, such as colour difference calculations and shade sorting can be carried out.

In general, colorimeters are used to compare colours, for example, the colour of a batch of dyed material against a standard colour. They may be used to measure the size of the visual colour difference between two similar colours, and also to specify the nature of the colour difference in terms of lightness, hue and saturation differences. Colorimeters are cheaper than spectrophotometers and are used mainly for monitoring colour control and for shade assessment. They are simple in operation and quick to use. The essential structural features of colorimeters are shown in Fig. 8.25.

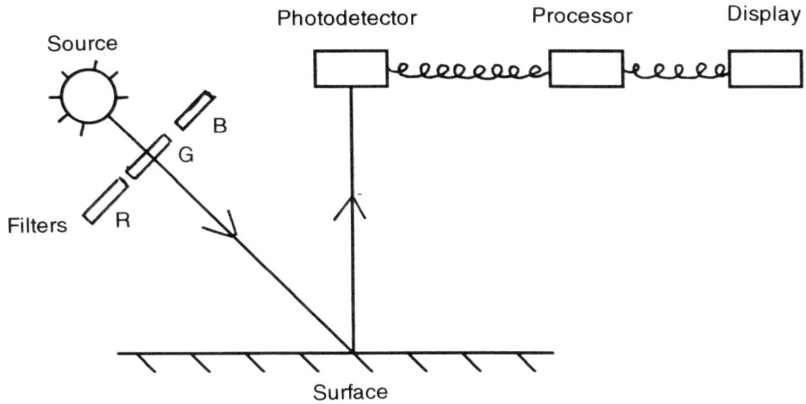

Fig. 8.25 Essential structural features of a colorimeter.

Spectrophotometers are used to measure the reflectance values of coloured materials and as such are more complex instruments. They are interfaced with dedicated microcomputers to enable them to calculate the tristimulus values. Modern spectrophotometers are capable of giving the tristimulus values of colours in as quick a time as colorimeters and in recent times have replaced the need for colorimeters. For those industries which only require a relatively cheap instrument for colour-difference evaluation, instruments are available which are described as spectrocolorimeters. These instruments are essentially a reflectance spectrophotometer interfaced to a microcomputer. The computer software is designed to perform those colorimetric analyses, such as colour-difference calculations, typical of colorimeters.

Spectrophotometers can carry out the same functions of colour assessment as colorimeters. Additionally they may be used for computer dye recipe prediction, the mathematics of which requires the reflectance values of the coloured material being matched. The microcomputer is then used to store the calibration data for the dyes and perform the recipe prediction calculations. The generally greater complexity of construction of spectrophotometers, together with the more comprehensive software for colour physics that accompanies them, means that they are the most expensive type of instrument on the market. Most of the spectrophotometers on the market are of the abridged type, which means that they provide the reflectance values of a surface colour at fixed wavelengths, instead of continuously across the spectrum. As an example, some spectrophotometers give reflectances at every 10 nm, between 380 and 710 nm, others give the values at every 20 nm, between 400 and 700 nm. The software which is provided with these instruments contains the appropriate weighting values $E_\lambda \bar{x}_\lambda$, $E_\lambda \bar{y}_\lambda$, $E_\lambda \bar{z}_\lambda$ for the spectral range and wavelength interval employed (see section 8.8.2).

The structural features of a typical spectrophotometer are illustrated in Fig. 8.26. The linear diode array detector system allows the reflectance at each wavelength to be measured simultaneously, a feature which facilitates the quick measurement of a sample. It will be noted that the sample measuring port is located on an 'integrating sphere', which gives a 0/diffuse illumination/viewing geometry. Some spectrophotometers have the opposite geometry, that is diffuse/0. These geometries are different from those employed in colorimeters or spectrocolorimeters, where a 0/45 or 45/0 illuminating/viewing geometry is usually used (see Fig. 8.25). The use of the 0/diffuse geometry with an integrating sphere allows the measurement of the light reflected from the surface in all directions. It also enables measurements to be made either including or excluding the specular (gloss) reflection. The various illuminating/viewing geometries and other variations in design, from one instrument type to another, mean that it cannot be expected that different instrument types will necessarily give the same reflectance readings (and hence X,Y,Z values) for the same sample.

Most commercially available spectrophotometers for colour measurement measure reflectance as the reflectance factor:

$$\frac{\text{amount of light reflected from the surface}}{\text{amount of light reflected from reference white similarly illuminated}}$$

instead of the simple ratio:

$$\frac{\text{amount of light reflected from the surface}}{\text{amount of light incident on the surface}}$$

The C.I.E. requires the reference white to be the 'perfect reflecting diffuser', that is a white surface which reflects all of the incident light at equal intensity in all directions in the hemisphere above it. In practice no material

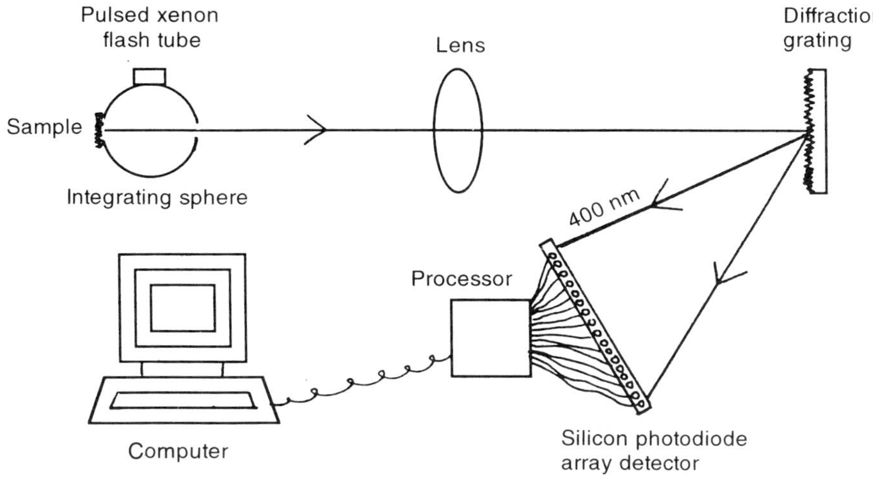

Fig. 8.26 Essential structural features of a typical single-beam spectrophotometer.

possesses such properties, though magnesium oxide and barium sulphate approach the requirement closely. These substances are not suitable for everyday use so the instrument manufacturers provide a white ceramic tile which is calibrated in terms of the perfect reflecting diffuser. Ceramic tiles have the advantage of stability over a long period of time, so long as they are carefully maintained.

It is essential, with both types of instrument, that regular standardisation is carried out, to ensure accurate measurements. Calibration of the instruments is easily achieved by the use of the calibration tiles provided with the instrument by the manufacturer. Also the accuracy can be checked with the calibrated standard tiles produced by the British Ceramic Research Association. These tiles are useful for monitoring the long-term stability of instruments and for detecting faults which may develop gradually.

8.10 Uses of colour specification

The uses of instrumental colour specification can be classified broadly into four areas:

(a) prediction of degree of metamerism;
(b) computer colorant formulation;
(c) shade passing and sorting;
(d) measurement of whiteness.

Each of these areas is discussed briefly below. They are in no particular order of importance but metamerism is discussed first since the original purpose of the C.I.E. system was to provide a method by which the influence of the light source on colour appearance could be specified.

8.10.1 Metamerism

Metamerism is the term used to describe the phenomenon whereby two coloured materials may appear the same under one type of light source (such as daylight), but appear quite different when viewed under a different type of source (such as tungsten light).

Two colours appear identical to the standard observer, under a given illuminant, if they have the same tristimulus values. This will be the case if the two coloured surfaces have exactly the same spectral reflectance values. However it is still possible for two surfaces, with different spectral reflectance curves, to have the same tristimulus values under one illuminant. For example, the two colours with the reflectance curves given in Fig. 8.27 have identical tristimulus values for Illuminant D_{65} and hence appear as the same colour. However, under different sources, for example tungsten or fluorescent sources, the two colours will not match, because their tristimulus values differ under these sources.

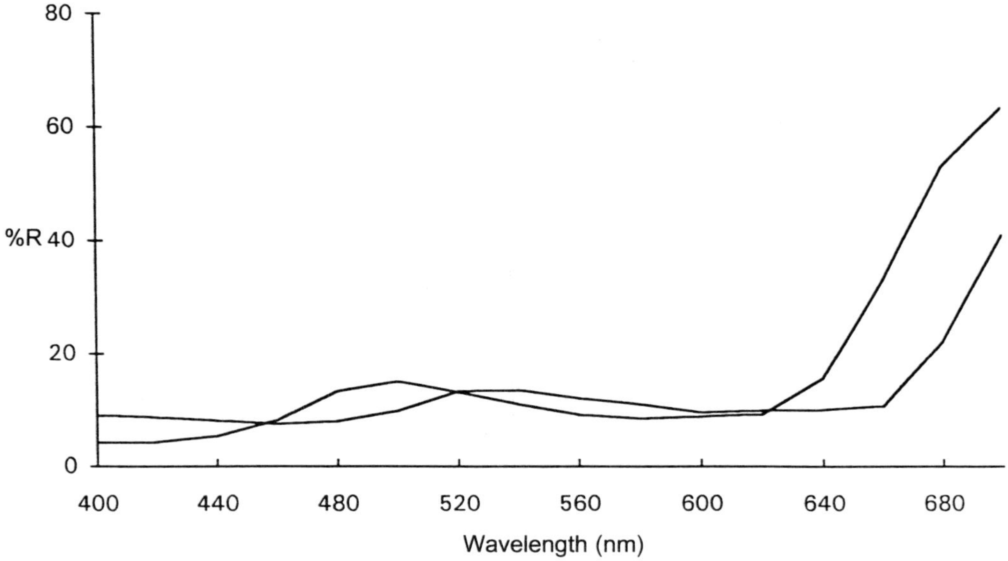

Fig. 8.27 Reflectance curves of two metameric colours.

Metamerism occurs between pairs of colours which have been coloured with different colorants. Often, different colorants have to be used, sometimes because those used to colour one substrate are unknown and the colorist has to use his own preferred ones. Also, for technical reasons, it may not be possible to use the same colorants anyway, as is the case when, for example, it is necessary to produce a colour match on polyester to a shade on cotton. Different application classes of dyes have to be used for the two fibre types.

In routine colour matching, metamerism is a very common occurrence. The fundamental problem is the difficulty in matching the reflectance curve of the standard exactly. Often the colorants used to dye the standard are unknown, so different colorants may be used to match it. Consequently an exact spectral match at each wavelength is not possible and the only solution is to find the match which is least metameric, with the colorants available. The difficulties are further compounded by the fact that a pair of samples which match under Illuminant D_{65} and show little metamerism under Illuminant A may still be very metameric under another source, for example a fluorescent source. It is necessary therefore to view the samples under as many sources as possible to check the constancy of the match.

Another property of colours and colour vision, closely related to metamerism is that of colour constancy. The two terms have quite distinct meanings in colour physics: colour constancy is a property of a single sample, whereas metamerism always refers to a pair of samples. Colour constancy is a property of the human colour vision system which enables colours to be perceived more or less consistently, as the type of illumination under which they are viewed changes. As illustrated in Fig. 8.13 the amount of light emitted from daylight and tungsten sources across the spectrum is very different, so the amount of light reflected to the eye from a coloured object will be very different, when it is viewed under these different sources. However, the colour of the object remains fairly consistent to us, so a form of compensation process must be taking place in the human colour vision mechanism. This process occurs when viewing other colours also and occurs sub-consciously so that we are unaware it is happening.

Although a colour will appear more or less the same when viewed under different types of light source, its tristimulus values will not be the same. This is a feature of colour vision which is not addressed by the C.I.E. system, which was developed only to indicate whether two colours would appear the same, when viewed by the standard observer under a standard illuminant. The development of a colour appearance model, which is an extremely complex task, is one which is the subject of much current research.

The relationship between metamerism and colour constancy is demonstrated in Fig. 8.28. Because metamerism has such commercial importance it is useful to be able to quantify the degree of metamerism. A metameric index (M.I.) is used for this purpose, and although many formulae for indices have been developed over the years, they are based on one of two methods. The first method is to compute the colour-difference, ΔE, between the pair of samples as seen under a range of different light sources (see section 8.10.3). From the magnitudes of the ΔE values obtained, an idea can be obtained about the degree of metamerism. The second method is based on the differences in the reflectance curves at each wavelength, (ΔR_λ) between the two samples because the greater the difference, the more highly metameric the pair of samples is likely to be. The index is obtained by summing the differences across the spectrum, but since the reflectance curves cross over each other (at least three times), positive differences at some wavelengths may be cancelled by negative differences

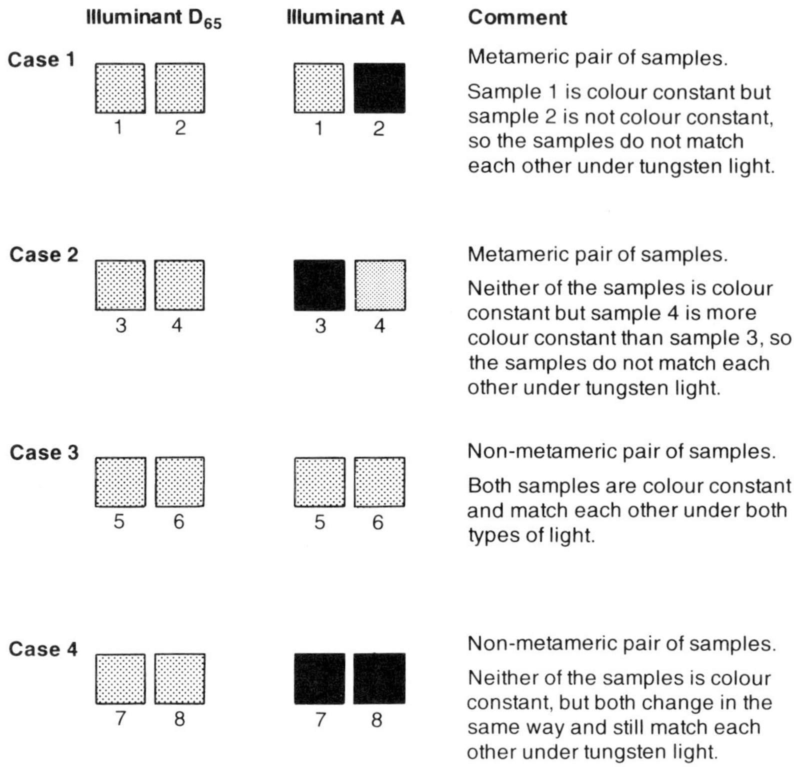

Fig. 8.28 Relationship between colour constancy and metamerism. In each case, the pair of samples match each other when viewed under Illuminant D_{65}, but not necessarily when viewed under a different type of light, such as tungsten.

at other wavelengths and it is necessary to square the differences before summing. The equation for this method is:

$$\text{M.I.} = \sum_{400}^{700} (\Delta R_\lambda)^2 \tag{8.12}$$

This simple formula ignores the fact that the eye is more sensitive to differences in reflectance in the middle region of the spectrum than at the two extremes (towards 400 and 700 nm). The general index of metamerism below addresses this aspect by weighting the differences at each wavelength according to the sensitivities of the red, green and blue cones.

$$\text{M.I.} = \left\{\sum \bar{x}_\lambda (\Delta R_\lambda)^2 + \sum \bar{y}_\lambda (\Delta R_\lambda)^2 + \sum \bar{z}_\lambda (\Delta R_\lambda)^2\right\}^{\frac{1}{2}} \tag{8.13}$$

8.10.2 *Computer colorant formulation*

The availability of spectrophotometers with dedicated microcomputers (see section 8.9) has led to the development of systems for the prediction of colorant recipes required to match standard shades. These systems have become widely used for the prediction of dye recipes to produce desired colours on textiles, and for pigment recipes for paints and printing inks.

The algorithms used to calculate the recipes require a mathematical relationship between the concentration of colorants in the substrate (e.g. textile fibre, paint, plastic, etc.) and the tristimulus values, X, Y, Z. Unfortunately there is no direct relationship and it is necessary to consider it as two separate relationships:

(a) relationship between tristimulus values (X,Y,Z) and spectral reflectance;
(b) relationship between spectral reflectance and concentration of colorant.

When combined, a relationship between X, Y, Z values and concentration exists. The first of the above relationships has been established by the C.I.E. (section 8.8.2), so once the reflectance values are known, the X, Y, Z values can be calculated, using Eqn 8.11.

The second of the above relationships is a more complex problem. For example, when increasing concentrations of dye are applied to a textile, the reflectance at each wavelength decreases. Gradually a limit is reached whereby increasing the concentration of dye still further does not reduce the reflectance any further. The graphs in Fig. 8.29(a) illustrate this behaviour. If one wavelength is selected and the reflectance at that wavelength plotted as a function of concentration, a curved relationship of the type shown in Fig. 8.29(b) is obtained. For the purpose of computer dye-recipe prediction, a linear relationship is required and this can only be obtained by plotting some mathematical function of reflectance against concentration. Many such functions have been developed, the simplest (and most widely used) of which is the Kulbelka–Munk equation. This equation relates the reflectance of a coloured substrate to the absorption (K) and scattering (S) coefficients of the colorants present. These coefficients are measures of how strongly a colorant absorbs and scatters light (see section 8.7.2). The Kubelka–Munk equation is:

$$\frac{K}{S} = \frac{(1-R)^2}{2R} \text{ or } R = 1 + (K/S) - \left[(K/S)^2 + 2(K/S)\right]^{\frac{1}{2}} \tag{8.14}$$

The K/S function of the coloured substrate is related to the specific absorption and scattering coefficients by the equation:

$$\frac{K}{S} = \frac{(1-R)^2}{2R} = \frac{K_s + c_1K_1 + \cdots c_nK_n}{S_s + c_1S_1 + \cdots c_nS_n} \tag{8.15}$$

where K_s and S_s are the absorption and scattering coefficients of the substrate.

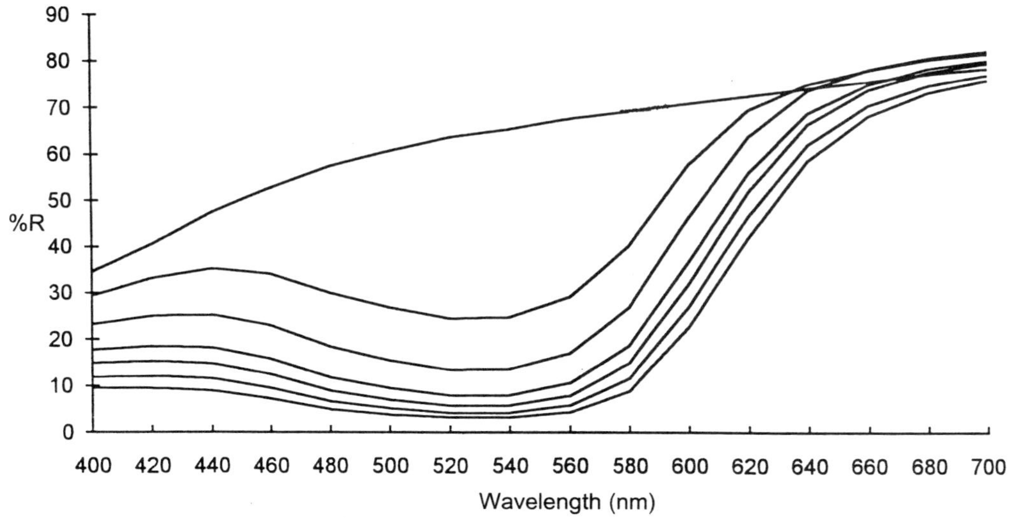

Fig. 8.29(a) Spectral reflectance curves of C.I. Acid Red 57 applied to wool fabric. 1, 0% (undyed wool); 2, 0.25% dye on weight of wool; 3, 0.50% dye on weight of wool; 4, 0.75% dye on weight of wool; 5, 1.00% dye on weight of wool; 6, 1.25% dye on weight of wool; 7, 1.50% dye on weight of wool.

K_1, K_n are the absorption coefficients of the colorants 1 ... n
S_1, S_n are the scattering coefficients of the colorants 1 ... n
c_1, c_n are the concentrations of the colorant present.

In the particular case of dyes on textile fibres, it can be assumed that the dyes are monomolecularly dispersed so that they are capable only of absorbing light, not scattering it. Therefore the only scattering of light

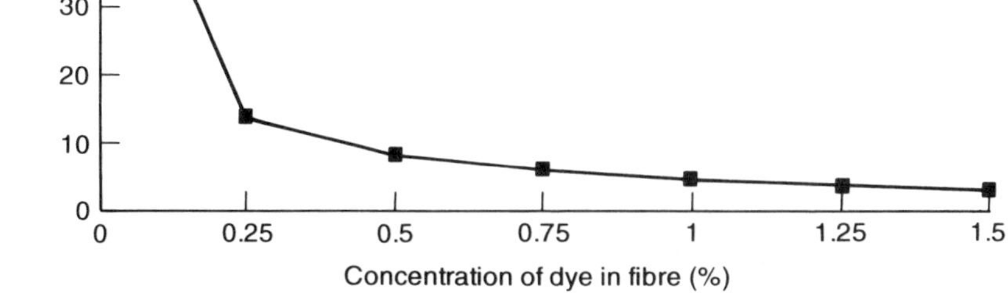

Fig. 8.29(b) Reflectance of C.I. Acid Red 57 at 520 nm plotted against concentration of dye in the fibre.

which occurs is that which is caused by the substrate (S_s) and Eqn 8.15 simplifies to:

$$\frac{(1-R)^2}{2R} = \left(\frac{K_s}{S_s}\right) + \frac{c_1 K_1}{S_s} + \cdots \frac{c_n K_n}{S_s} \quad (8.16)$$

or, in the case of a textile dyed with just one dye:

$$\frac{(1-R)^2}{2R} = \frac{(1-R_s)^2}{2R_s} + a.c \quad (8.17)$$

where R = reflectance of the dyed fibre
R_s = reflectance of the undyed fibre
a = absorption coefficient of the dye at a given wavelength
c = concentration of dye in the fibre

Eqn 8.17 has the form of the general linear equation $y = mx + c$, so a graph of $(1 - R)^2/2R$ plotted as a function of c should give a straight line, the slope of which is equal to the absorption coefficient, a. The values from the graph plotted in Fig. 8.28 are shown in this form in Fig. 8.30.

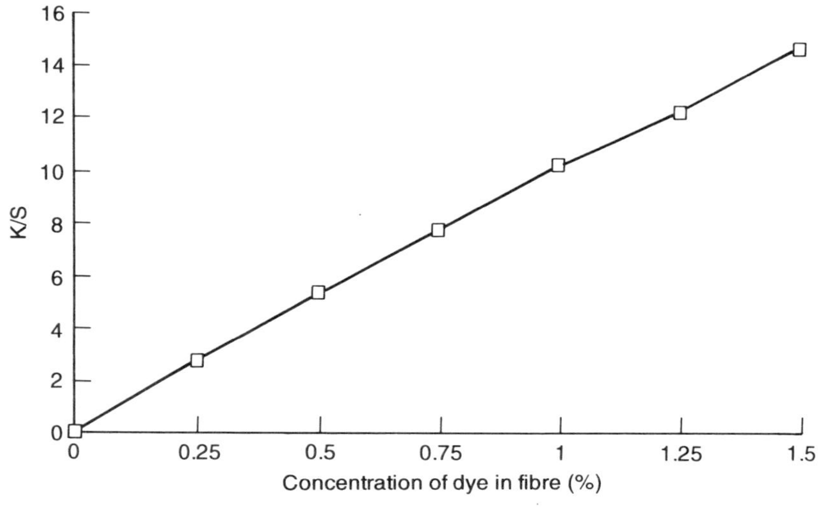

Fig. 8.30 Kubelka–Munk K/S function of reflectance (at 520 nm) plotted against concentration of dye applied to the fibre.

If a mixture of dyes is applied to a textile (as is usually the case in matching shades), the Kubelka–Munk equation can be written:

$$\frac{(1-R)^2}{2R} = \frac{(1-R_s)^2}{2R_s} + a_Y c_Y + a_R c_R + a_B c_B \quad (8.18)$$

where a_Y, a_R, a_B are the absorption coefficients of the dyes (e.g. yellow, red and blue)
c_Y, c_R, c_B are the concentration of the dyes

In the Eqns 8.14–8.18 the reflectance is used as a factor, that is, on a 0–1 scale, not a percentage scale. If the reflectance is used as a percentage the *K/S* function becomes $(100 - R)^2/200R$.

In a recipe prediction system, the reflectance values of the standard colour being matched are measured on the spectrophotometer and fed directly into the computer with which it is interfaced. The computer then calculates the tristimulus values of the standard. The combined C.I.E. and the Kubelka–Munk equations (Eqns 8.11 and 8.18 respectively) give a link between the concentration of dye on the fibre and the resulting X, Y, Z tristimulus values. Using the mathematical process of iteration, the amounts of the selected dyes which require to be applied to the textile material to produce the same colour as the standard are computed. The computation takes only a few seconds and it is possible for the computer to indicate also the degree of metamerism for the recipe. The computer can repeat the calculation with different combinations of colorants and print out, for each combination, the recipe, the degree of metamerism and the cost. The colourist can then select the recipe which gives the least metamerism and has the lowest cost.

Computer recipe prediction systems usually operate in the following sequence:

(1) measurement of spectral reflectance values of standard and substrate and reading of absorption and scattering coefficients of the colorants to be used;
(2) sequential selection of all possible (usually) three colorant combinations from the colorants selected;
(3) computation of the recipe, metameric index and cost, for each three colorant combination;
(4) output of all required information.

The dyer can select the recipe which best fits the requirements of cost, degree of metamerism, application conditions, fastness etc. The accuracy of the recipes predicted depends very much on the accuracy to which the absorption and scattering coefficients have determined. For example, the same dyeing method must be used to produce the calibration dyeings (from which the absorption coefficients are calculated) as will be used for dyeing the predicted recipes. In practice the calibration dyeings are normally prepared in the laboratory, which might not simulate exactly the bulk-scale dyeing process. To overcome this difficulty commercial recipe prediction software contains facilities to either correct off-shade batches or to re-calculate absorption coefficients on the basis of the colours actually obtained from the bulk scale production lots.

The computer system may be used to print the job ticket, giving such information as the weights of dyes and auxiliaries required for the amount of cloth, the volume of liquor required and procedure for dyeing. Also the computer can keep a record of dye stocks and estimate the amounts of dyes and auxiliaries required for the following week's production. Similar facilities are available for pigment based systems used for the production of paints and printing inks.

8.10.3 Colour difference assessment

In colour matching it is very difficult to produce an exact visual match to a standard colour and the colourist has to decide whether the colour of a batch is an acceptable match to the standard. Often it is not necessary that a perfect match be obtained, that is, where the tristimulus values of the batch are the same as those of the standard. Colorists aim to produce a 'good commercial match', but just what this is, in terms of how close the match is to the standard, depends on the end-use of the material. In the absence of any instrumental aids, the colorist has to decide whether a batch of material will be a good commercial match. Usually visual assessments are made, which although cheap and quick, are not necessarily very accurate. The human eye can detect very small differences in colour but it is difficult to quantify colour differences accurately. Judgements of colour differences are subject to statistical errors, so even experienced colorists will not agree with each other on occasions, when asked to assess the same pair of colours. Colorists may even change their decision when the same pair of colours is assessed on a later occasion.

If a batch of material, which is on the borderline of acceptability, is despatched to the customer, the customer may decide that it is unacceptable and return it with possible loss of confidence in the colorist. Alternatively, in order to avoid this difficult situation, the colorist may play safe and unnecessarily shade a batch to produce a closer match, which is time consuming and costly. Instrumental assessment can overcome these problems since modern instruments are able to measure tristimulus values accurately and reproducibly. A colour difference equation is then used, the function of which is to provide a single number which is proportional to the size of the visual difference between these two colours.

Most colour-difference equations are based on the principle of calculating the distance apart in the colour space between the standard and the batch, as illustrated in Fig. 8.31. If the colour space used is the three-dimensional X, Y, Z tristimulus space, the distance apart between the two colours (the colour difference, ΔE), is obtained by application of the Pythagoras theorem, which gives:

$$(\Delta E)^2 = (\Delta X)^2 + (\Delta Y)^2 + (\Delta Z)^2 \tag{8.19}$$

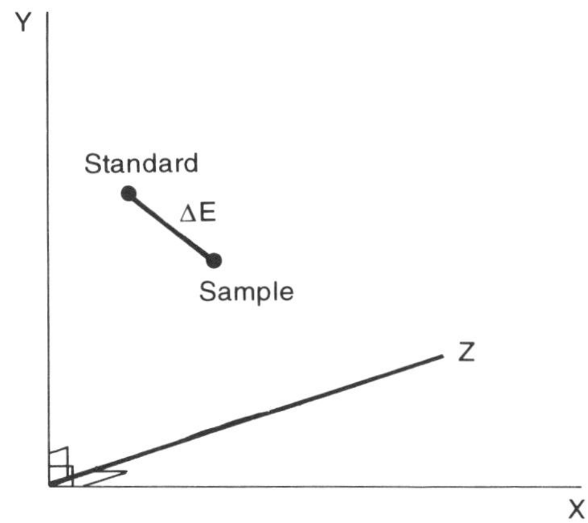

Fig. 8.31 Determination of colour difference.

This simple colour-difference equation is not satisfactory at all because it is found that pairs of colours with equal visual differences in colour do not have the same values for ΔE calculated using the equation. It is necessary to modify the X, Y, Z scales to produce a visually uniform colour space, which in practice, has proved a very difficult aim to achieve.

Consequently, many colour difference equations have been proposed over the years. In 1976 the C.I.E. recommended an equation called the CIELAB equation. This equation uses the LAB space, a visually uniform colour space derived from the XYZ space. The L^*, a^*, b^* values form the three perpendicular axes of this colour space and are obtained from the X, Y, Z values as follows:

$$L^* = 116\left(\frac{Y}{Y_n}\right)^{\frac{1}{3}} - 16$$

$$A^* = 500\left[\left(\frac{X}{X_n}\right)^{\frac{1}{3}} - \left(\frac{Y}{Y_n}\right)^{\frac{1}{3}}\right]$$

$$B^* = 200\left[\left(\frac{Y}{Y_n}\right)^{\frac{1}{3}} - \left(\frac{Z}{Z_n}\right)^{\frac{1}{3}}\right]$$

where X_n, Y_n, Z_n are the tristimulus values of the reference white under the source used. The colour difference, ΔE, between a sample and standard in this space is given by:

$$\Delta E = \left[(\Delta L^*)^2 + (\Delta a^*)^2 + (\Delta b^*)^2\right]^{\frac{1}{2}}$$

where $\Delta L^* = L^*_{\text{sample}} - L^*_{\text{standard, etc.}}$

In addition, the location of colours in this L^*, a^*, b^* space can be defined by their cylindrical co-ordinates, L^*, C^*, and h, where:

$$C^* = (a^2 + b^2)^{\frac{1}{2}}$$

and $$h = \arctan(b^*/a^*)$$

This space is illustrated in Figure 8.32.

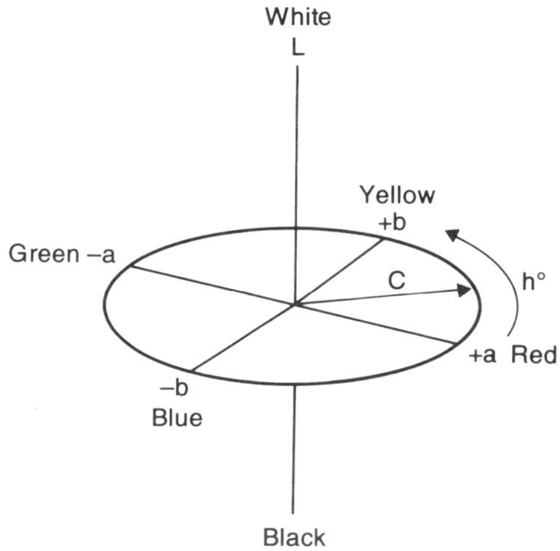

Fig. 8.32 CIELAB colour space.

The cylindrical co-ordinates have the advantage that they can be identified easily with the three attributes of colour vision, namely lightness, chroma and hue. A colour difference may then be split into its component lightness difference (ΔL^*), chroma difference (ΔC^*) and hue difference (ΔH^*) and the colour-difference equation written:

$$\Delta E = \left[(\Delta L^*)^2 + (\Delta C^*)^2 + (\Delta H^*)^2\right]^{\frac{1}{2}}$$

Individual tolerances can be set on each of these components of the overall colour difference (ΔE), which is called colour splitting. If the acceptability of a match is judged only on the overall ΔE value, it is termed 'single number shade-passing'.

The CIELAB equation does not perform consistently throughout the colour space and different tolerances have to be used according to the colour involved. Work is still being carried out at various research institutions to produce a better equation. The Society of Dyers and Colourists recommended another equation, called the CMC(*l*:*c*) equation, which is also based on the L^*, a^*, b^* space. This equation has the best overall performance of any published so far, and is suitable for single number shade passing. This equation became a British Standard (BS6923:1988) in 1988 and an ISO standard in 1996.

$$\Delta E = \left[\left(\frac{\Delta L^*}{l.S_L}\right)^2 + \left(\frac{\Delta C^*}{c.S_C}\right)^2 + \left(\frac{\Delta H^*}{S_H}\right)^2\right]^{\frac{1}{2}}$$

$$\text{where } S_L = \frac{0.040975 L_1^*}{(1 + 0.01765 L_1^*)}$$

unless $L_1^* < 16$ when $S_L = 0.511$

$$S_C = \left[\frac{0.0638 C_1^*}{(1 + 0.0131 C_1^*}\right] + 0.638$$

$$S_H = S_C(T.f + 1 - f)$$

$$f = \left\{\frac{(C_1^*)^4}{\left[(C_1^*)^4 + 1900\right]}\right\}^{\frac{1}{2}}$$

$$T = 0.36 + |0.4 \cos(h_1 + 35)|$$

unless h_1 is between 164° and 345°, when

$$T = 0.56 + |0.2 \cos(h_1 + 168)|$$

In this equation, L^*, C^* and h are determined using the CIELAB equation. L_1^*, C_1^* and h_1 refer to the standard. The values S_L, S_C and S_H are the tolerances of difference in the lightness, chroma and hue directions respectively and are calculated for the particular location of the colour standard in the L^*, a^*, b^* space.

The terms l and c are weighting values. When l and c are each unity, the equation is used to quantify the perceptibility of a colour difference. In this case, equal weighting is given to differences in lightness, chroma and hue. To predict the acceptability of a batch, it is recommended for the textile industry that $l = 2$ and $c = 1$. In this case the significance of the difference in lightness is halved, relative to the differences in chroma and hue. It has been recommended that other industries, such as leather, plastic, ceramic, etc. determine for themselves the most appropriate values for l and c.

A technical committee established by the C.I.E. in 1989 proposed a new formula which was subsequently recommended by the C.I.E. and is now known as the CIE94 formula. As with the CMC(l:c) formula, CIE94 is based on the CIELAB formula but uses less complex equations to compute the values of S_L, S_C and S_H. The formula enables the calculation of the visually perceived magnitude of the colour difference, ΔV, given by:

$$\Delta V = k_E^{-1} \Delta E_{94}^*$$

where k_E is an overall sensitivity factor, normally set to unity, and ΔE_{94}^* is given by

$$\Delta E^*_{94} = \left[\left(\frac{\Delta L^*}{k_L S_L}\right)^2 + \left(\frac{\Delta C^*}{k_C S_C}\right)^2 + \left(\frac{\Delta H^*}{k_H S_H}\right)^2\right]^{0.5}$$

The terms k_L, k_C and k_H are called parametric factors and are normally each set to unity. However in the case of textiles, correlation with visual assessments of differences is better if $k_L = 2$. The terms S_L, S_C and S_H are the lengths of the semi-axes of the ellipsoids (as in the CMC(*l*:*c*) formula), but in CIE94 they are computed using simpler equations:

$$S_L = 1, \quad S_C = 1 + 0.045C^* \quad S_H = 1 + 0.015C^*$$

where C^* is the value of the chroma of the standard. If neither of the samples of the pair can be easily identified as the standard, then the geometric mean of the chroma values is taken.

Whilst the computation of the values of S_L, S_C and S_H is obviously very different from that used in the CMC(*l*:*c*) formula, studies have shown the CIE94 formula to perform equally as well. However active research is still being conducted and it is likely that new equations for calculating their values will be recommended in future. It is also likely that recommendations for values of k_L, k_C and k_H other than unity will be made, as information on the effect of non-standard viewing conditions on visual assessments of colour differences becomes available.

8.10.4 Shade sorting

When many batches of material are produced as a match to a standard, it may be that the batches show small differences in colour from the standard. Rather than attempt to re-dye or shade each batch to produce a closer match to the standard, a cheaper alternative is to sort the batches into groups, so that all of the batches within any one group are an acceptable match to each other, though not necessarily to the standard.

Shade sorting can be carried out visually, but it is difficult to do consistently and is a monotonous task. It is also slow and expensive. A variety of numerical methods is available, in which the batches are sorted, on the basis of their L^*, A^*, B^* or L^*, C^*, h values, by computer. Typical of these methods are the '555' system, the truncated octahedron microvolume method and the clustering methods developed recently. The concept of the '555' system is illustrated in Fig. 8.33. Essentially an array of cuboids is defined in the colour space around the standard. The batches are sorted into the cuboids, according to the nature of their difference in colour (lightness, chroma and hue differences) from the standard. All of the batches sorted into the same cuboid are of an acceptable match to each other (and could be used together in a production run of garments), but not an acceptable match to the standard colour. As shown in Fig. 8.33, the cuboids are all coded by three digits, the first of which refers to lightness difference from

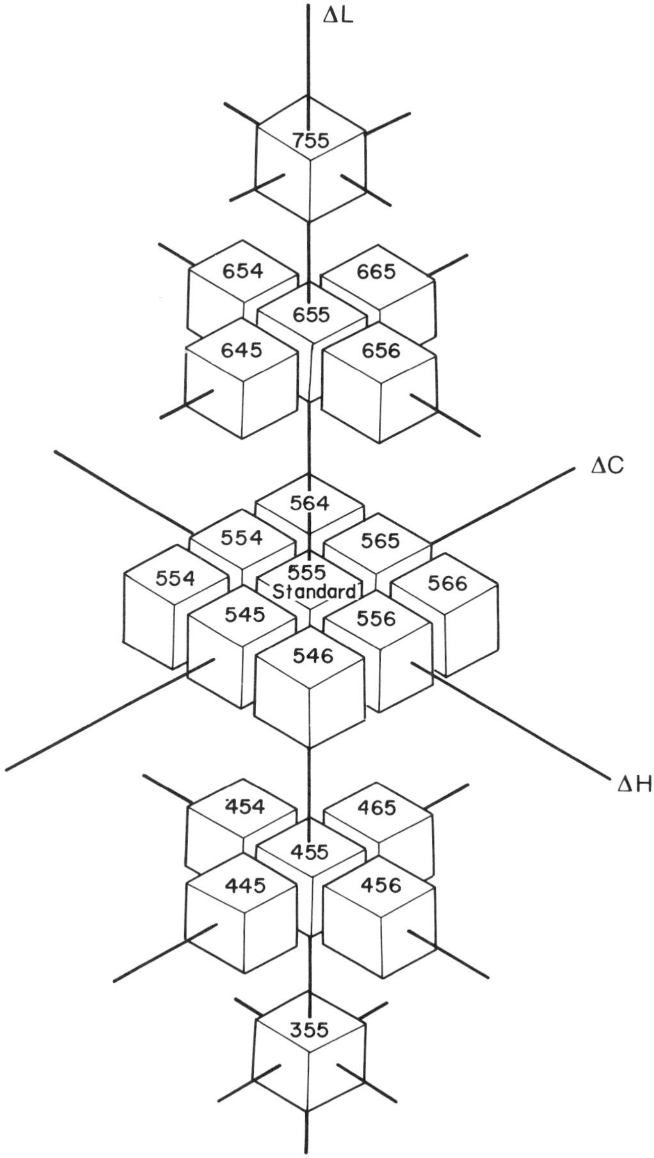

Fig. 8.33 The '555' system of shade sorting.

the standard, the second to chroma difference and the third to hue difference.

The cuboid shape is not the most efficient for sorting and it is found that batches are sorted into more groups than is really necessary. To improve the efficiency of sorting, the simple cuboid shape has been replaced by volumes with more than six sides, such as the twelve-sided rhombic dodecahedron and the fourteen-sided truncated octahedron shapes, which are available on some colour-measuring systems. More recently, a clustering method has been developed which sorts the batches into groups by pairing samples with their nearest neighbours in the colour space. The pairs are paired and so on, gradually building up the sizes of the clusters, until all the batches which are an acceptable match to each other have been included in the same cluster. This method is even more efficient than the truncated octahedron model, in terms of resulting in a minimum number of sorted groups.

8.10.5 Measurement of whiteness

Many articles are sold as whites, obvious examples being paint, paper and textiles. In many cases it is important that the quality of the white is the best that can be produced and the naming of some paints as 'brilliant white' or 'pure white' and even 'pure brilliant white' indicates how manufacturers wish to emphasise the quality of their product. It is also important that textiles, when sold as whites, are of as good quality white as possible, since any trace of yellowness is interpreted as poor quality of the articles. The measurement and specification of whites is a most important aspect for manufacturers, though it is an area of colour measurement which presents certain difficulties in carrying out accurately. The quality of a white is usually expressed by what is called an index of whiteness.

Whiteness is an attribute of colours of high luminous reflectance and low chroma. Many equations to quantify whiteness have been proposed since the C.I.E. system was agreed in 1931. The higher the lightness the whiter a sample will look, but chromaticity also has an effect. For example, if two whites have the same lightness but one appears slightly bluer than the other, then most people will consider it to appear a better white. Conversely, the presence of a yellow hue has quite the opposite visual impact. It is effects such as these which have to be accounted for by an index of whiteness. A number of factors can affect the accuracy of visual assessments:

(a) response of the observer's eyes to hue differences and any preference of the observer to the nature of the hue difference;
(b) intensity and spectral energy distribution of the incident light;
(c) reflectance characteristics of the substrate, such as opacity, matte (or lustre), total reflectance and colour of the substrate;
(d) geometry of viewing and the surrounding colour.

The use of commercially available viewing cabinets helps to standardise the viewing conditions, in particular the light source which should simulate average daylight, and the surrounding colour, which should be neutral grey. Such cabinets are always used by professional colorists in industry when making judgements about colours.

The yellowness in white textiles is due to the slight absorption of light in the blue region of the spectrum, as shown in Fig. 8.34(a). Before the advent of fluorescent whitening agents (fwa's), whiteness was enhanced by bleaching and blueing the textiles. Application of a small amount of a blue dye caused absorption of the longer wavelengths, with the result that the reflectance curve was flatter, so the textile had a more neutral appearance (Fig. 8.34(b)). However, although the yellowness was removed the 'white' was not very bright.

Fluorescent whitening agents absorb radiation in the ultraviolet part of the spectrum, (see section 8.2) just below the visible region, then emit this energy in the blue part of the visible spectrum. They are essentially

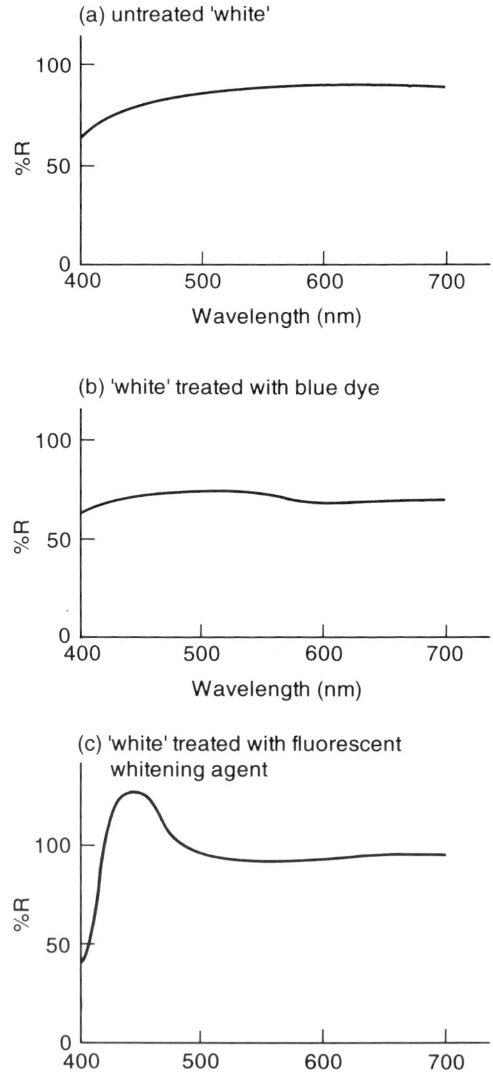

Fig. 8.34 Reflectance curves with white textiles. (a) Untreated 'white'; (b) 'white' treated with blue dye; (c) 'white' treated with fluorescent whitening agent.

colourless themselves because they absorb radiation outside the visible part of the spectrum. Their effect is to increase the amount of light radiated from the textile in the blue part of the spectrum. Since the amount of light radiated from the textile is now higher, the material appears whiter and brighter (Fig. 8.34(c)). In addition, the extra emission in the blue part of the spectrum gives the white a bluish appearance. A wide variety of fwa's is available, which emit their fluorescence in the violet, blue or even green regions of the spectrum. Consequently it is possible to produce enhanced whites which have reddish-blue, blue or greenish-blue tints to them.

Whites treated with fwa's can be evaluated visually by comparison of one white textile with another. The perception of whiteness and the exact nature of the white preferred by individuals may vary considerably, so any method of assessment must be related to the preferred white. Despite these complications, it is possible to measure whiteness instrumentally with reference to a preferred hue of white established from the opinions of a large number of observers. The light source in the instrument used is particularly important and must be such that it emits sufficient ultraviolet radiation to excite the fwa. The C.I.E. recommendation is that the source should simulate Illuminant D_{65}, but instruments achieve this requirement to different extents. Consequently the reflectance values of whites treated with fwa's may be different when measured on different instruments, even on different instruments from the same manufacturer.

The most appropriate white with which to compare the treated textile material, is the perfect white, that is, a surface which reflects 100% of all the incident wavelengths. However the use of fwa's can produce reflectances of greater than 100%, especially in the region 420–460 nm. Many attempts have been made to correlate instrumental measurements with visual assessments of whiteness and many whiteness scales have been proposed. To promote uniformity of practice, the C.I.E. in 1982 recommended the following index of whiteness:

$$W = Y + 800(x_n - x) + 1700\ (y_n - y)$$

where x, y, Y are the colorimetric values for the textile under D_{65} and x_n, y_n are the chromaticity co-ordinates of the light source (0.3137, 0.3309 respectively for Illuminant D_{65}).

With this formula, the perfect reflecting diffuser (perfect white) has a whiteness index of 100. However materials treated with an fwa may give values as high as 150. When increasing amounts of fwa are applied to a textile, its appearance becomes whiter but its lightness does not increase much. However the chroma changes and the whiteness index is therefore accompanied by a tint factor:

$$T = 900\ (x_n - x) - 650\ (y_n - y)$$

If T is near zero, the sample has a dominant wavelength of 466 nm. The more positive the value of T the greener the hue of the white; the more negative, the redder the hue of the white. For the perfect reflecting diffuser, $T = 0$.

The formulae for W and T are applicable only to samples whose values of W or T lie within the following limits:

$$40 < W < 5Y - 280 \text{ and } -3 < T < +3$$

Bibliography

Organic Spectroscopy, 3rd Edn, W. Kemp, MacMillan Education Ltd, London (1991).

Practical Absorption Spectrometry, A. Knowles and C. Burgess (eds), UV Spectrometry Group, Chapman & Hall, London (1984).

Colour Physics for Industry, 2nd Edn, R. McDonald (ed.), The Society of Dyers and Colourists, Bradford (1997).

The Colour Science of Dyes and Pigments, K. McLaren, Adam Hilger, Bristol (1983).

Measuring Colour, 2nd Edn, R.W.G Hunt, Ellis Horwood Limited, Chichester (1987).

The Physics and Chemistry of Color, K. Nassau, John Wiley, New York (1983).

Colour: Its Measurement, Computation and Application, G.J. Chamberlin and D.G. Chamberlin, Heyden & Son Ltd, London (1980).

Optical Radiation Measurements, Vol 2: Colour Measurement, F. Grum and C. Bartleson (eds), Academic Press, London (1980).

Human Colour Vision, R.M. Boynton, Holt, Reinhart & Winston, London (1979).

Colour: Art and Science, T. Lamb and J. Bourriau (eds), Cambridge University Press, Cambridge (1995).

Instrumental Colour Formulation, A Practical Guide, J. Park, The Society of Dyers and Colourists, Bradford (1993).

Chapter 9
Machinery for Dyeing Textiles

9.1 Introduction

A wide variety of machines are available for the coloration of textile materials, a consequence of the fact that textiles can be dyed at different stages of manufacture, either batchwise, semi-continuously or continuously. Additionally, the nature of the textile fibre influences the type of machine employed, an example being the need for high-pressure equipment for the dyeing of polyester at temperatures above 100°C. The dyeing process can be carried out at the loose fibre, tow or slubbing (parallel strands of untwisted fibres), yarn, fabric or garment stages. Additionally, synthetic fibres can be mass pigmented, whereby the pigment is added to the molten polymer immediately prior to its extrusion. The pigments used for this process must be stable to the high temperatures employed (about 150°C for polyolefins, 230°C for polyester) in extrusion.

Somewhat uniquely, acrylic fibres, which are wet spun rather than melt-spun, can be dyed by passing the fibres directly into a dyebath without letting them dry, so they are still in a swollen and in a highly receptive state for the diffusion of dye molecules. Dyeing at this stage of manufacture is termed 'producer dyeing'. The advantage of colouring synthetic fibres in this way is that the products have very high wet fastness properties because the dyes or pigments are locked into the polymer matrix and the molecules are insoluble in water anyway. The disadvantage is that the process does not easily facilitate the colouring of small amounts and large stocks of a fixed colour range may not sell if the fashion demands change.

The stages in the manufacturing sequence at which fibres may be dyed are shown in Fig. 9.1. The decision at which stage of manufacture to dye depends on various factors, probably the most important of which are cost and fashion considerations. In recent years there has been a noticeable shift towards late stage dyeing, to enable manufacturers to avoid over-production of unpopular colours and to respond quickly to repeat orders for those which are more popular. In particular, garment dyeing for this reason has increased significantly. Dyeing machines are available for dyeing loose fibres, yarns, woven or knitted piece goods, or partly or fully made up garments. Developments are continually taking place in textile dyeing machinery, as the coloration industry responds to market and environ-

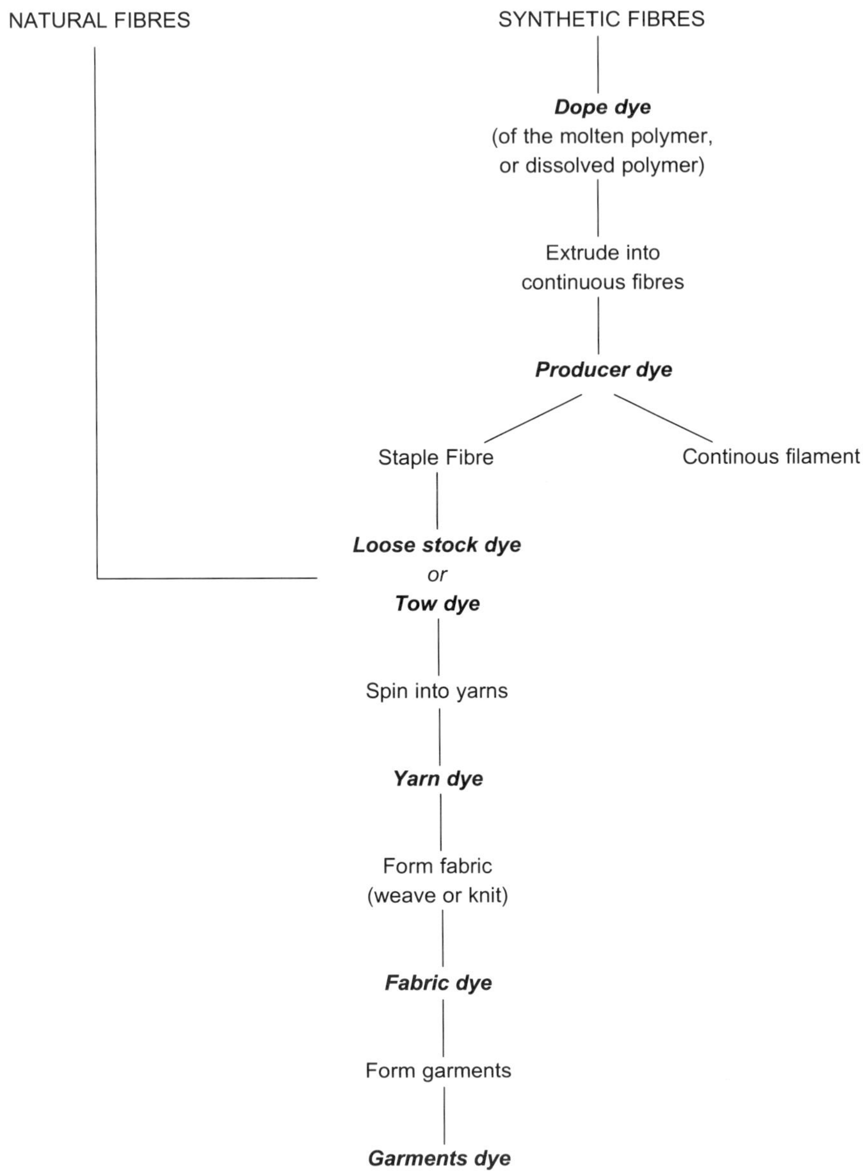

Fig. 9.1 Stages at which textiles can be dyed.

mental pressures. Environmental considerations have led machinery manufacturers to incorporate design features which minimise water consumption and have a greater degree of process control to increase batch-to-batch reproducibility and therefore production efficiency in the dyehouse.

The use of very low liquor-to-goods ratios to minimise water consumption has itself created additional difficulties for the dyer, because of the need to dissolve the dyes and auxiliaries in the small volumes of available water. It has been necessary to develop new dye application sequences, whereby the dye and the auxiliaries are not dissolved all together in the water, but instead are metered in gradually. An example of this type of approach is that devised for the application of reactive dyes to cellulosic fibres at low liquor-to-goods ratio. Here the fabric is run initially in a solution of electrolyte (either sodium chloride or the more soluble sodium sulphate) and the dye gradually added. Once the neutral exhaustion stage is complete the alkali is then metered gradually into the dyeing machine for the fixation stage to take place.

9.2 Basic principles

The fundamental aim of the dyeing process is to transfer the dye molecules from the dye liquor to the fibre in a uniform and efficient manner. The rate of uptake of dye by the fibre is significantly increased by the movement of the dye liquor relative to the fibre. Dyeing machines are designed to accomplish this transfer in one of three ways:

(a) the fibres are moved through the stationary dye liquor;
(b) the fibres are held stationary and the liquor pumped through them;
(c) the fibres and the liquor are moved together.

When a liquid moves past any solid surface, there is always a layer of stationary liquid adjacent to the surface (Fig. 9.2). This stationary layer, the hydrodynamic boundary layer, is significant in dyeing because dye molecules must diffuse across it in order to reach the fibre surface. The hydrodynamic boundary layer is defined as the distance from the fibre surface at which the velocity of the liquid has reached 99% of the main-stream velocity. In a similar manner a diffusional boundary layer is defined, this being the thickness of the layer in which the dye concentration changes from that at the fibre surface to 99% of that in the bulk liquor. It is an important concept because diffusion is a slow process, its rate being influenced by temperature, the size of the dye molecule (small molecules diffuse faster than large ones) and the concentration gradient. The concentration gradient is the rate of change of dye concentration with distance from the fibre surface. The greater the difference in concentration between dye solution at the fibre surface and that in the bulk solution, the greater the rate of transfer of dye. On reaching the fibre surface the dye molecules will be adsorbed into the surface and then diffuse into the interior of the fibre.

Three distinct stages in the dyeing process can be identified:

(a) transfer of dye from the bulk solution to the fibre surface;
(b) adsorption of dye on to the fibre surface;
(c) diffusion of dye from the surface into the fibre.

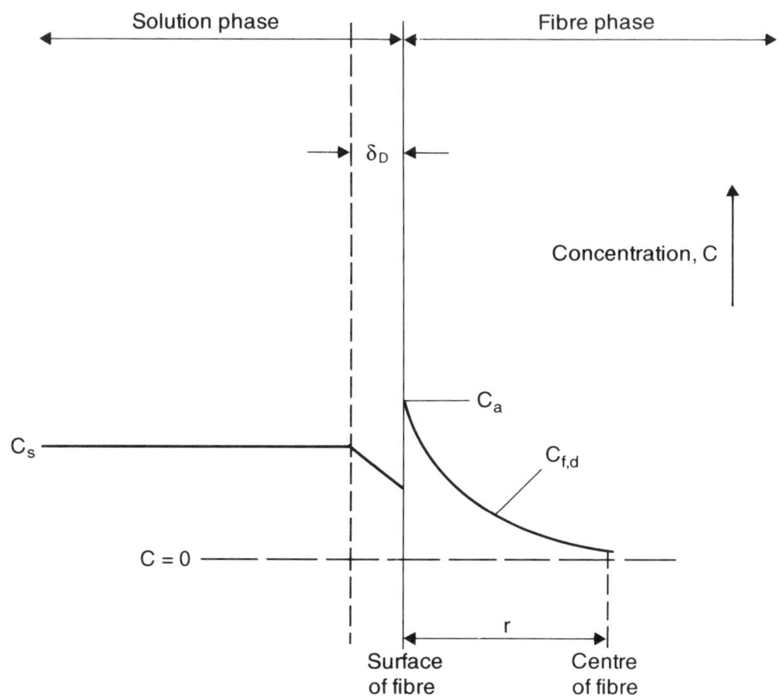

Fig. 9.2 Diffusion across the diffusional boundary layer into a fibre. δ_D = thickness of diffusional boundary layer; r = radius of fibre; C_S = concentration of dye in the bulk solution; C_a = concentration of dye adsorbed on the fibre surface; $C_{f,d}$ = concentration of dye in the fibre at distance d from the fibre surface.

Any one of these stages can be rate controlling but in practice the second step is instantaneous compared with steps one and three. The influence of the third step depends on the properties of the fibre and of the dyes being applied and can be influenced to some extent by the addition of auxiliary chemicals to the dyebath, such as carriers, retarders, or electrolytes.

It is the first of the stages which is most readily influenced by the design of the dyeing machine since in order to minimise the thickness of the diffusional boundary layer, an effective flow of liquor round the fibres is required. It is also important that the velocity of the fluid flow is consistent throughout the entire mass of fibres, in order that the rate of arrival of dye molecules occurs uniformly, if level dyeings are to be achieved.

Dyeing machines must be constructed of materials which will withstand the chemical and physical conditions used, not just in dyeing, but in pretreatments also, such as desizing, scouring and bleaching processes. For example, copper or iron components exposed to hydrogen peroxide bleaching liquors will cause catalytic decomposition of the peroxide, yielding free radicals which will attack and degrade the fibres. These metals may also cause colour changes of chelatable dyes. Stainless steel is the preferred material because it is inert to most process conditions and is easily cleaned.

The arrangement for heating the dye liquor must be such as to provide a uniform temperature distribution through the machine, though this is difficult to achieve. High pressure steam is the usual heating medium, which is passed through closed coiled pipes in the bottom of the machine. The coiled pipes give a large surface area and therefore effective heat exchange. Also there is no possibility of any contamination of the dye liquor by heavy metal ions from the boiler or pipework. It is necessary to incorporate a trap into the steam line so that any condensate can be removed. The coils can also be used for cooling the dye liquor by switching appropriate valves from the steam inlet from the boiler to cold water inlet. However a sudden change from heating to cooling can impose severe mechanical strain on the pipes and joints, leading to the possibility of cracks developing. Some machines therefore have separate heating and cooling coils. Dye liquors can be heated by blowing the steam directly into the dye liquor from perforated coils in the vessel, an arrangement which enables dye liquors to be heated very quickly. However, there is a possibility of contamination of the dye liquor by heavy metal ions and additionally, as the steam condenses in the dye liquor, the volume of liquor increases, so the liquor:goods ratio increases.

9.3 Dyeing in the loose fibre form

The main advantages of dyeing loose fibres ('loose stock') are the ease of circulation of the dye liquor through the fibres and the fact that any unlevelness in dyeing can be randomised during subsequent carding and spinning operations. The method has traditionally been used more for dyeing wool than for cotton or synthetic fibres, especially the application of acid milling dyes which have poor migration properties. However, even in the case of wool, the trend has been away from loose stock dyeing in recent years.

The traditional machines for loose stock dyeing are the conical pan and pear-shaped machines, in which fibres are loosely packed and the dye liquor pumped through them at relatively low pressures. In the conical pan machines the fibres are held in an inner container with a perforated bottom and covered at the top with a perforated lid (Fig. 9.3). Liquor is pumped upwards through the fibre mass, which has the effect of pushing the fibres against the conical walls of the inner vessel. The liquor flows out through the top plate and falls back to the base of the machine and then to the pump. The dye liquor is heated by steam coils in the base of the outer vessel.

The pear-shaped dyeing machine similarly contains the fibre mass between a perforated bottom and top plate. An impeller, at the side of the fibre container and partitioned from it, circulates the dye liquor from top to bottom through the fibres. In each type of machine the direction of flow of the liquor can be reversed.

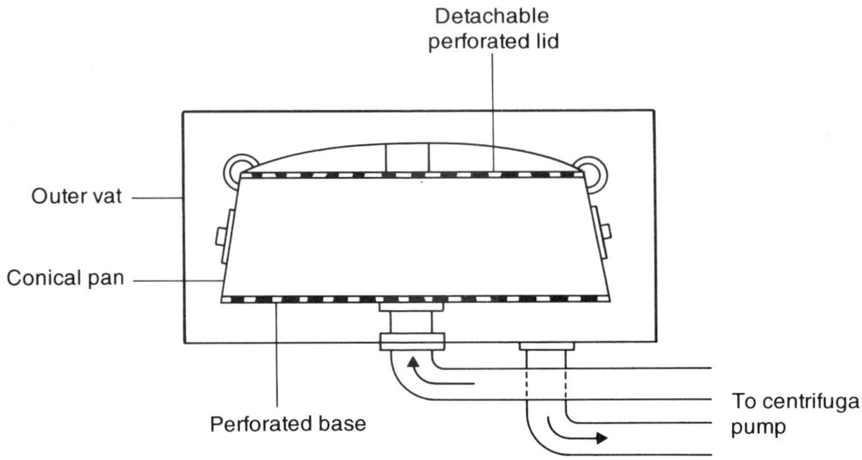

Fig. 9.3 Conical pan dyeing machine.

Another type of machine is the radial flow type, in which the fibres are held in a cage of approximately 150 cm diameter, with perforated sides and a central perforated column. The fibres are compressed into the cage by a top plate through which the dye liquor cannot flow. The liquor flows from the centre to the outside of the cage through the fibres. Because of the greater package density the productivity of this type of machine is greater than those of the pear-shaped or conical pan types. However, especially in the case of wool, the fibre damage caused is greater, since higher liquor flow rates are required causing the fibres to become compressed and physically 'set' into a distorted shape. This problem has led to the development of the WRONZ Soft Flo System in which the liquor flow is constantly monitored and maintained at a rate just sufficient to achieve level dyeing with minimal fibre damage.

9.4 Yarn dyeing

Yarns may be dyed in either hank form or as a package. Hank dyeing has always been carried out on yarns which are bulky in nature, whereas package dyeing is used for leaner yarns. For example, hank dyeing is used for hand-knitting yarns (wool, nylon, acrylic, etc.), carpet yarns (wool, wool/nylon, high-bulk acrylic) and high-bulk acrylic yarns for machine knitting. Package dyeing is used for weaving yarns (wool, cellulosic, acrylic, polyester), sewing threads (polyester or polyester/cotton core spun yarns) and regular acrylic yarns for machine knitting. However these are just general examples and the situation now is by no means so clear cut, for example high-bulk acrylic yarns can now be package dyed quite successfully.

9.4.1 Hank dyeing

This is a very long established method of dyeing yarn and, in spite of the advantages of package dyeing and the advances made in package dyeing machinery in recent years, is still an important method. Hank dyeing machines are relatively simple in construction, easily controlled and reliable. It is also considered that hank dyeing gives a product better bulk and handle (or 'softness to the touch') than can be achieved by package dyeing, for certain types of yarn.

With appropriate automation and accurate control of the dyeing conditions, a high degree of reproducibility is possible. The productivity of these machines is therefore good, although the throughput, in terms of the actual amount of yarn which can be dyed in one batch, is less than is possible with package dyeing machines.

The most traditional type of hank dyeing machine is the Hussong type (Fig. 9.4) in which the hanks are suspended from horizontal stainless steel 'sticks' in the dye liquor. The dye liquor is heated by either open or closed steam pipes, located underneath a false perforated floor in the machine. The liquor usually flows from the bottom to the top of the machines though the direction can be reversed if necessary. The upward flow of dye liquor tends to lift the hanks off the sticks, which prevents a resist on the yarn where the dye is unable to penetrate due to contact with the sticks. However the hanks tend to pack together at the top of the machine, which can impede the uniform flow of liquor through the fibres. A solid lid is clamped over the top of the machine during dyeing to prevent the hanks from floating.

To prevent the movement of the hanks, two-stick machines have been developed in which the hanks are supported both at the top and at the

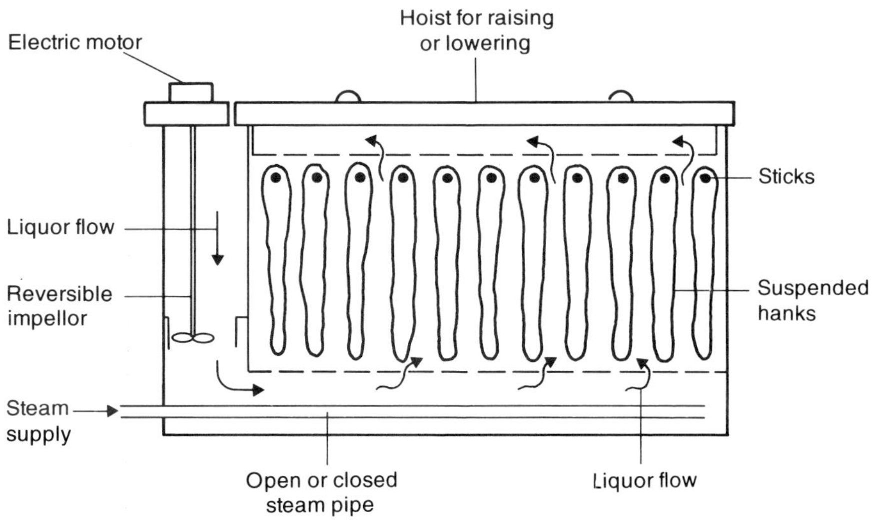

Fig. 9.4 Hussong type hank dyeing machine.

bottom by sticks. It is necessary in such machines to space the sticks at such a distance apart that the hanks are not stretched, which may cause resist marks. Also allowance must be made for the shrinkage of some yarns, such as high-bulk yarns, which can shrink by up to 25% during dyeing. The sticks can be adjusted to a certain extent but normally the length of the hanks produced in reeling is selected to suit the machine. Usually the hanks weigh up to about 0.5 kg and tie bands are fastened round them to prevent them from becoming tangled with each other in the machine. Recent developments have led to the dyeing of 'jumbo' hanks which, in the case of carpet yarns can contain up to about 10 kg of yarn. Such large hanks are sometimes wrapped in a stockinette of heat-set polyester to prevent tangling during dyeing.

The main problems with the Hussong type of machine arise through stick marking and the difficulty of achieving a uniform flow of dye liquor through the fibres of the hanks. In the modern machines the sticks are usually V-shaped and are perforated to reduce stick marking. It is usual practice to turn the hanks manually on the sticks so that different sections of the hank lie on the stick during the process. Rotating-stick machines have been developed which cause the hanks to rotate over the sticks during dyeing but these are little used now.

An alternative approach to minimise the problem of stick marking has been the development of a machine in which the dye liquor is circulated through and over the tops of the hanks and then down through them. A regulated proportion of the dye liquor is directed through the top sticks, the flow design causing the hanks to be lifted slightly off the sticks. The liquor flow through the hanks is gentle, minimising fibre damage and entanglement, but most importantly, it is uniform, so that temperature variation through the hanks is minimised and dyeings of better levelness are possible. The machines are usually the two-stick type, the sticks being easily adjusted to cater for hanks of different length, but some single stick machines are available. The machines can be pressurised for dyeing at temperatures up to 110°C.

Another development has been that of circular, multi-purpose machines, capable of dyeing loose stock or yarn in various forms, if necessary at higher temperature. For hank dyeing a circular frame with the sticks arranged in concentric circles on it is employed. However, loading the frames is a labour intensive process and the throughput of these machines is less than with the rectangular types. Given the higher capital cost of these machines, their success has not been very great.

Hank dyeing has retained its popularity because of the general simplicity and modest capital cost of the equipment. The method also gives reproducible results and high quality yarn. However the whole process is labour intensive because of the need to remove the yarn from the spinning machine and reel it into hanks. After dyeing, back-winding of the hanks on to cones is necessary and both of these processes can be costly because of the significant quantities of yarn which can be wasted. The actual loading and unloading of the hanks from the dyeing frames requires considerable care and thus takes time. However, by loading a spare set of frames with hanks, whilst a batch is being dyed, a quick change over can be made and the 'down-time' of the machine can be reduced. In addition, automated handling systems have been developed to address these handling problems.

9.4.2 Package dyeing

Prior to package dyeing, the yarn must be wound on to a support through which the liquor can flow. A variety of supports are available, either metal or plastic, cylindrical or cone shaped. Plastic supports are cheaper but are less robust and have a short life. Also they are prone to staining by some dyes, especially when heavy depths of shade are being applied and the dye liquors are fairly concentrated. This may cause contamination of dye liquors when they are used the next time in a dyeing operation.

The main types of packages used are cheeses, cones or cakes. Cheeses, which are the most commonly used type, are parallel sided and formed by winding the yarn on to cylindrically shaped formers, which can be either rigid or compressible. The rigid types are usually made of plastic. Formers are usually made of stainless steel for longevity, though some rigid types are made of plastic. The compressible types, which are usually made of stainless steel, allow the cheese to become compressed in the dyeing machine without any significant increase in the diameter of the cheese.

Cones are used for yarns which are difficult to unwind (back-wind) after dyeing, such as staple fibre yarns or yarns which swell when wet. The conical shape of the former allows the yarn to be moved more easily. However, cones have the disadvantages that they cannot be compressed in the dyeing machine and the flow of dye liquor through them tends not to be as uniform as with the parallel sided cheeses.

Cakes are formed in a similar way to cheeses, except that after winding the former is removed. The inner yarns, which are no longer supported, collapse into the void in the centre of the package and the whole structure becomes more relaxed. High-bulk acrylic yarns are often dyed in cake form, instead of by hank dyeing.

Packages are formed by winding the yarn at a constant speed on to the former, from a guide which traverses across the width of the rotating former at a fixed rate (Fig. 9.5). As yarn builds up on the former, the diameter of the package increases so that during one traverse of the guide, the number of wraps of yarn round the package will decrease. A stage is reached when the number of wraps round the package during a traverse of the guide actually coincides with the traverse repeat and the yarn is laid down almost in parallel in the layers. At this stage of winding a dense, close-packed section forms in the package, which will hinder the flow of dye liquor through it. Winding machines have the facility to adjust the rate of traverse of the yarn guide or the rate of rotation of the package to avoid this happening.

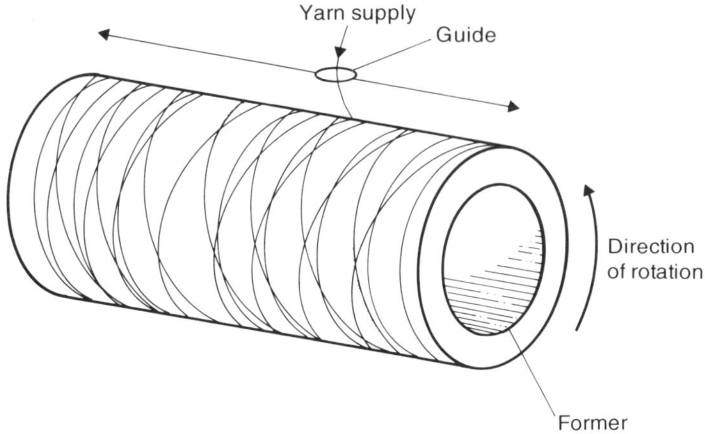

Fig. 9.5 Winding yarn packages.

Another feature of the winding process is that when the guide reaches the end of its travel and reverses direction back across the package, it is momentarily stationary. During this time the yarn still feeds through but because the guide is not moving, the yarn builds up parallel to the edge of the package. As a consequence, an excess of yarn can build up at each edge of the package. This feature is worst in the early stages of winding and with formers of small diameter. The greater the circumference of the package the more room there is available for this extra yarn at the edges. In modern precision winding this problem is overcome by controlling the extent of the traverse of yarn feed across the package.

It is essential that careful attention is paid to the winding of the yarn packages, otherwise both levelness and yarn quality will be poor. The packages must be wound in a uniform manner and to a suitable density, since channelling of the dye liquor through the package will occur if the package distorts under the pressure of the fluid flow. High density parallel sided packages are usually preferred because of their stability and superior fluid flow characteristics.

Developments in spinning machines have enabled the preparation of dyeing packages directly from the spinning frame, so that a separate winding operation is unnecessary. Staple yarns for example, are often delivered to the dyer on perforated plastic cones ready for dyeing. However this does mean that the dyer has no control over the quality of the winding of the packages in terms of their uniformity and density. For this reason many commission dyehouses prefer to rewind the packages prior to dyeing so that they can be sure of the quality. Also the use of cones as supports is not ideal for the dyer because the payloads possible are less than with cylindrical packages and expensive spacers between the cones are required in the dyeing machines.

One method of minimising channelling due to any variations in package density is to press pack the packages in the dyeing machine. About 8–10 packages are loaded by hand on to the dyeing spindles and then more packages added on top by means of a mechanical press, giving a compression of about 25–40%. In this way the packages are forced together to form a smooth vertical column of yarn in the machine, which greatly facilitates level dyeing.

Package dyeing machines may be open, in which case they are restricted to operating at atmospheric pressure, or enclosed, which allows dyeing processes to be carried out at high temperature (that is, temperatures above 100°C) and pressure. With open machines, the circulation of dye liquor from the inside of the packages to the outside is only possible because pumps are less efficient in the reverse mode where the action is by suction. Enclosed machines are preferred because the direction of the liquor flow can be reversed effectively. This facility is necessary because the surface area of the inside of the cheese is very much less than that of the outside. A given volume of dye liquor will therefore present more dye per unit area of cheese at the inner surface than the outer one, leading to unlevelness of dyeing. By alternating the direction of flow at regular intervals, usually about every three minutes, the potential for unlevelness due to this geometric property can be minimised.

In the machine the packages are held on hollow, perforated spindles, fixed vertically in a concentric arrangement to a circular base. The complete assembly is then hoisted into the cylindrically shaped dyeing vessel and seated carefully so that it aligns correctly with the outlet from the pump. An end-cap is screwed on to the end of each spindle to maintain the compression and to prevent dye liquor from spurting out from the top. The liquor therefore has to pass from the spindles through into the packages, or vice-versa. More recently, machines in which the spindles are mounted horizontally in the machine have been developed.

Package dyeing machines also contain an expansion tank, to take up excess dye liquor which expands as it is heated. The expansion tank is also used for making additions to the dye liquor during dyeing, for example the metering of alkali in the application of reactive dyes to cotton, after the initial neutral dyeing stage.

9.5 Dyeing of woven and knitted fabrics

A wide range of machine types are available for dyeing fabrics, the actual choice being governed by the nature of the fibre (e.g. wool, cotton, polyester, etc.) and the construction of the fabric. Also the quantity of cloth to be dyed will influence whether a batch or continuous process is employed..

9.5.1 Batch dyeing machines

The main types of machines in this category are winches, jigs, beam and jet dyeing machines, the latter being the most recently developed type.

Winch dyeing machines

This is a long established type of dyeing machine and is comparatively popular because of its low capital cost, simplicity of construction and operation. The machines basically comprise a deep vessel containing the dye liquor, above which are mounted two rollers (Fig. 9.6). The length of fabric is fed into the machine over the two rollers and the two ends stitched together so that a continuous 'rope' is formed. Depending on the width of the machine, many such ropes of fabric can be run in parallel over the rollers, each one kept separate from the next by a peg rail. The main roller (the winch reel) is located at the back of the machine and is motor driven to pull the ropes of fabric through the liquor. The smaller front roller (the fly roller) is free-wheeling and acts as a guide.

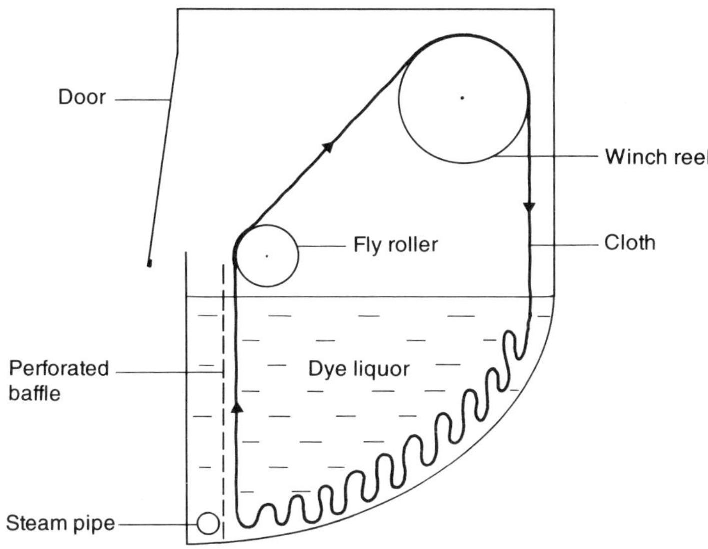

Fig. 9.6 Cross-section of winch dyeing machine (deep draught type).

At any one time the bulk of the fabric resides almost stationary, bunched in the dye liquor in the bottom of the vessel, though the liquor is constantly agitated due to the pulling of fabric out of the liquor and returning back to the liquor after passing over the rollers. In winch dyeing machines the steam for heating is usually blown directly into the liquor for rapid heating, which provides additional turbulence of the liquor.

Vessels of different cross-sectional shapes are available. For heavier woollen or cotton fabrics, a deep trough with a sloping back is preferred since the movement of piled-up cloth from the back to the front is then assisted. For lightweight fabrics, or fabrics made of thermoplastic fibres which may form creases as they pile up at the back of the machine, shallow draft winches are better. Generally, the mechanical action of the winch on the cloth is such as to develop fabrics with a thicker and fuller handle.

It is more difficult to maintain a uniform temperature in shallow vessels because the steam supply pipes are located at the front, separated from the fabric by a perforated baffle to prevent direct contact. The liquor at the back of the vessel is therefore not heated directly. The perforated baffle across the front of the machine enables the dye solution and solutions of auxiliary chemicals to be added gradually and uniformly across the full width of the machine. The winch reel is an important component of the machine, since it is responsible for the movement of the fabric rope. The reel itself is motor driven, but it is the downward force due to the weight of the wet rope which provides enough friction between the rope and the reel so that the rope moves over the reel. If any slippage occurs, chafing of the fabric surface will occur and spoil it.

Deep draught machines usually have circular, or only very slightly elliptical reels, which pull the fabric rope steadily out of the liquor, over the fly roller to the reel. Thereafter, the rope falls vertically back into the liquor with very little pleating action. Shallow draught machines have winch reels which are markedly elliptical in cross-sectional shape, the effect of which is to pull the fabric out of the liquor over the fly roller at an oscillating speed. After passing over the reel a folding action is imparted on the rope as it falls back into the liquor. Larger elliptical reels put less mechanical strain on the fabric and are better for delicate or lightweight cloths.

As the fabric is lifted vertically from the bath of a winch machine by the rotation of the reel, dye liquor runs back down the cloth towards the bath. Additionally a mild pressing action, due to the weight of the wet rope, is exerted on the fabric as it rotates over the two rollers, thereby removing more dye liquor. On leaving the reel and plunging back into the dye liquor, the cloth becomes associated with new liquor which is essentially trapped within the folds, so that this liquor more or less stays with that particular section of cloth as the rope gradually works its way to the front of the machine, to begin another cycle.

An inherent problem in winch dyeing is that fabrics tend to run through the machine as a 'rope' rather than in their full open width. As a result, running marks form along the length of the fabric, this unevenness being due to the fabric running too long in the same folded way. This problem can be overcome by manually opening the fabrics back to their full width as they pass over the fly roller. Another disadvantage of winch dyeing is that high liquor-to-goods ratios have to be used, causing expense in heating, dyes and auxiliaries. The dyebath exhaustion is lower under high liquor ratios, so the effluent is more contaminated than by other methods. Winch machines can also be used for desizing, scouring and bleaching processes and are especially useful for washing off, when large volumes of water are required for efficiency.

Jig dyeing machines

The principle of operation of jig dyeing machines, as of winches, is very simple. In jig machines the fabric is unwound from one roller, passed through a dye liquor and wound up onto a second roller (Fig. 9.7). When the transfer of cloth from one roller to the other is complete, the direction of drive is automatically reversed and the cloth transferred back again. During the transfer of cloth from one roller to the other, the fabric is kept in open width, so there are no problems caused by running marks. The rollers vary in size, but can be up to 2 metres wide and can accommodate up to 5000 metres of fabric, though this depends on the density of the cloth.

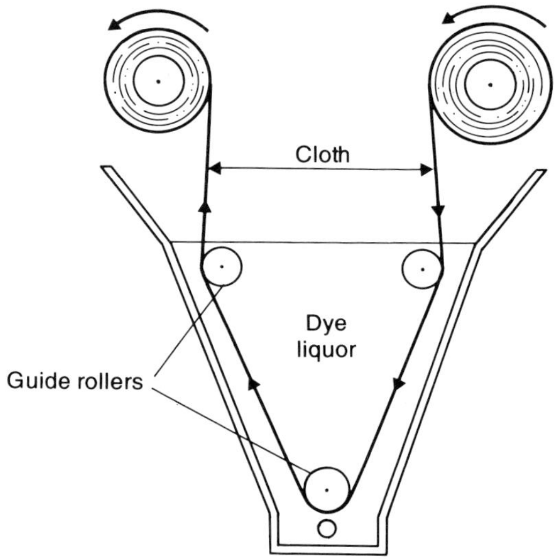

Fig. 9.7 Cross-section of a jig dyeing machine.

Jig dyeing machines are used for woven cloths which must not be creased during dyeing. They are ideally suited to applying vat dyes to cotton because the fabric spends most of the dyeing time wound tightly on a roll and has very little contact with air. In this way premature oxidation of the reduced dye is minimised and the need to replenish the dyebath with reducing agent avoided.

The fabric at the very ends of the roll cannot actually pass through into the dye liquor so a few metres of cheap 'feeder' fabric, of similar construction to the fabric being dyed, are stitched to each end. During its passage from one roller to the other through the dye liquor, the fabric experiences a warp tension, which may cause lengthways stretching and certain reduction in width. Considerable effort has been expended by jig manufacturers to develop a driving mechanism for the rollers which provides a reasonable fabric speed without exerting too much tension. The difficulty is that when the rollers have an equal amount of the fabric on them and their diameters are therefore equal they rotate at the same speed, but otherwise they have to rotate at gradually differing speeds. Also, on completion of the transfer, a jerking action on the fabric must be prevented as the direction of rotation is reversed.

The main advantage of jig dyeing machines is that they can be operated at very low liquor-to-goods ratios, as low as 1.5:1 in some cases. However the nature of the backwards/forwards transfer mechanism leads to an unlevelness effect known as 'ending'. After the fabric leaves a roller it passes through the dye liquor where it picks up interstitial liquor, which is liquor held loosely in the interstices of the fabric. The exhaustion of the dye liquor occurs very efficiently since the effective liquor-to-goods ratio at this point is only about 1:1. As the fabric winds on the take-up roller some squeezing of excess liquor from the fabric occurs. During the time the fabric resides on the roller, adsorption and diffusion processes take place. Each time the fabric passes back through the dye liquor, a proportion of the depleted liquor held in the cloth is exchanged for fresh liquor and the adsorption and diffusion processes continue. However, the fabric near the two ends of the roll has two passes through the dye liquor in quick succession, followed by a long dwell time on a roller, whereas the time intervals between immersions for fabric in the middle of the roll are more even. As a result, the fabric at each end of the roll does not have its interstitial dye liquor replenished as often, after the two times in quick succession, when the second immersion occurs before much exhaustion from the first has had chance to take place anyway. This effect gives rise to a gradual lightening of the shade towards the ends of the roll, due to a lower adsorption of dye.

Another unlevelness effect called 'listing', is a variation in shade across the width of the fabric. It occurs when the edges of the fabric cool whilst they are on the roller, causing lower adsorption and diffusion of dye. This effect can be reduced considerably by fitting the jig with a top cover, so that the temperature inside the machine is kept constant. The use of a top cover also minimises the exposure of cloth to air, useful for applying vat dyes, and prevents steam escaping into the atmosphere.

Beam dyeing

In this method, the fabric is wound, in open width, round a long perforated beam which is placed horizontally in the dyeing machine. The machines are usually pressure vessels, capable of operating at high temperatures, so that polyester can be dyed at 130–135°C. The largest vessels can be as long as 4.5 metres and have internal diameters of nearly 2 metres, and can accommodate beams containing something like 7000 metres of fabric. The actual amount of fabric which can be dyed depends of course on its density. Not all beam dyeing machines are as large and smaller ones are also used.

In a way, beam dyeing resembles the package dyeing of yarns because the fabric is wound on to a former, in this case the beam, through which the dye liquor is pumped from inside to out. The flow rate can be reversed as it can with enclosed package dyeing machines. However in beam dyeing fabrics, the out-to-in direction acts to compress the layers of fabric, thereby increasing the resistance to fluid flow. Also the pressure imparts a glazed appearance to the surface. Whilst flow reversal is used, the timing is adjusted so that the liquor flows in the in-to-out direction for a longer period than for the out-to-in direction. In beam dyeing there is less of a problem due to channelling of liquor flow than in yarn packages, but a perfectly uniform rate of flow through the beam is difficult to achieve because the flow tends to be the greatest at the end of the beam nearest to the pump.

The liquor-to-goods ratios typically used in beam dyeing machines are about 15:1, though in principle even shorter ratios are possible, so high efficiencies of dyeing can be achieved. As with enclosed package dyeing machines, beam dyeing machines have an expansion tank to take up the expanding dye liquor during heating. This expansion tank is used for making additions to the liquor during dyeing.

Pressure beams are particularly suited for dyeing lightweight delicate warp-knitted fabrics because there is no danger of creasing or stretching, as might occur in winch dyeing. Even so, fabrics to be beam dyed must have reasonable dimensional stability. For example, nylon fabrics need to be given a dry heat setting treatment prior to dyeing, otherwise they will shrink during dyeing, creating a pressure on the inner layers and compressing them. The flatter cloth from these inner layers on the beam will reflect light to a different extent than the cloth from the outer layers, giving an effect called 'watering'.

The beams used (beam barrels) are about 0.5 metres in diameter and contain thousands of holes, about 0.5 mm diameter, through which liquor can pass. The total area of the holes has to be as large as possible to permit efficient fluid flow. The large number of small holes allows uniformity of flow yet maintains the required rigidity of the beam. If the width of the fabric is less than the beam the exposed holes must be blocked off. This is usually achieved by fitting flexible stainless steel collars over the ends of the beam. The fit must be very good though, to prevent dye liquor from extruding from the extra holes near to the edge of the fabric and the collars are positioned so the edges of the fabrics just overlap them. However, the fabric must not overlap the collar too much or a resist effect at the edges will be obtained. Alternatively, if the collar is not close enough to the edge of the fabric, dye liquor will escape through, the increased flow rate in this vicinity causing the edges of the fabric to dye darker than the rest.

Prior to winding the fabric, the beam is covered with about ten layers of porous wrappers, usually cotton fabric, the purpose of which is to filter out any large particles (for example undissolved dye) and to produce an even flow of dye liquor as it enters through the holes in the beam. It is important to wind the fabric onto the beam with a uniform tension, sufficient to ensure the fabric will not hang loose when swollen, or be too tight to cause compression. The selvedge on woven fabrics can cause a bulge at the edges of the beam and winding must be arranged to try to smooth these edges.

For certain woven fabrics beam dyeing is very effective and produces more uniform results than can be achieved by jig or winch machines. The dyed fabrics from beams tend to be thinner and the compression effect mentioned above can give a slightly glazed appearance, or at worst, an embossed effect from fabrics with marked surface structures. The amount of fabric which can be dyed in one batch may be limited if the fabric swells much when wet, or if the fabric has a low porosity which restricts fluid flow through it.

Jet dyeing

Jet dyeing machines were developed in the late 1960s from the principles involved in winch dyeing. In jet machines, the fabric rope is moved by passing it through a strong jet of dye liquor, instead of by a reel as in the winch machine. Thereafter the fabric rope moves slowly in folds round the machine, much as in a winch, before passing through the jet to commence another cycle. In the short time during which the rope passes through the jet, an exchange of dye liquor occurs, then during the rest of the cycle, the 'passive' phase, dye adsorption from this liquor and diffusion of dye into the fibre takes place. A complete cycle of the rope takes about one minute but the time taken for a particular section of fabric to pass through the jet is less than one second. One of the earliest types of jet machines was the Gaston County machine developed in the USA (Fig. 9.8).

Jet dyeing machines are somewhat different from the other types of dyeing machines described in that they move both the fabric and the liquor. Jet dyeing machines of different designs have been marketed, though all

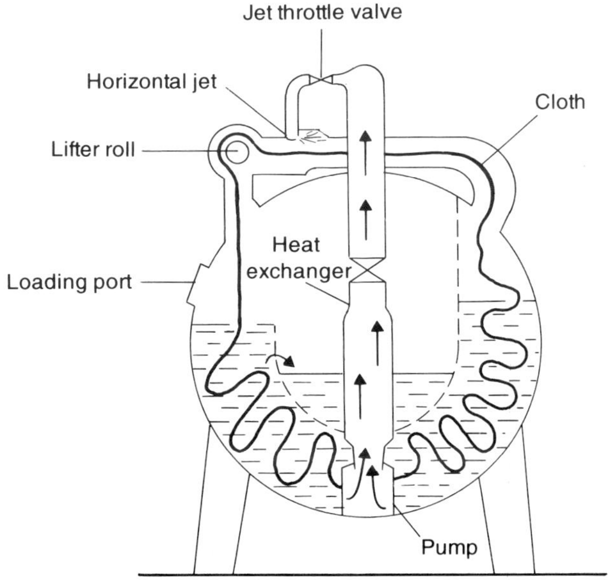

Fig. 9.8 Gaston County jet dyeing machine.

work on the same basic principle. One of the features of the jet of liquor is that it creates creases longitudinally in the fabric, but these creases disappear after the fabric leaves the jet. The high degree of turbulence due to the continuous fluid motion causes the fabric to balloon out. This feature makes jet machines especially useful for dyeing knitted textured-filament polyester at high temperature and indeed it was the difficulty encountered in dyeing these fabrics to heavy depths on a winch which led to the development of jet machines.

The high degree of liquor turbulence in jet machines gives a faster rate of dyeing and good levelness. The machines can be operated at liquor-to-goods ratios of about 10:1 so exhaustion is good and the process economical in terms of heating. A wide variety of fabric types can be dyed but the force of the jet of liquor can be too severe for some delicate fabrics.

More recent developments in jet machines have seen the introduction of fully-flooded and the soft-flow types. In the former, the fabric is kept completely immersed in the dye liquor throughout the entire cycle. Longitudinal creasing of the fabric as it passes through the jet is prevented by feeding it through the jet at high speed and then along a narrow tube to a much wider section, where it moves more slowly. On reaching the end it transfers again to a narrow tube and back through the jet. The diameter of the jet can be adjusted to suit fabrics of different weights. Fully-flooded machines are particularly suitable for delicate fabrics.

Soft-flow machines, developed in the early 1970s, are so called because the dye liquor in them moves more slowly than in earlier jet dyeing machines, giving a gentler mechanical action on the fabric. These machines comprise the main dye vessel (Fig. 9.9), which is tubular and contains the bulk of the fabric. At one end a motor driven circular reel lifts the fabric from the dye liquor and feeds it into the vessel. The other end of the vessel is connected to a narrow transportation tube located either over the top or underneath the main vessel, which delivers the fabric back to the reel. Soft-flow machines are more versatile than fully-flooded machines, cause less

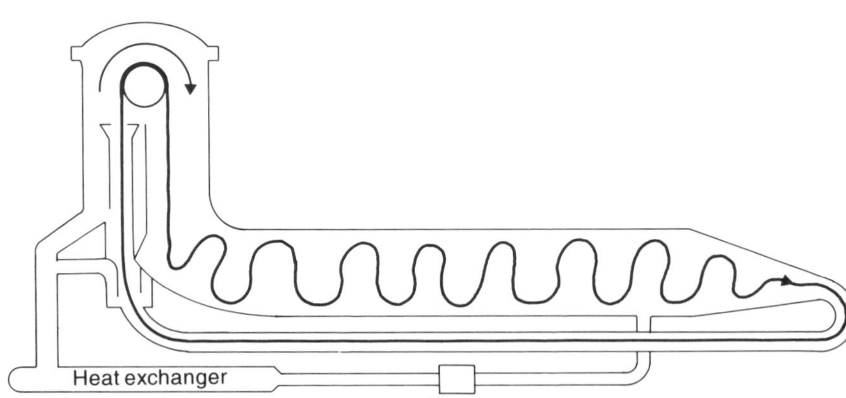

Fig. 9.9 Principle of the soft-flow jet dyeing machine.

creasing in delicate fabrics and put less strain on them than a conventional winch reel.

The trend towards the use of lower liquor-to-goods ratios to minimise water consumption and the use of auxiliaries and maximise dyebath exhaustion, has led to the development of low liquor reel jet dyeing machines. Essentially these machines are a combination of the jet and soft-flow principles, in that the fabric is lifted from the dyebath by a motor driven reel and fed across the top of the machine to the jet. Thus the movement of fabric is brought about mainly by the reel, but is supplemented by the jet. The jet is fed with dye liquor which is pumped to it from the bottom of the dyebath via a heat exchanger.

9.5.2 Continuous dyeing of fabrics

Continuous dyeing has always been of greater importance in the USA than in Europe. In the USA the market has demanded much longer runs of fabric of the same shade, whereas in Europe shorter runs, something of the order of 10 000–15 000 metres are the norm and this practice is likely to continue. The main advantage of continuous dyeing is that a better quality of product, in terms of levelness, is achieved and the repeatability is greater than can be obtained by batch dyeing.

The type of fabric which is dyed by the continuous method is mainly either 100% cotton or a blend of polyester and cotton. Blends of polyester and viscose are often dyed continuously also. The dye classes used for cotton or the cotton component in blends are reactive, vat, sulphur, direct and azoic types, whilst disperse dyes are used for the polyester. In the case of blends, the dyes can be applied either together, when they must be compatible with each other and with the auxiliaries in the dyebath, or separately.

Continuous dyeing of 100% cotton fabrics basically involves passing the cloth through the dye liquor (padding), drying and steaming (for fixation). If reactive, vat or sulphur dyes are being applied, the cloth is passed through another pad liquor of auxiliary chemicals, for example to bring about reduction (of vat or sulphur dyes), prior to steaming. The final stages of the process are then the application of any after treatments, followed by washing and drying.

In blends containing polyester, the polyester component is dyed using the thermofixation process, the important feature of which is the treatment of the cloth after padding, at 200–220°C for 30–60 seconds, to bring about diffusion of the disperse dye into the polyester. The full process sequence is then pad (dye), pre-dry, dry, thermofix, cool, pad (chemical), steam, after treatment, wash, and dry.

Padding involves feeding the fabric under a slight tension through a narrow bath containing the dye liquor and any auxiliaries. The fabric is guided through the bath by rollers, which maintain the state of tension and prevent any creases from forming. On emerging from the liquor, the fabric passes between two nip rollers, whose function is to squeeze excess liquor from the cloth. From this description, padding appears to be a fairly straightforward operation, but in fact it is highly complex because there are many variables which can influence the process. Many different arrangements of the rollers for the process have been devised, some of which are illustrated in Fig. 9.10.

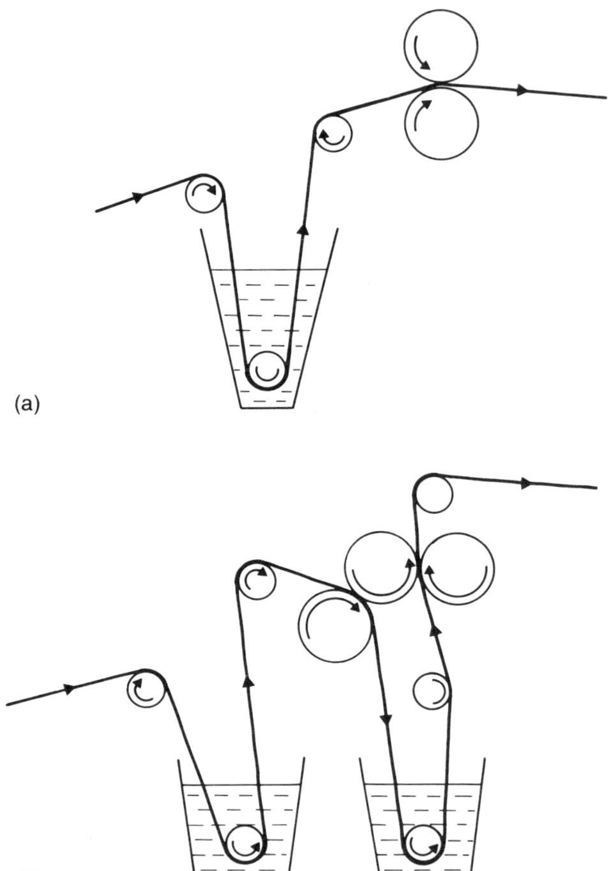

Fig. 9.10 Typical arrangements of pad-mangle systems. (a) Two-bowl mangle; (b) three-bowl mangle.

The duration of immersion of the fabric in the liquor, the time of contact with the nip rollers and the pressure these rollers exert on the cloth are all important aspects. The last of these, the pressure, influences the liquor pick-up, that is, the weight of dye liquor picked up as a percentage of the weight of the fabric. The pick-up needed for a certain fabric is governed by the type of fabric and the depth of shade required. An optimum diameter of the nip rollers has to be selected, because if it is too high, there will be a large area in contact with the fabric under pressure and there will be a greater deformation of the roller surfaces. However, larger diameter rollers can be used for high speed running and there is less deflection of the rollers. The rollers themselves are covered with a rubber surface to allow some deformation to take place under the pressures used. The harder the rubber, the less the deformation and the smaller the area in contact with the cloth. Particular care has to be taken with the rubber surfaces since they may develop cracks in their surfaces and have to be re-ground. Also it is vitally important that the diameter of each roller is exactly consistent along its length.

As the fabric passes through the dyebath, which takes only about 0.5 seconds, it becomes wetted with the dye liquor. Dye then immediately begins to adsorb into the fabric surface so that the excess liquor which is squeezed by the nip rollers back into the trough will be depleted in dye content. As fabric continues to pass through the dyebath, the volume of dye liquor in the bath steadily decreases, but the concentration of dye in it decreases also. When applying mixtures of dyes, this decrease in dyebath concentration becomes apparent because some dyes in the mixture pick up at a faster rate than others. This leads to a variation in shade (tone) as the dyeing process proceeds, an effect known as 'tailing'. By using a padding trough which is deep and has low volume and replenishing it with fresh dye liquor frequently (about every 2–3 minutes), the tailing effect can be minimised.

After the padding operation, which applies dye liquor evenly to the fabric, it is necessary to dry the fabric to avoid any migration of the dye. This is particularly important with polyester or blends containing polyester. It is achieved firstly by pre-drying the fabric, which involves passing it through a bank of infra-red heaters. About 50% of the water is removed quickly and from both sides of the fabric simultaneously, which minimises migration. The fabric is then given a conventional drying until just about all moisture is removed. The separate thermofixation process is then used for polyester blends. Under the high temperature of this treatment the disperse dyes, which after padding, somewhat surprisingly tend to reside more in the cotton component of a cotton/polyester blend, transfer to the polyester component through the vapour phase. Diffusion of the dye into the polyester also takes place at this stage.

After the steaming stage, to bring about diffusion of the cotton dyes, after-treatments may be necessary. For example, if reactive dyes have been used, the fabric will need to be neutralised and washed well to remove any unfixed dye. In the case of vat or sulphur dyes, an oxidation treatment will be needed, again followed by a thorough washing off. The final stage then is drying, achieved by passing the cloth over steam heated cylinders, at a speed which will produce a normal moisture regain in the fabric.

9.6 Garment dyeing

Garment dyeing is an increasingly important method of coloration, having the advantage that the decision about the colour to be dyed can be delayed until the very end of the production sequence. The garments can therefore be dyed to order, with repeat orders being made for the most popular colours. The lead-time for production involving garment dyeing is only a few days, whereas the production of garments manufactured from fabric dyed and finished in the traditional way can take weeks. The types of garments most usually dyed at this stage are knitted garments, such as socks, tights, stockings, hats and sweaters. In addition garments made from woven fabric are used. In particular cotton, cotton/polyester and viscose garments for leisurewear have seen significant growth in recent years. Design and fashion have become important features for this sector which has led to the need for quick response in manufacture.

Obviously only solid shades can be produced and depending on the construction of the garment, the possibility of cross-staining must be considered if the garment contains more than one fibre type. Care has to be taken to avoid damage to the garments by the machine components and all surfaces which come into contact with the garments must be perfectly smooth. Also the fluid flow must not be too vigorous to cause damage. Shrinkage of the garments may also occur, varying from about 2% for woven fabrics to 10% in the case of knitted garments. It is common practice to make the garments up oversize to allow for shrinkage.

Clearly all of the components of these garments must be compatible with the dyeing process. Different types of machines have been developed to handle the various shapes and sizes of garments, most designed to operate at atmospheric pressure, but high-temperature machines for dyeing garments containing textured polyester yarns have been developed.

9.6.1 Side paddle machines

These are very simple in construction, comprising basically an oval-shaped bath with a central (oval-shaped) island (Fig. 9.11). A paddle, whose axle is mounted just above the normal level of the dye liquor, rotates, so that the blades enter the dye liquor and move the liquor and the garments in the channel around the central island. There are usually five blades on the paddle and the rate of rotation is about 10 revs per minute. The larger machines stand nearly 1 metre from the floor and can hold up to about 200 kg of garments. It is necessary however to keep the garments moving freely in the circulating dye liquor and to achieve this large volumes of liquor are needed. Consequently the liquor-to-goods ratios which have to be used are no lower than about 25:1, otherwise there are penalties in dyebath exhaustion and steam consumption.

The walls of the central island are perforated so that dye liquor can flow in and out, but the garments cannot. The water and steam inlets are situated within the island, though in the larger machines, the steam is blown directly into the liquor in the main channel, for faster heating. The central island

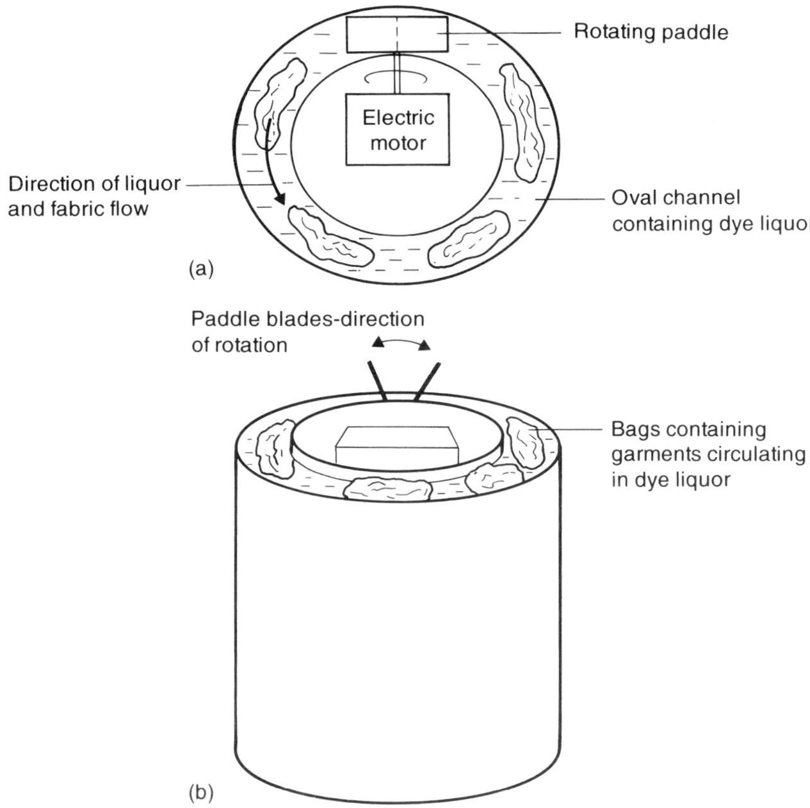

Fig. 9.11 Principle of the side paddle machine. (a) Top view; (b) side view.

acts as a convenient point for making additions of dye liquor or auxiliaries, which are then gradually moved out into the main channel by the fluid motion.

The garments are often placed in open mesh bags made of polyester or polypropylene, to protect them from becoming tangled. The garments can be grouped in the bags, for example all garments of a certain size located within one bag, so that identification of the garments after dyeing is made easy. The bags need to be packed so that the garments within them have room to move to avoid creasing and the dye liquor around them can be regularly replenished. Conversely, if the garments are packed too loosely in the bags they will rub each other and cause surface damage. The machine is designed such that it imparts fluid flow as well as movement of the material being dyed. The main stimulus to both is brought about by the action of the paddle, which creates a very high turbulence of liquor around the garments as they pass under it. Thereafter, the garments, or the bags containing them, make their way round the channel, most of the liquor entrapped in the folds and creases of the garment staying there until the next pass under the paddle.

The unloading of these machines after dyeing can be awkward. The bath must first be cooled by running in cold water, with the excess hot water running away through the overflow pipe. The machine is then filled again with cold water and the garments circulated for a while to remove dye liquor trapped within the garments and this water then drained away. Finally though, the wet garments have to be lifted out of the machine manually, which is a messy and laborious task.

9.6.2 Rotating drum machines

Side paddle machines have the disadvantage of using large volumes of water and machine manufacturers have responded by developing more sophisticated low-liquor rotating drum machines. These machines consist of a perforated drum, which is mounted horizontally in the main vessel, which just encloses it. The garments are loaded into the drum and the drum then rotated slowly in the dye liquor contained within the main vessel (Fig. 9.12). As the drum rotates, the garments, or the bags containing them, ride up its inner wall and then fall back to the bottom under gravity, so that the liquor agitation is very effective.

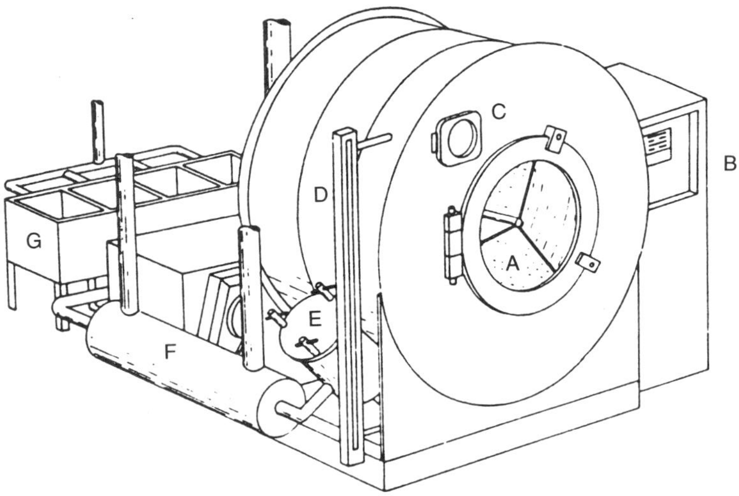

Fig. 9.12 Diagram of a typical rotary drum garment dyeing machine. A, Perforated drum (with Y-pocket); B, control panel; C, sampling port; D, liquor level indicator; E, in-line lint filter; F, heat exchanger; G, addition tanks.

Some machines consist of two such drums, which rotate side-by-side within the main vessel. Another variation is to segment the drum (called the 'Y' pocket) which has a gentler action with brushed fabrics. Rotating drum machines are mainly used for dyeing hosiery goods.

9.6.3 Tumblers

Again, the garments, either loose or contained within open mesh bags, are front loaded into a perforated circular drum, though in tumbler machines the drum rotates on a horizontal shaft at the back of the machine. The design is very similar to that of a domestic front loading washing machine, though obviously on a much larger scale. The drum rotates at about 10 revs per minute in the dye liquor, which is contained within the main outer vessel.

After dyeing the liquor can be run out of the machine and the drum rotated at a higher speed to remove imbibed liquor by centrifugal force. The goods are then rinsed, spun again and finally dried. The machines can be used for scouring or milling woollen fabrics as well as dyeing, the rate of rotation of the drum being selected to suit the particular process.

A method has been developed in which the dye liquor is sprayed cold as a foam onto the garments in the drum. An anionic surfactant is added to the concentrated dye liquor to produce the foam. The drum is then rotated for about 15 minutes, during which time the foam distributes itself uniformly throughout the garments, but, because the dye liquor is cold, no fixation occurs. Steam is then blown directly into the drum, causing rapid heating and bringing about the fixation of the dye. After the fixation stage, the garments are rinsed, centrifuged by rotating the drum at high speed and finally dried. The process has the advantage that dyeing can be carried out at liquor-to-goods ratios as low as 1.5:1.

9.6.4 High-temperature garment dyeing

Garments containing polyester need to be dyed at high temperature if the use of carriers, with their residual odours, are to be avoided. For this purpose, a machine has been developed in which dye liquor is pumped up the centre of the vessel against a mushroom shaped inverted baffle. The liquor flow is directed outward to the edges of the vessel, upwards to the top and then back down to the bottom in the central region of the vessel (Fig. 9.13). The liquor then falls through a perforated false base back to the pump. The circular flow of the liquor in the vessel has sufficient energy to move the garments, thereby giving the necessary agitation.

The heating and cooling coils are mounted below the perforated false base. The lid on the top of the vessel can be pressure sealed to enable the high pressures and the temperatures of 130–135°C required for dyeing polyester, to be achieved.

9.7 Future developments

The increasing awareness about the environment, ecology and pollution issues have influenced political thinking with the result of the introduction

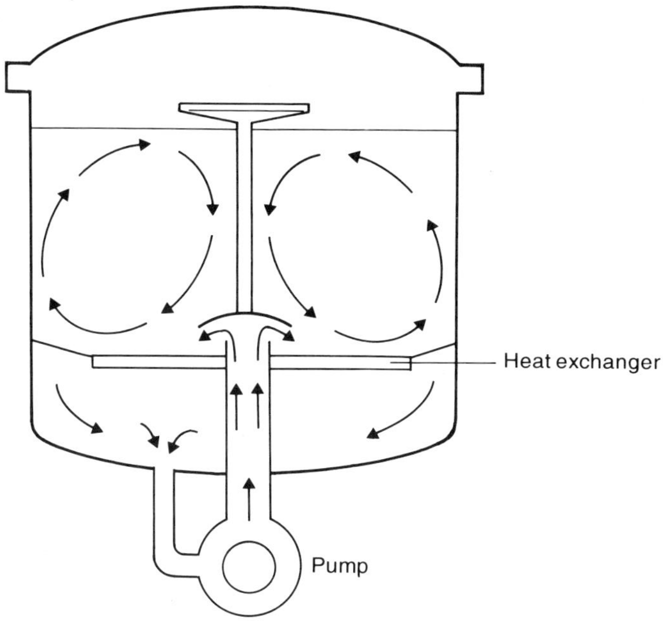

Fig. 9.13 Principle of a high temperature (Pegg toroid) garment dyeing machine.

in recent years of increasingly rigorous legislation. The textile industry, as a major user of water and energy has had to respond to such regulations as discharge of effluent into community sewerage systems and the restrictive exhaust emission controls, which are especially challenging. The manufacturers of dyeing machines have responded by aligning design to economical use of water and power, with process cycles fully automated. In this way the textile processing industry can save costs and meet the strict legislative requirements by savings on labour, power, water and chemicals.

Bibliography

Engineering in Textile Colouration, C. Duckworth (Ed.), The Dyers Company Publications Trust, Distributed by the Society of Dyers and Colourists, Bradford (1983).

Batchwise Dyeing of Woven Cellulosic Fabrics, G.W. Madaras, G.J. Parish and J. Shore, The Society of Dyers and Colourists, Bradford (1993).

A Practical Introduction to the Dyeing and Finishing of Wool Fabrics, I. Bearpark, F.W. Marriot and J. Park, The Society of Dyers and Colourists, Bradford (1986).

A Practical Introduction to Yarn Dyeing, J. Park, The Society of Dyers and Colourists, Bradford (1981).

Chapter 10

Pigment Dispersion Technology

Pigments are insoluble colouring materials which are incorporated into their application media by a dispersion process. They are used mainly in paints, printing inks, and plastics but also to a certain extent in an extremely diverse range of miscellaneous applications, including the coloration of paper, textiles, elastomers, glass, ceramics, cosmetics, crayons, and building materials such as cement and concrete. Generally, although not always, the medium into which the pigment is dispersed is a liquid, for example a wet paint or ink or a molten thermoplastic material. Subsequently, the pigmented medium converts into a solid form and the pigment particles are then 'fixed' since they are held mechanically in a polymeric matrix.

The distinction between dyes and pigments as colouring materials has been presented in some detail in Chapter 2, together with a discussion of how, for selected cases, organic pigment molecules are designed to meet certain application requirements. The physical and chemical principles underlying the application of pigments including a description of the structures and properties of the most important inorganic and organic pigments are given in Chapter 6. This present chapter deals with the principles of the industrial technology used in the manufacture of paints and printing inks and in the compounding of plastic materials, with particular reference to those aspects of the manufacturing processes used to incorporate the pigments. In this sense, it complements the discussion of the manufacturing technology used in textile dyeing which the reader will find in Chapter 9. The chapter contains an outline of the mechanism of the dispersion process involved when a pigment is incorporated into its application medium, followed by a description of the principles of some selected examples of the milling equipment used industrially to carry out this dispersion process.

10.1 The mechanism of pigment dispersion

The dispersion process is a critical stage in the manufacture of a paint, printing ink or plastic material. The purpose of the dispersion process is to distribute the insoluble pigment particles through the application medium and to reduce the pigment to as finely-divided a form as possible. This section describes the mechanisms by which this is achieved in practice. For an appreciation of the physical and chemical principles of pigment dispersion, it is of fundamental importance firstly to consider the nature of pigment particles, and in particular the way in which they form clusters.

10.1.1 *The nature of pigment particles*

Pigments are finely divided solid materials consisting of clusters of particles. Primary particles are the smallest pigment particles not normally further subdivided. However, during the manufacture of pigments, and especially at the stage where the pigments are dried to produce a powder, particles tend to cluster together. Two different types of particle clusters can be considered: aggregates and agglomerates. Aggregates are considered to be groups of primary particles joined at faces. Agglomerates are considered to be groups of aggregates or primary particles joined at edges and corners. These types of particles are illustrated in somewhat idealised form in Fig. 10.1. Because of the nature of the particle contact, agglomerates are much less tightly bound together than aggregates and are therefore easier to break down by dispersion.

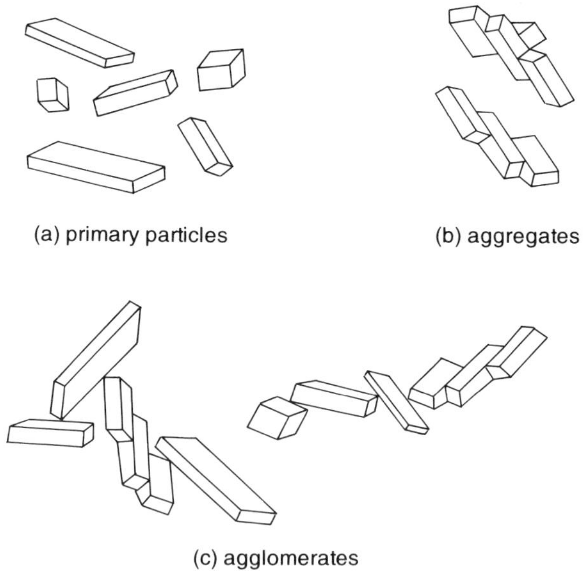

Fig. 10.1 Types of pigment particles. (a) Primary particles; (b) aggregates; (c) agglomerates.

10.1.2 *The dispersion process*

It is essential to achieve a good dispersion of pigment in its application medium for a variety of reasons. Control of dispersion is important in ensuring that the pigment produces optimised optical properties in its application. For example, the colour strength obtained from an organic pigment increases as the particle size is reduced during the dispersion process (see Chapter 6) and for obvious economic reasons it is desirable to obtain the maximum colour intensity from the minimum amount of pigment. In addition, it is generally important to achieve a smooth texture in the final product by ensuring the absence of oversized particles. Large numbers of coarse undispersed particles can give rise to a number of problems in application. In paint films, for example, large particles may protrude through the surface of the dry film giving rise to a reduction in the gloss of the surface. In traditional printing inks, coarse particles can scratch printing plates leading to poor print quality. In ink-jet printing, which has the potential to become a dominant large-scale printing method, water-soluble dyes are currently preferred to pigments because of the tendency of oversize pigment particles to block the nozzles of the jets which deliver the ink. In mass-pigmented textile fibres, large particles can produce weak points leading to breakage of the fibres. The degree of dispersion also influences the flow properties of wet paints and inks, which are of importance in the way that these products are applied (see Chapter 4).

The dispersion of a pigment into a liquid medium may be separated conceptually into three stages: *wetting*, *mechanical disruption* and *dispersion stabilisation*. These three stages are not necessarily sequential but it is convenient to discuss them separately and in this order. The wetting of the pigment particles by the liquid medium is the replacement of the solid–air interface, which is present originally in the dry pigment powder, with a solid–liquid interface, and is an important early feature of the dispersion process. Efficient wetting will be achieved if there are strong intermolecular forces between the surfaces of the pigment particles and the constituent ingredients of the medium into which the pigment is being concorporated. The disruption phase is that part of the overall process in which the clusters of pigment particles are reduced to a much more finely-divided form. It involves the mechanical breakdown of agglomerates and, to a certain extent, aggregates into smaller aggregates and primary particles. The nature of the forces used to effect this mechanical disruption are discussed later in this chapter. A third stage of the process is dispersion stabilisation. A dispersion of fine solid particles in a liquid possesses high surface energy and is thus thermodynamically rather unstable so that the particles will have a tendency to come together again, or flocculate, unless the dispersion is stabilised in some way. Generally, in paints and printing inks this is achieved by careful formulation of the medium, for example by an appropriate choice of polymeric film-former, by the use of specific additives where necessary and by careful control of the viscosity of the medium. The main mechanism by which pigment dispersions are stabilised is referred to as *steric stabilisation*. In this mechanism, polymer molecules, for example of the film-formers in paints and inks, are adsorbed in layers at the surface of pigment particles, forming a barrier to the flocculation process and thus preventing pigment particles from coming together.

The factors affecting the colour strength of a pigment in its application have been discussed previously in Chapter 6. In general, for a typical organic pigment there is an increase in colour strength of the pigment as the particle size is reduced, especially at particle sizes less than 0.1 μm. In the case of many inorganic pigments, a different type of behaviour may be observed with an increase the colour strength reaching a maximum at around 0.2–0.4 μm. Fig. 10.2 illustrates how, for the dispersion of two samples of a pigment, (a) and (b), into an appropriate application medium, the colour strength varies with the time of dispersion. It may be seen that dispersion of the pigment causes a progressive increase in colour strength as the particle size is reduced. There is initially a pronounced increase in colour strength as weak agglomerates are broken down, followed by a more gentle change as stronger clusters of particles are disrupted. Finally, the colour strength approaches a limiting value at the point when no further particle size reduction is occurring as any remaining aggregates are too tightly bound. This effect may be demonstrated experimentally by visual comparisons of films drawn down from samples of the ink or paint during the dispersion process and the method may be used to assess the degree of dispersion, for example for quality control purposes.

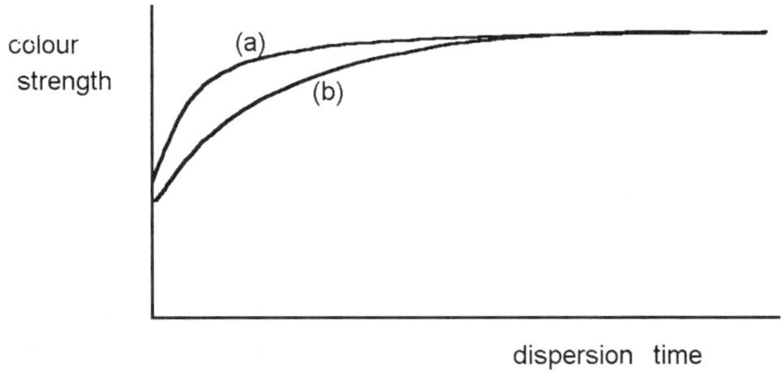

Fig. 10.2 Colour strength development during the dispersion of pigment samples (a) and (b) into an appropriate application medium, sample (a) showing better dispersibility than (b).

Dispersibility is an important parameter in pigment application technology. It is defined as the ease with which a pigment powder may be dispersed into a medium to give the optimum degree of dispersion. Dispersibility is a difficult property to quantify and is usually considered in qualitative terms as illustrated in the following example. Fig. 10.2 shows the outcome of the dispersion process, assessed by colour strength development, for two pigment samples (a) and (b). Ultimately, the two samples of pigment are capable of the providing the same colour strength but sample (a) reaches the maximum more quickly as a result of its better dispersibility. The dispersibility of a pigment is frequently improved by specific treatments applied to the surfaces of pigment particles (see Chapter 6). The surfaces of untreated pigment particles often tend to be highly polar, or hydrophilic. This may lead to poor wetting by organic media, such as those into which the pigments are incorporated in printing inks, paints and plastics. Additionally, the nature of these surfaces produced during the manufacture of the pigments can lead to aggregation, especially at the stage when they are dried, by removal of water, to give powders. The pigment particle aggregates which are produced in this way are tightly bound and difficult to disperse. Appropriate treatments applied to the pigments can dramatically alter the nature of their surfaces and lead to improved dispersibility. For example, this may be achieved when certain organic pigments are treated with surface-active agents. The molecules of these agents may be considered as consisting of a polar 'head' , often a sulphonic or carboxylic acid group or an amino group, and a 'tail', usually a long alkyl chain. When the pigment is treated with the surface-active agent, the polar group attaches itself to the polar pigment surface. As a result, the alkyl chains provide the pigment with a new surface which is non-polar or hydrophilic and hence more compatible with an organic application medium. In addition, because of the steric barrier presented by these alkyl chains, the pigment particles are kept apart and are thus less susceptible to aggregation. Also, those aggregates which are formed are more easily broken down by the dispersion process.

10.2 Forces involved in pigment dispersion

There are two principal types of force acting in the pigment dispersion process to bring about mechanical disruption of particle clusters: crushing and shearing. These are illustrated diagramatically in Fig. 10.3. In most industrial dispersion equipment (commonly referred to as *mills*), both types of force will operate to a certain extent although depending on the action of

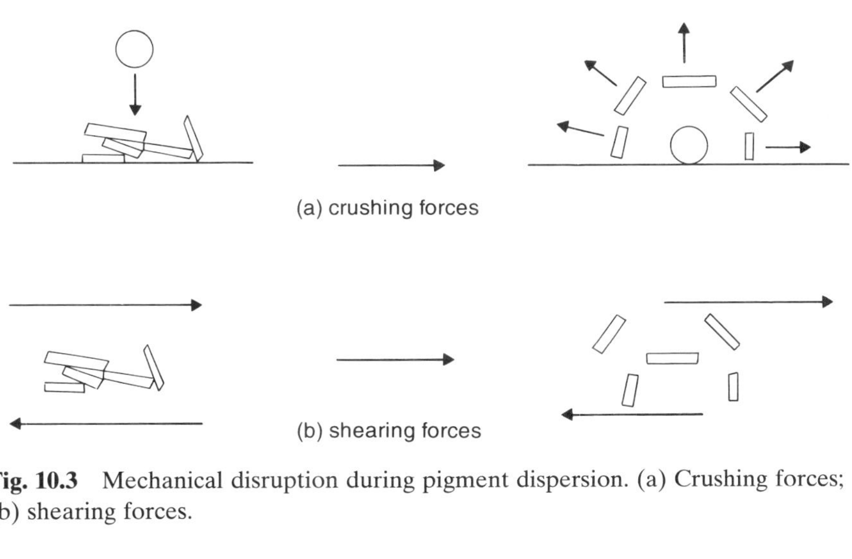

Fig. 10.3 Mechanical disruption during pigment dispersion. (a) Crushing forces; (b) shearing forces.

the particular mill, either one or the other may dominate. Crushing involves impact forces, often due to projectiles (e.g., balls or beads), moving at high speed and giving rise to disintegration of pigment agglomerates. These forces are of particular importance in milling equipment such as ballmills, sandmills and beadmills. These mills are commonly used for dispersion of pigments into relatively low viscosity media such as paints and liquid inks, where sufficiently high impact velocities may be achieved. In contrast, shearing forces result from stresses exerted within a high viscosity fluid as the fluid flows, and these in turn cause mechanical breakdown of the pigment agglomerates. For efficient shearing, the flow pattern within the fluid must be *laminar*, a type of flow which is characterised by a regular linear flow pattern. The chaotic movement of fluid which is caused by *turbulent* flow conditions gives efficient mixing of the ingredients in the formulation but a poor dispersion results due to the low shearing forces which are generated under these conditions. Shearing is the principal force involved in dispersion of pigments into high viscosity media such as paste printing inks, for example those used for lithographic printing, and into molten thermoplastics.

10.3 Industrial dispersion equipment

Early industrial methods for dispersing pigments generally used milling machines capable of applying powerful forces, both crushing and shearing, in order to achieve appropriate size reduction of the pigments available at the time. A number of these long-established mills are still used today in the manufacture of printing inks, paint and plastics. They include heavy-duty mixers, such as the Banbury mixer for plastics, roll mills, such as the two roll mill for plastics and the triple roll mill for paste inks, and the ballmill for paints and liquid inks. However, for many applications, particularly for paints and inks, these have been superseded by modern milling equipment such as the highly efficient beadmills and sandmills which are capable of continuous as well as batchwise operation. In addition, the high speed impeller mill has become more extensively used. This is a mill which is simple in its construction and is of only moderate efficiency, but it is capable of producing good dispersions because most pigments are nowadays available in highly dispersible grades.

There is currently an immense range of types of dispersing equipment available. The selection of a mill for a particular process will depend on a variety of factors. One of the most important of these is the viscosity of the medium or vehicle into which the pigment is to be dispersed. Dispersion takes place as a result of a combination of crushing and shearing forces, the former being most important at low viscosities and the latter at high viscosities. The diagram shown in Fig. 10.4 provides a guide to the dispersion mechanism involved for some of the important mills in use industrially. An outline of the principles of a selected range of these mills follows.

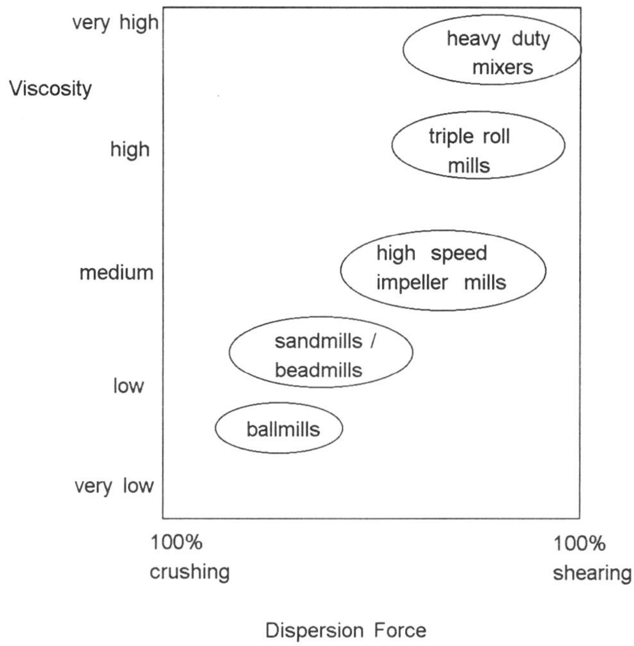

Fig. 10.4 Crushing and shearing forces in industrial dispersion equipment in relation to the viscosity of the application medium.

10.3.1 The Banbury mixer

The Banbury mixer (Fig. 10.5) is one of the most versatile and widely-used mills for the large-scale incorporation of additives into plastics. This heavy-duty mixer provides the high shear conditions required for the dispersion of pigments into the highly viscous polymer melt and in addition has the facility for good temperature control. The device consists of a jacketted chamber whose cross-section is in the shape of a horizontal 'figure of eight'.

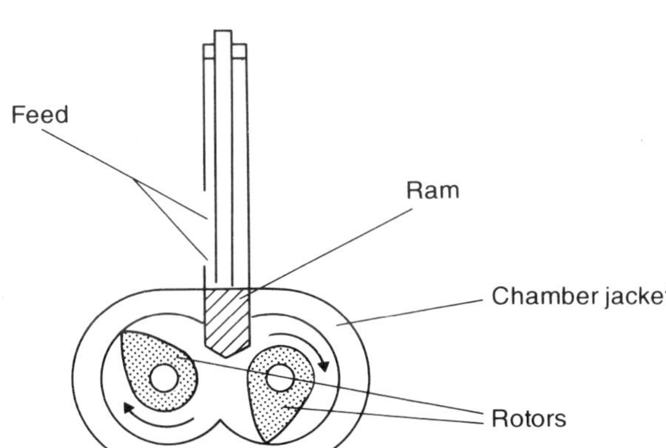

Fig. 10.5 The Banbury mixer.

In this chamber, two rotors rotate at slightly different speeds inducing a kneading action and conditions of high shear between the rotors and the chamber wall. The jacket and the core of the rotors give provision for heat transfer by means of a circulating fluid. The heat transfer will involve heating in the case of thermoplastics in order to ensure that the polymer is at an appropriate processing temperature. In the case of thermosets, cooling may be required as the shearing action generates heat which might induce premature polymerisation.

10.3.2 The triple roll mill

The triple roll mill (Fig. 10.6) is a long established mill used mainly in the manufacture of highly viscous paste inks such as those used in offset lithographic and letterpress printing. The mill consists of three cylinders, a feed roll, a centre roll and an apron roll. These rolls rotate in the directions indicated in the diagram and at different speeds, the feed roll slowest and the apron roll fastest. The hydraulic pressure between the rolls is adjustable and they are hollow to permit heat transfer by water-circulation, either heating or cooling whichever is desired.

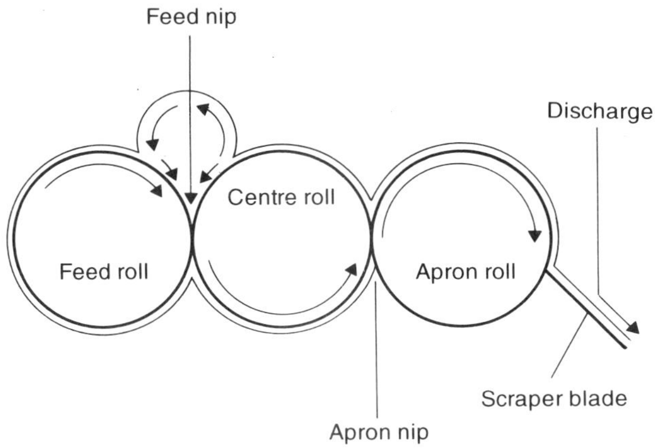

Fig. 10.6 The triple roll mill.

In the manufacture of a paste printing ink using a triple roll mill, it is essential that a degree of premixing of pigment with the ink vehicle is carried out. This ensures efficient wetting of the pigment and the breakdown of very large agglomerates. Often, this is carried out using a high speed impeller mill. The premix is then fed to the triple roll mill and the mixture passes from roll to roll via the points of contact of the rolls known as 'nips'. Dispersion results from the shearing action in the nips due to the different rotation speeds of the rolls and crushing forces as a result of the hydraulic pressure between the rolls. Finally, the ink is removed from the apron roll by a scraper blade. The triple roll mill remains of some importance in the manufacture of paste inks and is capable of producing high quality dispersions even from pigments with poor dispersibility. It does, however, operate on a batchwise principle with relatively low production rates and the need for constant attention by the operator. For these reasons, it has been largely superseded, where appropriate, by beadmills which are capable of continuous operation and are effective with the modern range of highly-dispersible pigments.

10.3.3 High speed impeller mill

This type of mill is referred to variously as a high speed impeller mill, mixer, disperser, shear mill or cavitation mill. The mill (Fig. 10.7) consists of a motor driving a shaft at the end of which is a disc impeller which may be raised or lowered into the vessel in which the dispersion is to be carried out. There are a number of different types of disc impeller used, although the most common type involves a 'saw-tooth' design. The impeller rotates in the paste and dispersion of the pigment in the medium results as a result of a combination of shearing and crushing forces, but largely from the shearing forces acting in the flow beneath the impeller. Effective use of the high speed impeller mill requires careful selection of a number of variables, the most important of which are the viscosity and volume of the dispersion medium, the design of the impeller, its location in the vessel and its speed of rotation. The dispersion procedure is controlled carefully so that laminar flow conditions occur beneath the impeller in order to maximise the shear

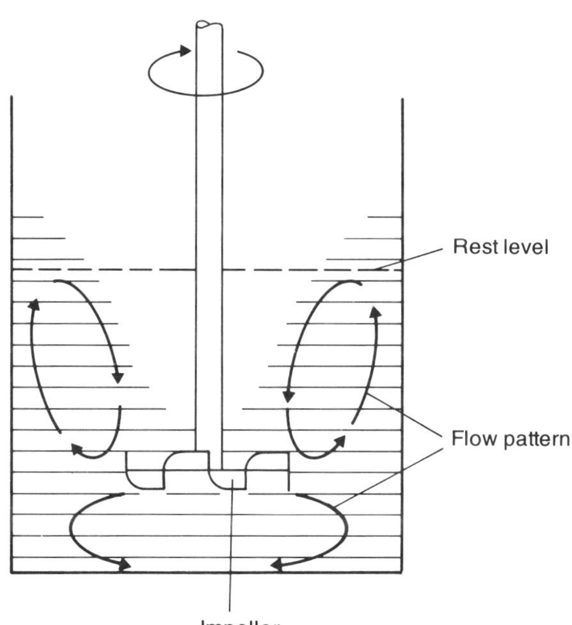

Fig. 10.7 The high speed impeller mill.

forces. This mill is particularly important in the manufacture of high volume decorative paints but, because of its only moderate efficiency, requires the use of pigments which show good dispersibility. It is also used for the premixing of pigments into printing inks prior to dispersion using a sandmill, beadmill or triple roll mill.

10.3.4 The ballmill

The ballmill was for many years the most widely-used type of mill for paint and liquid ink manufacture and it remains of some importance, although it has been replaced for many applications by continuous sandmills or beadmills. The ballmill (Fig. 10.8) consists of a horizontally-mounted cylindrical vessel, half-filled with steel or ceramic balls together with the dispersion mixture, which is rotated about its axis causing the contents to roll around inside. In general, the mill is rotated at an approriate speed such that the balls are carried up to a point 20–30° short of the zenith, so that they cascade down the sloping mass. In this way the whole mass of balls is in constant motion, each rolling over or rubbing against each other.

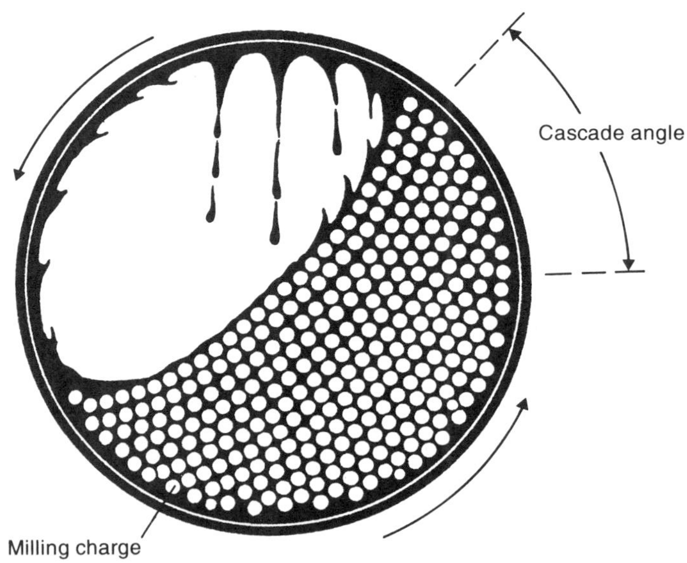

Fig. 10.8 The ballmill.

A combination of shearing and crushing forces operate in the ballmill as illustrated in Fig. 10.9. In particular, the cascading effect of the balls produces impact forces which are effective in producing good dispersions even from pigments with lower dispersibility. There are also shearing forces within the fluid arising from the rotation of the balls with respect to one another and to the wall of the mill. The ballmill generally works best with relatively low viscosity media so that the balls are capable of attaining high

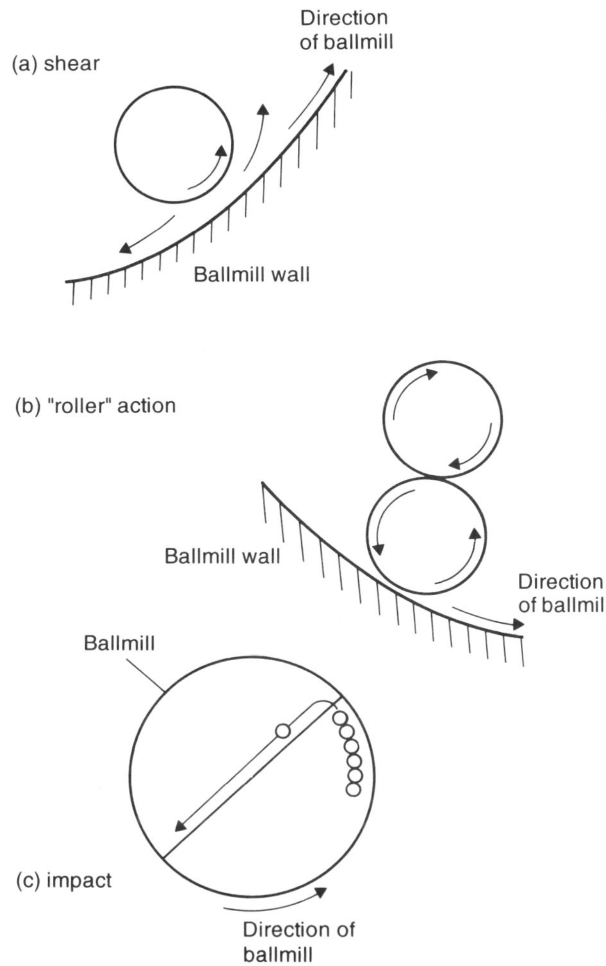

Fig. 10.9 Dispersion forces in the ballmill. (a) Shear; (b) 'roller' action; (c) impact.

impact velocities. However, in order to ensure the optimum combination of impact and shear forces, the consistency of the medium should be as high as will permit satisfactory movement of the ball charge. The ballmill is a versatile and effective mill for pigment dispersion, and requires little maintenance or attention during operation. However, it operates batchwise and production rates are low and it is mainly for these reasons that it has been largely superseded by more modern devices such as sandmills and beadmills.

10.3.5 Sandmill/beadmill

Sandmills and beadmills may be considered as having developed historically from the ballmill. Both types of mill use very small grinding media in an attempt to improve efficiency by increasing the number of contact points. In practice, the sandmill uses graded sand of 0.5–1 mm diameter as the grinding charge. In contrast, the beadmill uses graded glass beads of 1–3 mm diameter. Both mills use an impeller to agitate the medium and the design of the impeller is crucial to the effectiveness of the mill. Mills of a great variety of designs are in use, including batch types and continuous versions and they may be designed to operate either vertically or horizontally. The diagram shown in Fig. 10.10 serves to illustrate the principles of a vertically-operating mill. The mill consists of a stationary water-cooled cylindrical container and an impeller, consisting of a number of discs mounted regularly along a shaft, which rotates about its axis. The chamber is partly filled with the grinding charge. A premix of pigment and the paint or printing ink vehicle is fed to the bottom of the milling chamber and the dispersion is achieved by continuous passage through zones between the discs in which the grinding elements circulate continually. As with the ballmill, a combination of crushing and shearing is involved. A screen with an appropriate pore size allows the finished paint or ink to pass through while retaining the grinding charge. These mills, particularly those operating continuously, are capable of producing dispersions of high and consistent quality, quickly and cheaply. On the other hand, as with the triple roll mill a degree of premixing of the pigment with the medium is required.

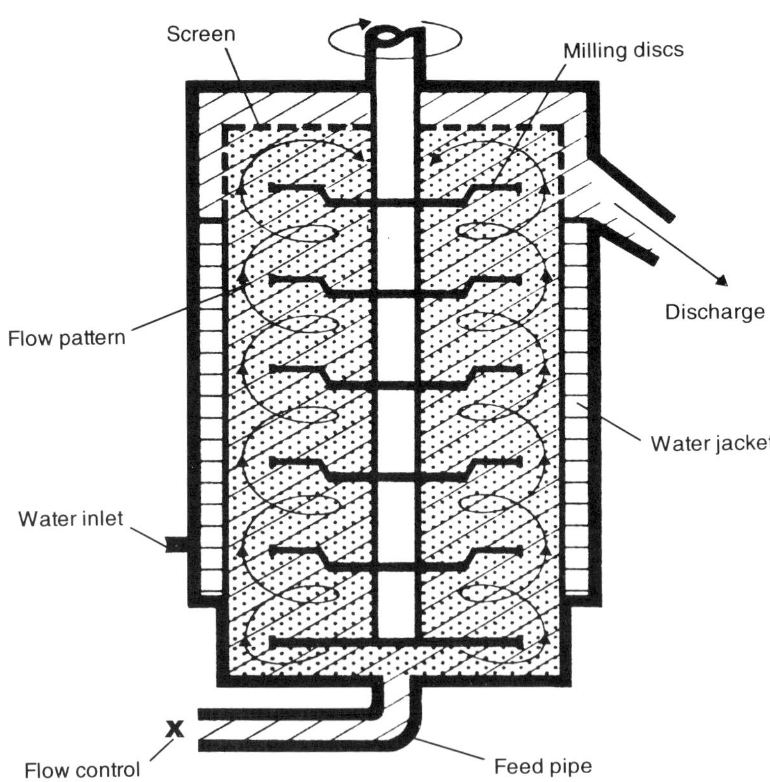

Fig. 10.10 The sandmill/beadmill.

Bibliography

Outlines of Paint Technology, Vols 1 and 2, W.M. Morgan, Griffin, London (1984).

Surface Coatings, Vols 1 and 2, 2nd edition, Oil and Colour Chemists' Association, Australia, Chapman & Hall, London (1983).

The Printing Ink Manual, R.H. Leach (ed.), 5th edition, Blueprint, London (1993).

Physical-chemical Aspects of Pigment Applications, J.T. Guthrie and L. Lin, Oil and Colour Chemists' Association, London (1994).

Paint Flow and Pigment Dispersion: A Rheological Approach to Coating and Ink Technology, T.C. Patton, John Wiley & Sons, Chichester (1979).

Dispersion of Powders in Liquids: With Special Reference to Pigments, G.D. Parfitt (ed.), Applied Science Publishers, London (1981).

Dispersion and Dispersion Equipment, R. Stanfield, OCCA Student Monograph No. 5, Oil and Colour Chemists' Association, London (1993).

Chapter 11

Other Applications of Colorants

11.1 Introduction

So far attention has been focussed on just some of the many applications of dyes and pigments, yet both types of colorant have many other practical uses. Most of these can be broadly assigned, as shown in Table 11.1, into three groups, photographic, biological and electrical/electronic. In addition, dyes may serve as titration indicators, food colorants and hair colorants.

Table 11.1 Examples of other applications of colorants.

Photographic	Biological	Electrical/electronic
Dye-forming process	Histology	Liquid crystal displays
Dye-bleach process	Affinity chromatography	Electrophotography
Image-diffusion process	Biochemical assay	Solar energy conversion
	Drugs	Organic conductors

Some of the uses are well established, as in colour photography, food colouring and histology, although developments in these areas are continuing. The expansion of colorants into other areas is more recent: these applications are mostly electronic, but also embrace developments in biological techniques of analysis, such as affinity chromatography and enzyme assay. Dyes, in particular, are even becoming valuable as probes for the location of cells in living organisms and the destruction of cancer cells.

In this chapter, it is clearly impossible to cover all these other uses of colorants. An overview, covering the chemistry underlying some of these applications must necessarily suffice.

11.2 Colour photography

Colour photography provides an important commercial outlet for dyes. Colour photographs can be obtained by the well established dye-forming process or by the more recently developed dye-bleaching and 'instant' dye-diffusing processes.

11.2.1 Dye-forming process

The dye-forming process involves the use of a solution of colour developer which reduces photosensitive silver halide in the film to silver. The oxidised form of the developer so produced then immediately reacts with a suitable colour coupler to form a dye. Whereas hydroquinone is used to develop black and white film, the most usual type of colour developers are provided by the *p*-phenylenediamines. During reduction of the silver halide to silver, the *p*-phenylenediamine is oxidised to a semiquinone diamine, as illustrated in Fig. 11.1. The semiquinone diamine is then either further oxidised to a quinonediimine (Route 1) or alternatively disproportionates to give the quinonediimine and the original *p*-phenylenediamine (Route 2). The quinonediimine rapidly reacts with a colour coupler to form a yellow, magenta or cyan dye, i.e. one of the three subtractive primary dyes.

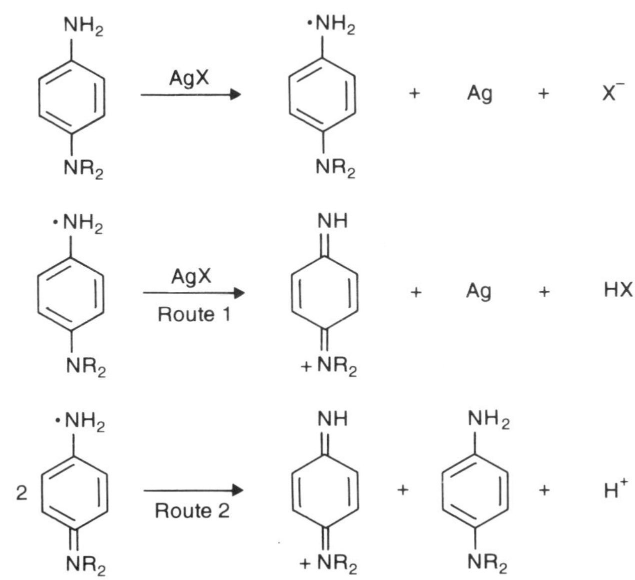

Fig. 11.1 Routes for the conversion of a *p*-phenylenediamine to a quinonediimine (AgX represents a silver halide).

Prominent amongst yellow couplers are those β-ketoamides of general formula:

$$R—CO—CH_2—CO—NH—Ar$$

where R represents Ar′— or $(CH_3)_3C$—, and Ar and Ar′ represent aryl ring systems. The methylene (—CH_2—) groups are thus activated by the electron-withdrawing carbonyl (>C=O) groups each side and readily lose protons to form carbanions. The anionic form reacts with the quinonediimine to give a colourless compound, which on oxidation by further silver halide forms a yellow azomethine dye (Fig. 11.2).

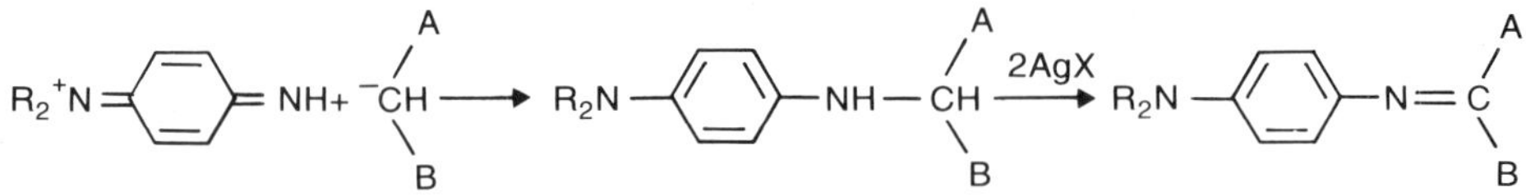

Fig. 11.2 Formation of yellow azomethine dyes.

In other yellow couplers, —CH_2— is replaced by —CH(Y)—, where Y is a good leaving group, often called a coupling-off group. This group is usually halogen, acetate or sulphonate. For these couplers, the coupling reaction alone gives the required yellow dye, without the need for the subsequent oxidation step. Less silver halide is, therefore, needed.

Magenta couplers are also active methylene compounds, though mostly pyrazolones of the type **1** where R′ is usually —NH.CO.R or —NH.R. The formation of magenta dyes is, therefore, similar to that of yellow dyes. Moreover, some pyrazolone couplers contain coupling-off groups instead, for example, thiocyanate.

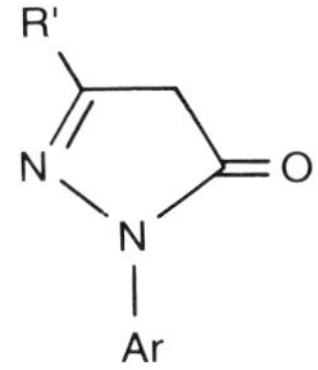

1

Cyan couplers are almost all phenols and naphthols, many containing a carbonamido (—CO.NH.R) group at the 2-position. A coupling-off group, Y, usually —Cl or —SO_3H, can also be incorporated at the position of coupling with the quinonediimine. Thus, the coupling reaction for phenolic couplers can be represented as in Fig. 11.3.

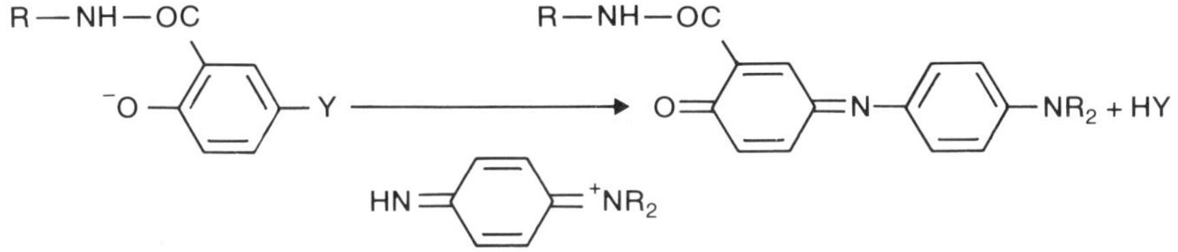

Fig. 11.3 Formation of cyan dyes.

11.2.2 Dye-bleach process

The dye-bleach process is much less widely utilised than the dye-forming process. Generally, a hydroquinone developer is used, and the photosensitised silver formed is able to reduce yellow, magenta and cyan azo dyes already present in the film:

$$4Ag + R{-}N{=}N{-}R' + 4H^+ \rightleftharpoons 4Ag^+ + RNH_2 + R'NH_2$$

A positive image is thus formed directly. To push the equilibrium in the desired direction, the silver ions formed are removed either by precipitation as a halide or by the formation of a silver complex. The process is catalysed, usually by quinoxaline.

11.2.3 Image-diffusion process

The image-diffusion process is notable for its speed, only about one minute from taking the photograph to viewing the print. The best-known example is probably the Polaroid process. Dye and hydroquinone developer are present in the same molecules, separated from each other by alkyl chains, commonly —$(CH_2)_2$—. Development is triggered by the release of alkali as soon as the photograph is taken (Fig. 11.4).

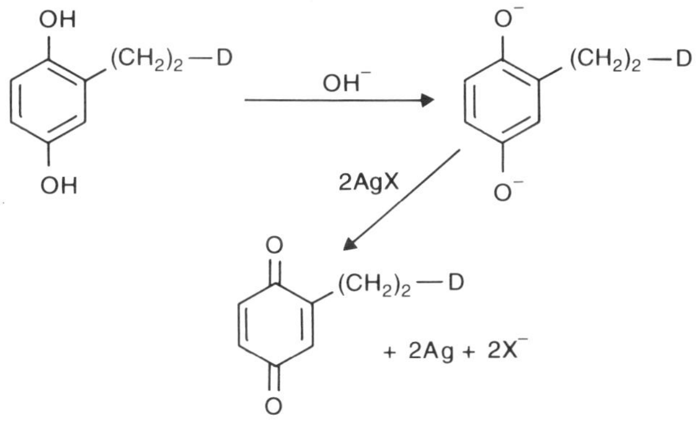

Fig. 11.4 Formation of a quinone in the image-diffusion process (D represents dye).

The soluble anions formed diffuse towards a light-sensitive layer. If, as shown in Fig. 11.4, they encounter silver halide in the layer, they are oxidised to a quinone, which is precipitated. If, on the other hand, no silver halide is encountered, the ions diffuse through the layer to a receiving layer, where they are mordanted to render them insoluble. They, therefore, become trapped in the receiving layer, and a positive image is directly formed. Also during the process, a so-called timing layer is gradually degraded by the alkali. Once this layer is completely degraded towards the end of the process, the alkali can reach acid contained on a polymer and be neutralised. More recent developments of the Polaroid process have incorporated metal-complex dyes on account of their superior light-fastness.

11.3 Food colorants

A variety of dyes are used to enhance the attraction of food to the consumer, for consumers often use colour to gauge the desirability of a particular food product. In this respect, the nature of the light source, as well as the colorants used, is an important consideration in the presentation of food. In addition, the public perception of the appropriate colour for a given type of product may also change over a period of time.

Natural and synthetic colours are both used as food colorants, although natural colours in this context often embrace synthetic analogues of natural materials, in addition to natural extracts. Such synthetic analogues are known as *nature identical* food colours. Moreover, chemically modified natural colours may be used, a notable example being the greater stability conferred on chlorophyll, the green colorant in photosynthetic plants, when its constituent magnesium is replaced by copper (II).

The most widely applied food chemicals are caramels. They are used in a variety of drinks, sauces, liquorice and confectionery, for example, to enrich their brown colour, and in addition they confer particular flavours. Caramels are produced by heating carbohydrate solutions to *c.* 160°C, or to 130°C in the presence of alkali, ammonia or ammonium sulphite. Four different classes of caramel have been described, according to the manufacturing process used. These classes differ in colour strength, shade and electrical charge and have different applications. Thus, beer requires a positive caramel, because beer contains positively charged components derived from the malt; the use of a negative caramel would be liable to precipitate these components. By contrast, soft drinks such as colas require negative caramels.

Many of the red, orange and yellow food colourants are carotenoids. These are available from natural sources but, in practice, are often synthesised on a production basis. The most widely used is the yellow food colorant, β-carotene (**2**).

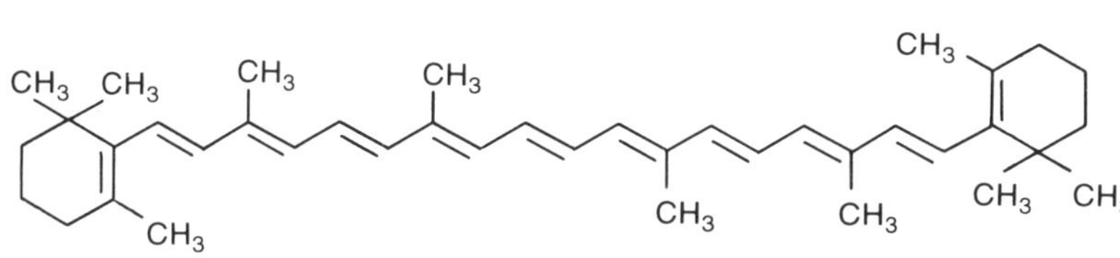

2

Natural red and purple anthocyanins, which occur in fruits and flowers, are also widely used, but their colour is more sensitive to pH, temperature and oxygen. Green colour is often supplied by chemically modified chlorophylls. A number of other natural or nature identical colorants are also supplied.

The most important class of synthetic food colorants are azo dyes, such as

Tartrazine and Amaranth. They provide mainly yellow, orange, red and black colours. However, there is increasing concern over the toxicity of azo colorants, and a number of them are now prohibited in many countries. Some coloured inorganic pigments, such as titanium dioxide and iron oxides, also find use as food colorants.

One approach to countering concerns over the toxicity of traditional food dyes has been to develop food dyes which are not absorbed into the bloodstream but pass straight through the gastro-intestinal tract. These non-absorbable food dyes are polymeric forms of more conventional dyes, and their molar mass often exceeds 20 000 g mol^{-1}. In many cases, the coloured molecules are attached as side-chains to a conventional polymeric backbone. An example is **3**.

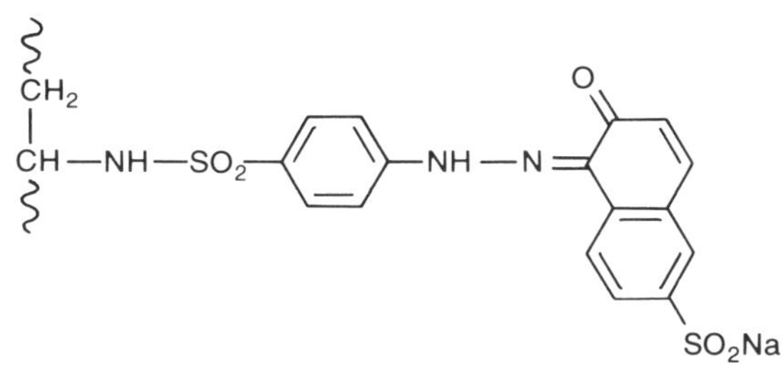

3

Azo groups, however, are often avoided in these polymers, for it has been found that azo groups in traditional food dyes can be reduced in the intestine to amines, which are potentially toxic.

11.4 Hair colorants

Hair colouring has been practised for many thousands of years. Indeed, the oldest known hair dye, henna, used traditionally in the Indian subcontinent and the Middle East, still finds use today. Hair colorants can be conveniently classified as permanent, semi-permanent and temporary colorants, and all three types are subject to considerable competition between the world's major manufacturers. Hair colorants based on metal salts also exist.

Permanent hair colorants are produced from the formation of dye within each hair fibre. A so-called primary intermediate, usually a *p*-phenylenediamine or a *p*-aminophenol, is oxidised to a quinonediimine, which then reacts with a coupler (also known as a secondary intermediate) to give a diphenylamine. The diphenylamine is then further oxidised to produce the dye. It will be noted that the process bears some resemblance to the dye forming process in conventional colour photography (outlined in section 11.2.1). The oxidising agent is normally hydrogen peroxide. Some of the commonly used colour couplers are listed in Table 11.2, along with the colours they produce on reaction with *p*-phenylenediamine.

Table 11.2 Colour couplers.

Coupler	Colour produced
m-Phenylenediamine	Purple-blue
m-Aminophenol	Red-brown
Resorcinol	Green-brown
l-Naphthol	Magenta

Several other means of achieving hair coloration have also been tried. These include the use of mordant dyes, and the addition of a diazonium salt to hair already treated with a coupling component. Moreover, formulations have also been patented which make use of reactive dyes and cationic polymeric dyes.

Semi-permanent hair colorants are designed to provide a colour which will last about 5–10 shampooings. Dye is applied directly to the hair fibres and can diffuse into them. No oxidation process is involved. When the hair is washed, the dye tends to diffuse out of the fibres.

Semi-permanent colorants tend to be of low molar mass and only sparingly soluble in water. The commonest dyes are aminoanthraquinones, nitrophenylenediamines and nitroaminophenols. Sometimes azo dyes are used too. The colorants are applied as foams or lotions containing a thickener, such as carboxymethyl cellulose, and ethanolamine, to assist diffusion of dye into the hair fibres. A few examples of semi-permanent hair colorants are given in Table 11.3.

Table 11.3 Semi-permanent hair colorants.

Colorant	Colour
2-Amino-4-nitrophenol	Yellow
2-Nitro-*p*-phenylenediamine	Red
1, 4-Diaminoanthraquinone	Violet

Temporary hair colorants are designed to be washed out completely when the hair is shampooed. The dyes are deposited on the surfaces of the hair fibres, and do not penetrate at all into the bulk of the fibres. They are used mostly for toning and highlighting. Temporary hair colorants are applied as 'leave-in' products: the dye solution is applied at pH 3–4, for example, by combing, spraying or as a mousse, and then allowed to dry.

Dyes of quite high molar mass are used, in order to prevent diffusion of colour into the hair fibre bulk. Most temporary hair colorants are acid and cationic dyes.

Colorants based on metal salts have been used for several centuries. These colorants include salts of silver, copper, nickel and iron, but nowadays commercial products are generally based on formulations containing lead ethanoate (acetate) and sulphur. The development of colour in the hair, however, requires daily application and often takes over a week. The colours are limited to browns and yellows. The production of colour is probably through the formation of insoluble oxides and sulphides, which once formed are very difficult to remove.

11.5 Biological applications

11.5.1 Histology

Histology concerns the study of biological tissues. Dyes can stain specific tissues and thus assist their characterisation. Indeed, successful selective staining depends on the same types of interaction (and sometimes rates of interaction) between dye and tissue as govern the application of dyes to textiles and other goods.

One of the earliest uses of staining was in classifying bacteria. In 1884, Christian Gram discovered that Crystal Violet colours some bacteria irreversibly but is removed from other bacteria by washing with alcohol. These types of bacteria are known, respectively, as *Gram positive* and *Gram negative*. The contrasting behaviour arises from differences in the structures of the bacterial cell walls.

Another well established technique is the use of Schiff's reagent, a deep-red solution of rosaniline which has been decolourised by sulphur dioxide. On contact with aldehydes, Schiff's reagent restores the magenta colour, but the mechanism of the reaction is obscure. Schiff's reagent can be used to identify deoxyribonucleic acids (DNA) in cell nuclei. The pentose rings in the polymeric chains of DNA are first opened by mild acid hydrolysis to form aldehyde groups, which are then exposed to Schiff's reagent.

Nowadays, there is great interest in the use of dyes for identifying nucleic acids. Cationic dyes, for example, stain nucleic acids rather than proteins in the pH range of *c.* 2–5. In this pH range, proteins carry an overall positive charge (see e.g. wool keratins, section 1.3), whilst nucleic acids are anionic. An interesting aspect of this behaviour is seen in the antiseptic properties of such cationic fluorescent dyes as Acridine Orange and Acriflavine. These dyes have a lethal effect on many bacteria, which arises from interaction with some of their nucleic acids.

Some dyes can even show specificity for DNA or RNA. The cationic dye, Pyronine G (red with a yellow fluorescence), selectively stains RNA, and the cationic dye, Methyl Green, stains only DNA. This difference has, for example, even been exploited in the treatment of tumours.

11.5.2 Fluorescent stains

Fluorescent compounds are being increasingly used for staining, for in some applications they have greater sensitivity than conventional dyes (see Chapter 2). A number of these fluorescent compounds are reactive dyes, such as the vinylsulphone dye, Lucifer Yellow VS (**4**) which reacts with cell proteins. Moreover, these dyes have no toxic effects so can be used to stain living tissues. They have become especially useful for staining neurons (nerve cells), whose cross-section may be as low as 1 μm.

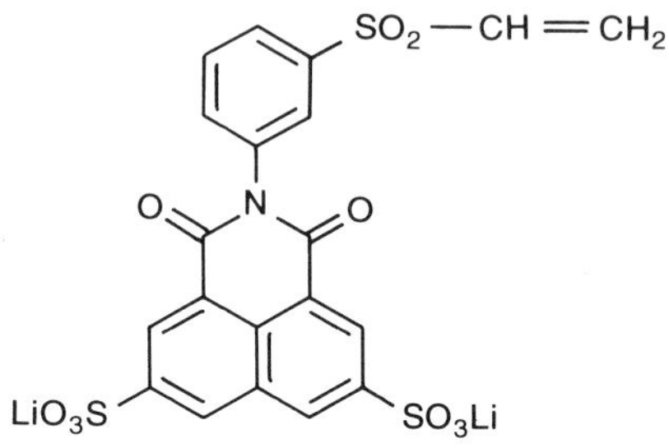

4

Fluorescent dyes can additionally be used for monitoring the activity of neurons. Nerve impulses arise from changes in electrical potential in nerve cell membranes. In the past, these changes have been followed with the aid of microelectrodes, but one microelectrode can monitor the activity of only one neuron, whereas in practice an impulse arises from very many neurons operating together at the same time. More recently, however, an alternative technique has been developed utilising some types of fluorescent dye. Changes in electrical potential in the membranes of neurons, stained with one of these dyes, are accompanied by changes in the dye's absorbance or fluorescence. The types of fluorescent dye which have the greatest sensitivity to changes in potential include merocyanine dyes (e.g. **5**).

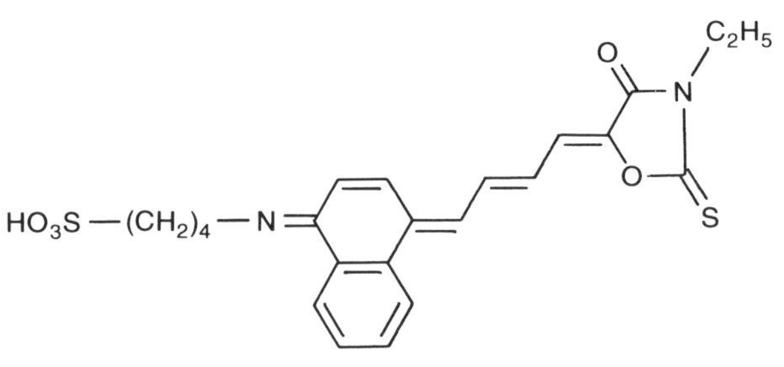

5

Fluorescence staining is also useful in textile microscopy, for example in examining the behaviour of cross-linking agents applied to cellulosic fabrics. Many of these agents are designed to cross-link the hydroxyl groups of neighbouring cellulosic chains and confer a crease-resistant finish to a fabric. Movement of these agents within the individual fibres will, therefore, affect the fabric's physical properties in use. A highly sensitive method for following the movement of cross-linking agents is to stain them with the fluorescent cationic dye, Rhodamine B, or even with a combination of

Rhodamine B and an anionic fluorescent brightening agent. The cationic dye stains the cross-linked parts brown and the non-cross-linked cellulose, to which the fluorescent brightening agent is attracted, exhibits a blue fluorescence.

11.5.3 Affinity chromatography

Affinity chromatography is a rapidly developing separation technique making use of reactive dyes and is particularly important for the purification of proteins, especially of enzymes, whose function is to catalyse reactions in biological systems. Thus, each enzyme has a specific affinity for another biologically active compound or for a functional group in a compound. The compound is known as a *bioligand*. If a bioligand is retained in a chromatographic column, through which a biological extract is then passed, the enzyme in the extract which is attracted to the bioligand will be adsorbed onto the column. The rest of the extract will pass through the column. Indeed, the process can be refined to separate a series of proteins on a column, depending on different affinities of the proteins for the bioligand.

Some reactive dyes can confer the properties of a bioligand on polysaccharides such as dextran and agarose, both of which are extensively used in chromatographic columns. Prominent amongst these dyes is the anthraquinone dye, C.I. Reactive Blue 2, which assists the purification of a number of enzymes catalysing dehydrogenation reactions. Other reactive dyes which have been successfully used are the azo dye, C.I. Reactive Red 120, and the phthalocyanine dye, C.I. Reactive Green 5.

11.5.4 Biochemical assay

Various methods have been devised for utilising dyes in the assay of proteins and nucleic acids. Assay refers to the determination of the content of biochemical substances in organisms. One method for assaying an enzyme depends on the rate at which it catalyses the release of a reactive dye covalently bound to insoluble material. This method finds use, for example, in detergent and brewing technologies. A method for assaying the protein, albumin, in serum makes use of a so-called affinity electrode. This electrode consists of a titanium wire with a layer of the dye, C.I. Reactive Blue 2, covalently bound to the titanium dioxide present on its surface. When albumin becomes linked with the dye, there is a change in electrical potential. A method for estimating the content of some forms of DNA depends on the use of particular cationic dyes. Reference has already been made to the attraction of cationic dyes for nucleic acids (section 11.5.1).

11.6 Electrical/electronic applications

11.6.1 Dyes for liquid crystal displays

Liquid crystal displays have made a great impact commercially over the past ten years or so. Many everyday items of equipment contain liquid crystal displays: digital watches, calculators, display panels in cars, petrol pumps and numerous other items. The attractiveness of coloured displays is perhaps self-evident, but the presence of colour also considerably widens the angle of vision through which the display may be viewed.

A number of organic compounds are capable of existing in phases which exhibit character intermediate between the crystalline solid and normal liquid phases. Such compounds are termed mesogenic, or liquid crystalline. The intermediate phases are fluid, yet highly anisotropic, in that many of their physical properties vary according to direction. Liquid crystallinity results from long-range ordering of molecules into particular orientations, such that the ordering is sufficient to confer a few solid-like properties. An important consequence of this fine balance is that liquid crystals are sensitive to external changes, for example in electric fields. This property is the basis of liquid crystal displays. When an electric field is placed across a liquid crystal, the orientation of molecules is altered or even destroyed. The path of light through the liquid crystal is consequently affected.

Mesogenic compounds consist of elongated, rod-like molecules. Thus, the anisotropic nature of the molecules is translated throughout the entire liquid crystal medium. A measure of their degree of alignment is the order parameter, which has a value between 0 and 1. For completely random molecular orientation, as in true liquids, the order parameter is 0. For perfect alignment of molecules, as might be found in a crystalline solid, the order parameter is 1. In liquid crystals, a high degree of molecular alignment is generally a desirable requirement, i.e. an order parameter nearer to 1.

For the successful incorporation of dyes to give coloured liquid crystal displays, a number of stringent conditions must be satisfied. The dye must readily dissolve in the host liquid crystal medium and, like the liquid crystals themselves, have a high order parameter to promote good alignment of dye molecules with liquid crystal molecules. The dye should, moreover, be nonionic and possess a high state of purity. The contrast observed in a display arises from the different intensities of colour, leading to light and dark display areas.

Because of the rod-like structures of mesogenic molecules, dyes of similar molecular structures would seem to be clear choices for incorporating into liquid crystals; and indeed a number of azo dyes have been successfully applied. However, the lightfastness of these azo dyes is poor. More surprising has been the discovery that, although their molecules are not rod-like, some anthraquinone dyes too can be very successfully used in liquid crystal displays. Anthraquinone dyes have much better lightfastness properties, and several have now been developed with high order parameters and good solubilities in liquid crystals. Suitable dyes in a number of colours are now available, and some examples are given in Fig. 11.5. More recently, fluorescent perylene dyes have also been developed for use in liquid crystal displays. The structure of one of these dyes is also included in Fig. 11.5.

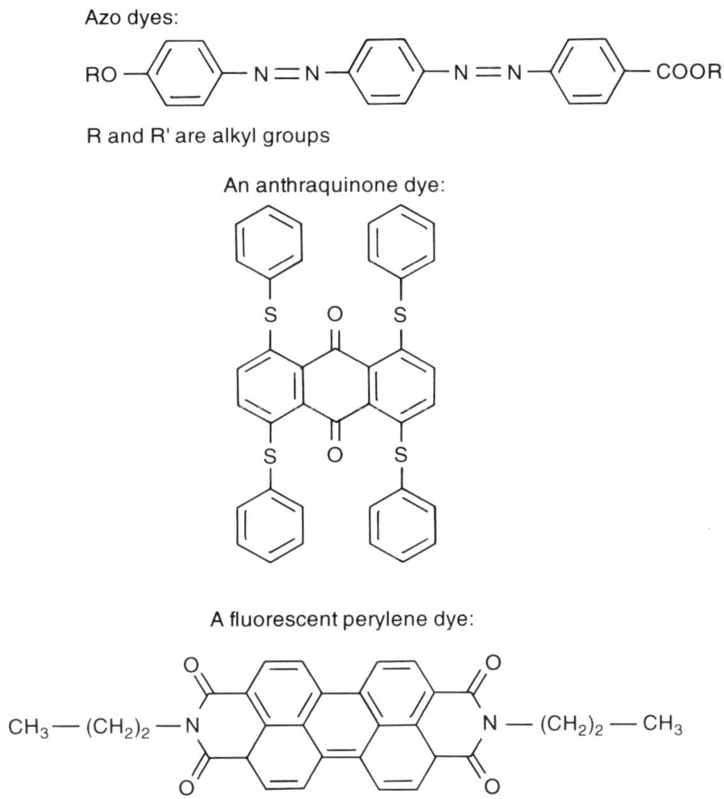

Fig. 11.5 Examples of dyes for liquid crystal displays.

11.6.2 Electrophotography

As discussed in section 4.2.1, the most widely known electrophotographic process is probably photocopying, which like conventional photography is based on the generation and subsequent development of a latent image. However, the mechanisms underlying photography and photocopying are very different. Whereas a photographic image is developed by chemical means, a photocopied image is developed electrostatically.

Photocopying is still largely a black and white process, utilising carbon black pigments, but the use of colour photocopying is increasing rapidly. Three pigments, yellow, magenta and cyan, are required and, as in three-colour printing, these pigments are applied in succession. To date, the pigments used are those already well established in the more traditional areas of pigment technology, discussed in Chapter 6. They include bisarylamide yellows, such as C.I. Pigment Yellow 12, quinacridones, such as C.I. Pigment Red 122 and copper phthalocyanine (C.I. Pigment Blue 15).

The photocopying process uses a plate consisting of an electrode separated by an insulator from a photoconductive layer of selenium or selenium alloy. The use of selenium for the photoconductive layer has, however, been challenged by the advent of a number of organic photoconductors, many of which are pigments. These pigments have the advantages that they are cheaper than selenium and are not toxic. They are being applied more and more to small-scale colour copying equipment and represent a wide variety of pigment types, notably azos, phthalocyanines and perylenes. Coloured compounds, therefore, play two important roles in photocopying: as pigments in image development and in the photoconductive layer.

11.7 Titration indicators

Dyes can be used as titration indicators, whereby the end-point of a titration is determined by a change in the colour of the indicator. Even though other techniques for determining end-points enjoy wide appeal, such as conductimetric and potentiometric methods, the use of coloured indicators is still widespread. The change in colour is derived from the donation or acceptance of entities such as protons (acid–base titrations) or electrons (redox titrations).

For acid–base titrations, the pH at which a colour change is observed varies widely amongst indicators and indeed influences the choice of indicator for a particular titration. For example, for titration of a strong acid against a weak base, a colour change at acidic pH is required. A commonly used indicator for such titrations is Methyl Orange, whose colour changes from red to yellow with a rise in pH from 3.0 to 4.4. Alternatively, where a colour change at alkaline pH is required (weak acid against strong alkali), phenolphthalein is commonly used as an indicator. This indicator changes from colourless to red from pH 8.3 to 10.0. Another phthalein, Cresol Red, changes colour over *two* intervals of pH: red to yellow from pH 0.2 to 1.8, and yellow to purple from pH 7.0 to 8.8.

In redox titrations, the indicator gains electrons on reduction or loses them on oxidation. The choice of dyes is very wide and, not unexpectedly, includes derivatives of vat dyes, such as indigo disulphonic acid. Other types of redox indicator include coloured metal complexes.

Bibliography

High-technology Applications of Organic Colorants, P. Gregory, Plenum Press, New York & London (1991).

Color Chemistry: Syntheses, Properties and Applications of Organic Dyes and Pigments, H. Zollinger, VCH, Cambridge (1991).

Chapter 12

Environmental and Toxicological Implications of Coloration

The manufacture and usage of colorants amounts to hundreds of thousands of tonnes per year worldwide. Thus, any waste colour arising from incomplete application, even at a very small percentage, may present an environmental and toxicological hazard. It is perhaps a paradox that whilst colour is so widely used to brighten our environment, its application may possibly endanger the environment. Equally, the aesthetic qualities of colour, designed to enhance our way of life, can, if applied without care, present severe toxicological problems. It is, therefore, not surprising that strict rules have been progressively developed for premises where colour is applied, to safeguard the health of personnel involved, and that well defined maxima are being introduced to limit the level of colorants and auxiliaries in effluent. Indeed, arguably the biggest challenge to today's colour industry is the challenge of meeting stringent environmental controls and the demands of health and safety. Considerable legislation has been introduced into many countries, as public perceptions of these dangers have been heightened, but a discussion of this legislation is outside the scope of this book.

Important factors which must also be taken into account are hazards arising from starting materials and intermediates in the synthesis of colorants, for traces of these materials may persist into the final coloured product. Hazardous compounds may also be produced during correction procedures involving stripping colorant from a substrate.

Moreover, as is manifest from the preceding chapters, colourless compounds are also deliberately incorporated during colour application. The impact of these compounds on our health and environment must also be taken into account. Prominent among those compounds associated with the application of dyes are salts (e.g. sodium chloride and Glauber's salt), acids and alkalis, chromium compounds (e.g. with chrome mordant dyes), organic compounds used as carriers or other types of auxiliaries, and surface-active compounds, such as those used as levelling agents. Commercial pigments often contain additives to assist dispersion into the chosen application medium. As discussed in Chapter 6, these additives include natural resins, pigment derivatives and so on. The presence of many of these compounds is undesirable once the colour has been applied, and unless they can be readily recycled for further use, they have to be disposed of. In this respect, they too may present an environmental or toxicological hazard.

The products used by dyers and printers which have the greatest toxicological and environmental implications include some types of carrier used to assist the application of disperse dyes to polyester, some chromophores and some types of reactive dye. Insect-proofing agents commonly applied to wool during dyeing and many halogen-containing compounds are also increasingly being put under close scrutiny. In the application of reactive dyes to wool and cellulosic textiles, the quest for complete adsorption and fixation of dye is being urgently pursued. At present, as much as 50% of the dye may be lost to effluent, as hydrolysed and unreacted dye. Extensive research is under way both to develop new reactive systems and to devise fibre pretreatments which then enable complete fixation of existing reactive dyes.

The use of many metals in the colorant industries also presents severe problems. For example, this is an important consideration for the inorganic pigments industry. Mercury and cadmium are metals which are under attack on account of their toxicity, persistence and accumulation in biological organisms. The importance of cadmium sulphide pigments in the red and yellow shade range has already been highlighted in Chapter 6. Mercury is used to catalyse the sulphonation of anthraquinone compounds, to render them soluble as dyes. The removal of mercury from the effluent of sulphonation reactions is critically important, and indeed resins for this purpose are now commercially available.

Many other metals used in the coloration industry, whilst not considered as hazardous as mercury and cadmium, nevertheless require environmental quality standards to be set for them. Amongst these metals are chromium, lead, copper and iron, all of which are important in the pigments industry. Specific examples include lead chromate, iron oxide and copper phthalocyanine pigments. Some of these metals, notably chromium, also feature extensively in the dyeing industry. As discussed in Chapter 5, processes using chrome mordant dyes are extremely important for dyeing wool, especially deep shades; so too are processes for the the application of metal-complex dyes, particularly 1:2 metal-complex dyes. In the application of chrome mordant dyes, chromium is added to wool fibres as Cr(VI), either as dichromate or chromate, but subsequently reduced to Cr(III). Very strict concentration limits in effluent are set on both forms.

The continuing use of chrome mordant dyes has resulted from technological developments in their application. The improved methods of application depend on the use of optimum pH conditions, often in the presence of selected reducing agents, and generally involve the addition of lower levels of Cr(VI) and lower dyeing temperatures. It is noteworthy that these milder conditions usually reduce damage to the wool fibres as well.

Dichromate can also be used to oxidise the leuco forms of vat and sulphur dyes after their application to cellulosic fibres. However, dyers are now progressively turning to hydrogen peroxide and, if possible, air as safe oxidising agents.

Copper too is beginning to come under scrutiny. It is, of course, a constituent of the major organic pigment, copper phthalocyanine, and is used in the 'coppering' of direct dyes, to increase their wetfastness in cellulosic textiles. In addition, copper is used as a catalyst for reacting aromatic amines with bromoanthraquinone. Legislative controls on the level of copper in industrial effluent are likely to become widespread.

One major factor which can render a colorant environmentally or toxicologically undesirable arises from the very ease with which colour is discernible to the naked eye. Even concentrations as low as a few parts per million are usually detectable. Thus, the degree of solubility becomes an important consideration. Organic pigments, for example, have low solubilities in nearly all naturally occurring aqueous and fatty media, a factor which serves to reduce their environmental and toxicological impact. By contrast, dyes generally exhibit a far greater impact. High solubility implies that they can be readily distributed through the environment, via rivers for example, and can accumulate in particular plant and animal organisms. The extent of colour loss to effluent depends markedly on the type of application process. Table 12.1 gives an indication, for example, of the extent to which dyes are commonly lost in effluent. The table reveals that there is a clear variation amongst the various dye application classes.

Table 12.1 Loss of dyes to effluent.

Dye class	Proportion of dye applied which is generally lost to effluent (%)
Acid	7–20
Cationic	2–3
Direct	5–20
Disperse	1–20
Premetallised	2–5
Reactive	20–50
Sulphur	30–40
Vat	5–20

The chemical structures of colorants would, therefore, seem to have bearing on their toxicological and environmental impact, yet the relationship is seldom straightforward. In some cases the toxic risk is less apparent because the dye is not intended for continuous close contact with the user. In other cases, as in food and cosmetic dyes, the risk may be much greater. However, a few simple examples relating chemical structure to potential for hazard may be cited. Thus, cationic dyes may have greater potential for hazard, if they contain quaternary ammonium groups (see Chapters 2 and 5), for such groups can be biologically active. Moreover, as already described in Chapter 11, azo dyes if ingested may be reduced to potentially harmful primary amines.

It is clear, therefore, that during the application of colour, as in its manufacture and end use, proper care has to be taken to ensure that the colorants and auxiliaries used present the minimum of hazard. Moreover, there must be as little waste as possible. In the case of dyes, complete adsorption of dye from the dyebath is highly desirable, as is the recycling or safe disposal of auxiliaries. The choice of colorants and the methods by which they are applied has, therefore, to be strongly influenced by environmental, toxicological and legislative considerations.

Bibliography

Colour in Dyehouse Effluent, P. Cooper (ed.), Society of Dyers and Colourists, Bradford (1995).

Color Chemistry: Synthesis, Properties and Applications of Organic Dyes and Pigments, H. Zollinger, VCH, Cambridge (1991).

Index

absorbance spectrum, 183
absorbance, 182
absorption coefficient, 213
  of dyed textiles, 215
absorption of radiation, 10, 178
  quantitative aspects, 180
acid dyes, 19–21, 98–102, 111–112, 118, 120
  application of 99–102, 111–112, 118, 120
  chemical structures,19–21
  properties of, 98–99
acrylic fabric printing, 153
additive colour mixing, 189–190
alginates for thickening agents, 148,149
alkyd resins, 68–69,
anthraquinone dyes, 15
arylcarbonium ion dyes and pigments, 17
auxochromes, 10
azo coupling, 13
azo dyes, 13–15
azo pigments, 13–15, 28–29, 136–137, 138–139
azoic dyes, 2, 92–94
  application of, 93–94
  properties of, 92–93

ballmill, 262–263
Banbury mixer, 259–260
basic dyes *see* cationic dyes
Batik, 156
beadmill, 263–264
beam dyeing, 241–243
Beer's law, 180
  use for quantitative analysis, 184–185
Beer-Lambert law, 181–182
binders for pigment printing, 167,168
biological applications of dyes, 273–275
  affinity chromatography, 275
  biochemical assay, 275
  fluorescent stains, 273–275
  histology, 273
blind spot, 196
block printing, 157

cadmium sulphide pigments, 135
calibration dyeing for recipe prediction, 216
calibration of colour measuring instruments, 209
camouflage, 1
carbon black pigments, 134
carbonyl dyes, 15–16
carbonyl pigments, 15–16, 139–141
cationic direct dyes
  application of, 119
cationic dyes, 27–28, 112–114, 119
  chemical structures of, 27–28
  application of, 112–114, 119
  properties of, 112
cellulose acetate fibres
  dyeing of, 108–111
  properties of, 40
cellulosic fibres
  dyeing of, 84–97
  dyes for, 22–26
  pretreatment of, 84
  properties of 36–37
cement, 49–50
ceramics, 49
channelling in dye packages, 236–237
chroma, 219
chromaticity coordinates, 204
chromaticity diagram, 204
chrome mordant dyes, 21–22, 102–105
  chemical structures of, 21–22
  application of, 102–105
  properties of, 102
chromophores, 10
CIE *see* Commision Internationale de l'Eclairage
CIE94 colour difference formula, 220–221
CIELAB colour difference formula, 218
CIELAB colour space, 219

CMC(l:c) colour difference formula, 219–220
colour constancy, 211
colour coordinates, 204–206
  cylindrical coordinates, 219
colour difference, 217–221
  assessment 217
  CIE94 formula, 220–221
  CIELAB formula, 218
  CMC(l:c) formula, 219–220
Colour Index, 12–13
colour matching using lights, 197–198
colour measuring instruments, 206–209
colour mixing, 189–190
colour perception, 177
colour photography, 266–269
  dye-bleach process, 268–269
  dye-forming process, 267–268
  image-diffusion process, 269
colour quality, 173
colour solid, 190
colour space, 190
colour splitting, 219
colour temperature of light sources, 193
colour vision, 1
  factors affecting, 191
colour, relationship with molecular structure, 10–11
colours, attributes of, 219
Commision Internationale de l'Eclairage, 196
complementary colours, 189
computer aided design in textile printing, 169–171
computer colorant formulation, 212–217
concrete, 49–50
cone cells in the eye, 195
conical pan dyeing machine, 232
continuous dyeing of fabrics, 245–247
copper phthalocyanine, 16, 124–125, 137–138
covalent bonding, 7
cylindrical colour coordinates, 219

D series illuminants, 202
diazonium salt, 14
diazotisation, 13
diffuse reflection, 208
diffusion properties of dyes
  importance for dyeing, 229
  importance for printing, 147
diffusional boundary layer, 229
dipolar forces, 8
dipole-induced dipole forces, 8
direct dyes, 22–23, 84–88, 118
  aftertreatments of, 86–88
  application of, 85–86, 118, 120–121
  chemical structures of, 22–23
  properties of, 84–85
direct printing, 150–154
  of acrylic fabrics, 153
  of cotton fabrics, 150–151
  of polyamide fabrics, 153–154
  of polyester fabrics, 152–153
  of wool fabrics, 151–152
discharge printing, 154–155
disperse dye printing, thickeners for, 149
disperse dyes, 2, 26–27, 109–111
  chemical structures of, 26–27
  application of, 109–111
  properties of, 109
dispersibility, of pigments, 131, 256–257
dispersion forces, 9
dispersion of light, 177
dominant wavelength, 204
drying of printed textiles, 166
dye mixtures, application of, 121–122
dyeing fibre mixtures, 114–115
dyeing machines, 227–252
  beam, 241–243
  conical pan, 232
  consideration in design of, 229
  Gaston County, 243
  heating of dye liquors in, 231
  Hussong, 233–234
  jet, 243–245
  jig, 239–241
  materials for construction of, 230
  package, 235–237
  Pegg Torroid, 252
  rotating drum, 250
  side paddle, 248–250
  soft flow, 244
  winch, 238–239
dyeing, stages of, 229
dyes
  affinity for textile fibres, 11–12
  application classification, 11, 18–28
  chemical classification, 11–18
  distinction from pigments, 2, 5–7
  fastness properties, 12, 83

levelness of, 83–84
mixtures of, 121–122
selection of, 83–84

elastomers, 47
electrical/electronic applications, 276–278
dyes for liquid crystal displays, 276–277
electrophotography, 77–78, 277–278
electromagnetic radiation, 174–178
wave model, 175
electromagnetic spectrum, 176
electron spin, 186
electronic energy, 179
electronic transitions, 10, 180
electrophotography, 77–78
emulsion thickening agents, 168
ending, in jig dyeing, 241
energy level diagram, 181
energy of radiation, 177–178
environmental aspects, 279–282
dyeing, 227–228, 252
pigments, 132
printing, 168, 170
epoxy resins, 69–70
equi-energy spectrum, 198
excitation purity, 206
eye, 194–196

fabric batchwise dyeing, 237–245
fabric continuous dyeing, 245–247
fibres, mass pigmentation, 227
relationship with plastics, 48
film-formers, in paints, 57
flash-age printing, 166
flat screen printing, 159–160
fluorescence, 185–187
fluorescence emission, 186–187
fluorescent lamps, 192
fluorescent whitening agents, 185, 224–225
flushed pigments, 144
foam, printing using, 149
food colorants, 270–271
fovea, 195
frequency of radiation, 176

garment dyeing, 247–251
high temperature, 251
glass transition temperature, significance in paints, 64
glass, 50
gloss, 127
guar gum thickening agent, 148
gum tragacanth thickening agent, 148

hair colorants, 271–273
hank dyeing, 233–235
high speed impeller mill, 261–262,
high temperature garment dyeing, 251
hue angle, 219
hue circle, 190
hue difference, 219
humectant for print pastes, 148
hydrodynamic boundary layer, 229
hydrogen bonding, 9

illuminants
distinction from sources, 201
standard, 201–202
illuminated discharge printing, 155
imaginary primaries, 198
incandescent light sources, 191
indigo, 15
infrared radiation, 175
ink-jet printing, 75–77, 163–164
inorganic pigments, 18, 133–135
instruments for colour measurement, 206–209
ion-dipole forces, 8
ionic forces, 8
iron oxides, 134–135

jet dyeing machines, 243–245
jig dyeing, 239

Kubelka-Munk equation, 213–216

Lambert's law, 180
laser printing, 78
latex paints, 65–66
lead chromate pigments, 135
leather
dyeing of, 119–121
properties of, 43–44, 119
tanning of, 43–44
leuco form of vat dyes, 2
levelling agents, 115–117
modes of action, 115–116
types of, 116–117
light, 176

sources, 191–194
speed of, 176
lightness, 206
liquid, in paints, 57
liquor circulation in dyeing machines, 230
liquor to goods ratio, 228
listing in jig dyeing, 241
locust bean gum thickening agents, 148
loose fibre dyeing, 231–232

mercerisation, preparation for printing cotton, 150–151
metal complex dyes *see* premetallised dyes
metameric indices, 211–212
metamerism, 210
relationship with colour constancy, 212
methine dyes, 17
molar extinction coefficient, 181
molecular energies, 179,181
molecular size of dye molecules for printing, 147
mordant dyes *see* chrome mordant dyes
multiplicity, 186

natural thickening agents for printing, 148–149
Newtonian and non-Newtonian fluids, 60–63
nitro dyes, 17

opponent colours diagram, 182
optic nerve, 1, 196

package dyeing, 235–237
padding, 245
paints, 51–69
classification of, 53–55
composition, 55–59
decorative, 54
industrial, 54–55
nature of, 52–55
principles of film formation, 63–69
purpose of, 52–53
rheology of, 59–63
paper
dyeing of, 117–119
properties of, 42–43
particle size, influence on colour of pigments, 125, 256
Pegg Toroid garment dyeing machine, 252
perfect reflecting diffuser, 208
phosphorescence, 188
photocopying, 77–78
photon, 177
phthalocyanine dyes and pigments, 16
pigment dispersion, forces involved, 257–259
industrial equipment, 258–264
mechanism, 253–257
technology, 253–264
pigment printing on textiles, 167–169
pigments, 2, 28–30
application of, 123–145
chemical classification, 11–18
chemical resistance, 131
factors influencing colour, 124–126, 129
factors influencing opacity, 126–127, 129
fastness properties, 12, 129–131
heat stability, 130
high performance organics, 138–140
lightfastness properties, 129–130
manufacture, 142–144
particulate nature, 254
relationship between chemical structure and properties, 132–141
requirements for printing textiles, 167
role in paints, 56
solvent resistance, 28, 130–131
surface treatments, 143, 256–257
Planck's constant, 177
Planck's radiation law, 191
Planck's relationship, 10
plastics, 44–48
coloration of, 48
polyacrylate thickening agents, 149
polyamide fabric printing, 153–154
polyester fabric printing, 152–153
polyester/cellulose continuous dyeing, 245
polyethylene, 45–46
polymers
fibres, 33, 48
glass transition temperature, 32
introduction, 31–33
molar mass, 31
polymorphism, of pigments, 124–125
polypropylene, 46
polyvinyl chloride, 46
premetallised dyes, 21–22, 105–107, 112, 121
application of, 105–107, 112, 121
chemical structures of, 21–22
properties of, 105

primary colours, 189
printing inks, 69–81
drying mechanisms, 80–81
nature of, 79–81
printing of carpets, 164–165
printing pastes, 147–150
typical composition, 147
printing processes, 70–81, 157–164
printing styles, 150–156
printing, use of computer aided design, 169–171
of textiles, 146–172
colour processes, 78–79
discharge, 154–155
flexographic, 73
gravure, 73–74
ink-jet, 75–77
letterpress, 74–75
lithographic, 71–72
resist, 155–156
roller, 157–158
rotary screen, 75, 160–161
toner, 77–78
transfer, 161–163
producer dyeing, 227
protein fibres
dyeing of, 108
dyes for, 19–22
pretreatment of, 98
properties of, 38–40
Prussian blue, 135
PTFE, 46

quality of colour, 173
quantitative analysis of coloured solutions, 184–185
quinacridone pigments, 15, 28–29, 139–140

reactive dyes, 2, 24–26, 95–97, 107–108
application of, 97, 107–108
chemical structures of, 24–26
properties of, 95–97
recipe prediction for dyed textiles, 216–217
reflectance, 194
measurement of, 208
typical curves, 194
resist printing, 155–156
retina, 194
rheology of surface coatings, 59–63
rod cells of the eye, 195
roller printing, 157–158
rotary drum garment dyeing machines, 250
rotary screen printing, 160–161
rotational energy of molecules, 179
rubber, 47

sandmill, 263–264
saturation of colour, 206
scattering coefficient, 213
screen printing, 75, 159–161
secondary colours, 189
shade sorting of dyed textiles, 221–223
side paddle garment dyeing machine, 248–250
single number shade passing, 219
singlet state, 186
soft flow jet dyeing machines, 244
spectral energy of sources, 191
spectrophotometers, 206–209
spectrum locus, 204
spectrum, electromagnetic, 176
specular reflection, 191
spin angular momentum of electrons, 186
standard illuminants, 201
standard observer, 197–199
starch thickening agents, 148,149
Stokes shift, 186, 187
sublimation, of disperse dyes for transfer printing, 161
subtractive colour mixing, 190
sulphur dyes, 17, 90–92
aftertreatments of, 92
application of, 91–92
properties of 90–91
surface coatings, 51–82
drying with chemical reaction, 66–69
drying with no chemical reaction, 64–66
synthetic fibres
dyeing of, 108–114
dyes for, 26–28
properties of, 40–42
synthetic thickening agents, 149

textiles, 33–42
classification, 35
properties of, 33–34
thermoplastics, 44–47
thermosets, 47
thickening agents, 148–149, 168

thixotropy in surface coatings, 61–63
three line fluorescent lamp, 192
tiles, for calibration of colour measuring instruments, 209
tint, in whites, 225–226
titanium dioxide, 126–127, 133
titration indicators, 278
toxicological issues, 132, 279–282
transfer printing, 161–163
transmittance, 182
triphenylmethane dyes, 17
triple roll mill, 260–261
triplet state, 186, 188
tristimulus colorimeters, 207
tristimulus values, 198, 199–201
tumblers for garment dyeing, 251
tungsten lamps, 191

ultramarine blue, 135
ultraviolet radiation, 175

van der Waals forces, 9
vat dyes, 2, 23–24, 88–90
  application of, 89–90
  chemistry of, 23–24
  properties of, 88–89
velocity of light, 176
vibrational energy of molecules, 179
viewing geometry of colour measuring instruments, 191
visco-elastic properties of thickening agents, 148
visible spectrum, 176–177
VOCs from paints, 58

washing off, in textile printing, 166–167
watering, in beam dyeing, 242
wavelength, 175
  of maximum absorbance, 183
weighting values for computing tristimulus values, 203
white, perfect reflecting diffuser, 208
whiteness, 223–226
  assessment, 223
  index, CIE formula, 225
winch dyeing, 238–239
winding, of yarns for package dyeing, 235–236
wool printing, 151–152
Wright and Guild colour matching experiments, 198

yarn dyeing, 232–237
yarn winding, 235–236
yellowness in textiles, 223